For Doug,
  Thank you for your interest in the Plants of Deep South Texas.

  Ken King
  Al Richardson

PLANTS OF DEEP SOUTH TEXAS

Perspectives on South Texas
Sponsored by Texas A&M University–Kingsville
Timothy E. Fulbright, General Editor

A&M nature guides

Plants of Deep South Texas

*A Field Guide
to the Woody
& Flowering Species*

Alfred Richardson
& Ken King

Texas A&M University Press
College Station

Copyright © 2011
by Alfred Richardson and Ken King
Manufactured in China
by Everbest Printing Co.
through FCI Print Group
All rights reserved
First edition

This paper meets the requirements of
ANSI/NISO Z39.48–1992
(Permanence of Paper).
Binding materials have been chosen
for durability.

Library of Congress
Cataloging-in-Publication Data

Richardson, Alfred, 1930–
  Plants of deep south Texas : a field guide to the woody and flowering species / Alfred Richardson and Ken King.—1st ed.
    p. cm.—(Perspectives on south Texas)
Includes bibliographical references and index.
ISBN-13: 978–1-60344-144–5
(pb-flexibound : alk. paper)
ISBN-10: 1–60344-144–1
(pb-flexibound : alk. paper)
1. Plants—Texas—Lower Rio Grande Valley—Identification.   I. King, Ken, 1960–   II. Title.
III. Series: Perspectives on south Texas.
QK188.R528   2010
581.9764—dc22
2010005066

A. Berlandier's fiddlewood, *Citharexylum berlandieri,* on a clay dune or loma near the coast. In times of drought, the leaves turn yellow to orange in color.
B. Golden wave, *Coreopsis tinctoria,* and Dakota vervain, *Glandularia bipinnatifida,* in a fallow field, Cameron County.
C. Spanish moss, *Tillandsia usneoides,* in the Rio Grande floodplain, Santa Ana National Wildlife Refuge.
D. Yellow lotus, *Nelumbo lutea,* in a wetland.
E. Gulf cordgrass, sacahuista, or needle grass, *Spartina spartinae,* occurs in almost pure stands in saline soils at or near the coast.
F. A field of Texas bluebonnet, *Lupinus texensis,* Willacy County.
G. Square bud primrose, *Calylophus serrulatus,* and southern Indian blanket, *Gaillardia pulchella,* at the coast.
H. Sabal Palm Audubon Sanctuary, preserving one of the few remaining stands of *Sabal mexicana* near the Rio Grande.
I. Laguna Madre, with black mangrove, *Avicennia germinans,* in the foreground. The vertical "dead sticks" in front of the mangroves are pneumatophores that arise from the roots.

DEDICATED TO
Christina Mild & Mike Heep

CONTENTS

Acknowledgments	ix
Map of the Lower Rio Grande Valley	xii
Introduction	1
Plant Species Accounts	9
Algae	11
Ferns	12
Gymnosperms	14
Monocots	16
Dicots	48
Appendix: Butterflies and Moths	429
Glossary	431
Bibliography	435
Index	439

ACKNOWLEDGMENTS

This book is the result of the generosity, cooperation, and encouragement of a large number of people. Identification of difficult species was provided by Dr. Tom Wendt, Dr. Paul Fryxell, Ann Vacek, Dr. Billie Turner, Dr. David Austin, Dr. David Rosen, Dr. Robert Lonard, Dr. William Ocumpaugh, and Dr. Andrew McDonald.

In order to photograph many species, it was necessary to enter a number of private ranches in the Lower Rio Grande Valley. Fidencio Guerra (Guerra Ranch), Hardin Gwin (Gwin Ranch), Karen and Dr. Philip Hunke (Tecolote Ranch), John and Audrey Martin (The Hills), James and Georgiana Matz (Matz property), Lowry and Jessica McAllen (Las Colmenas Ranch), Raquel and Carlos Olivo (La Sagunada Ranch), Betty Perez and Susan Thompson (Perez Ranch), Benito and Toni Treviño (Treviño Ranch), Santa Margarita Ranch, Armando Vela, Chris Wattenpool, and Kate Cervone and were all generous in allowing us to enter their properties.

Photographs were taken, with permission, at the Valley Nature Center and Frontera Audubon Thicket in Weslaco, Laguna Atascosa National Wildlife Refuge, Quinta Mazatlán in McAllen, the North American Butterfly Association (NABA) Butterfly Park in Mission, the Edinburg Scenic Wetlands, and Sabal Palm Audubon Sanctuary at Brownsville. Lisa Williams, at that time director of the Tamaulipan Thornscrub Project of the Nature Conservancy of Texas, gave us access and guided us through Las Estrellas Preserve. Ellie Thompson guided us through various parts of the Laguna Atascosa National Wildlife Refuge. Chris Hathcock gave us access and led us through Resaca de la Palma State Park.

Tom Patterson was especially helpful in guiding us and sharing with us his knowledge of plants on Padre Island and in Starr County. He also made many valuable suggestions regarding the manuscript. Mike Heep was helpful and generous with information regarding locations of various plants and also in allowing us to photograph plants at his nursery in Harlingen. John Pemelton and Lindley Lentz gave us locations of important rare plants. Christina Mild was enthusiastic in reporting new plants she had found, sometimes guiding us to the localities. Ann Vacek generously shared with us her knowledge of the local plants. Martin Hagne guided us to the localities of several species. James Everitt gave us localities for several species. J. Rodney Sullivan was generous with advice and aid when the computer became stubborn and refused to cooperate.

Members of the local Audubon societies, especially Master Naturalists Diane Ballesteros and Frank Wiseman, were encouraging and helpful in pointing out plants for which photographs were needed. Rick and Patricia Choate were generous in calling a species to our attention and allowing us to photograph it at their residence. Katie and Violet Springman generously allowed us to photograph plants at their residence.

The reviewers made suggestions that greatly improved the manuscript. Christina Mild, John Pemelton, and Ann Vacek read it and offered many valuable suggestions. Special thanks go to John Pemelton, who prepared the map of the Lower Rio Grande Valley.

PLANTS OF DEEP SOUTH TEXAS

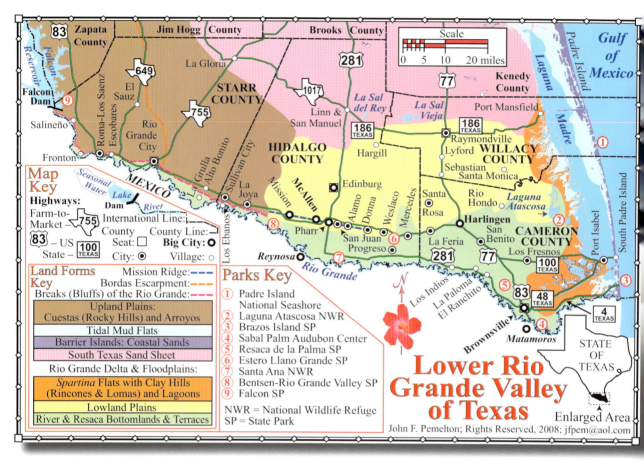

Map of the Lower Rio Grande Valley by John Pemelton.

INTRODUCTION

The purpose of this book is to provide, through pictures and short descriptions, a simple means of identifying plants that grow naturally in Cameron, Hidalgo, Willacy, and Starr counties, known locally as the Lower Rio Grande Valley. This includes native plants, locally introduced naturalized plants, and a few invasive introduced plants. Also included are some species that are known to grow a few miles from the borders of the counties mentioned but that have not yet been found within the specified area. Except where noted, photographs were taken by the authors. Most were taken in the field, but a few species were photographed in cultivation.

This is the most comprehensive photo-illustrated flora of the Lower Rio Grande Valley. It includes undescribed species, newly described species, the first published photographs of many species, and range extensions of many species, using the latest available taxonomy. There are photographs of more than 800 species and varieties, about 77 percent of the total number (excluding the grasses) reported for this area. Books on selected wildflowers, woody plants, and broad-leaved plants are available, but they cover a much broader geographic area. Richardson's *Plants of the Rio Grande Delta* (1995) combines the technical with more "user friendly" color photographs, and it includes only Cameron, Hidalgo, and Willacy counties. Starr County has not been studied as much as the other three and almost all the information available for that county is in technical manuals. The grasses have not been included for several reasons: Dr. Robert Lonard has already produced *Guide to the Grasses of the Lower Rio Grande Valley, Texas* (1993), with drawings of the grasses. Another book on the region's grasses is now being prepared by James Everitt, Robert Lonard, Christopher Little, and Lynn Drawe. A third book, *Weeds in South Texas and Northern Mexico* by James Everitt, Robert Lonard, and Christopher Little (2007), contains some beautiful pictures of grasses.

A brief description is given for each species pictured in the present volume, along with other information such as blooming time and distribution of the species. In our climate, the blooming time can vary considerably. Leaf size also varies widely. In seasons of abundant rain, plants produce larger leaves. In drier times, they may produce leaves that are highly reduced in size. Notations on the distribution of a species may fail to include a county where it actually exists, simply because it has not yet been reported.

Voucher specimens were not collected for species that are well known to be present in this area. For others, especially those that might be questioned, voucher specimens were collected and are stored in the herbaria at the University of Texas at Austin and at the University of Texas at Brownsville.

Cameron, Hidalgo, Willacy, and Starr counties, the four southernmost counties in Texas, are collectively referred to as the Lower Rio Grande Valley (LRGV). Together they comprise 2,986,240 acres. Compared with the area of the whole state, this is not a huge area. The *Manual of the Vascular Plants of Texas* reports almost 5,000 species and varieties of plants for the state. Excluding the grasses, 1,058 plant species and varieties have been reported to occur in the Lower Rio Grande Valley.

There are two major vegetational areas in the Valley: the Rio Grande Delta, and part of the Hebbronville Plain. The Rio Grande Delta includes Cameron and Willacy counties and most of Hidalgo County. The Gulf of Mexico forms the eastern boundary. The soils are various mixtures of clay and sand, with pockets of almost pure clay and almost pure sand. The second area includes western Hidalgo County and Starr County. According to Elzada Clover (1937) they are part of the Hebbronville Plain. The topography becomes hilly, and the soils are caliche and gravel. The Bordas Escarpment forms the western boundary of the Hebbronville Plain. Each of these areas has its own plant associations but with some plants common to both.

Within these two general areas, the Valley enjoys a number of different habitats for plants and animals. There are plants found here that occur nowhere else in the United States. At the coast are areas of almost pure sand. Farther inland are mixtures of clay and sand. Because of proximity to the Gulf and relatively low rainfall, there are large areas of salt flats that may be interrupted by clay hills (usually called clay dunes, or lomas). Low, salty areas may extend several miles away from the coast. La Sal del Rey and other salt lakes occur in Hidalgo County, and there are saline soils in Starr County.

A narrow band of "river bottom" vegetation follows the Rio Grande on the southern border. The river was once navigable, with steamboats being a common sight. As the population grew and agriculture expanded, demands on water grew accordingly. Dams were built along the river. Now there is not enough flow to allow large boats to navigate the river. In periods of extreme drought, the flow slows and even stops. During the last drought period, in the years 2001 and 2002, sand bars gradually built up at the mouth of the river, entirely blocking its connection with the Gulf of Mexico.

The sand deposits on the northern boundary, referred to as the South Texas Sand Sheet, provide a different habitat. In the spring of the year this area is popular for wildflower tours. The sand deposits, which extend down to the bedrock, are seen in northern Hidalgo County and in portions of Starr and Willacy counties. This area is different from the coastal sands. It does not have the high salt content, and the sand grains are smaller.

Where eastern Cameron and Willacy counties border the Gulf of Mexico, that large body of water has a moderating effect on temperatures. Accordingly, areas nearest the Gulf are slightly cooler in summer and warmer in winter than inland zones, with higher humidity. As one moves westward, temperatures become more extreme (hotter summers, colder winters) and the humidity becomes lower. Elevation is not a major factor in producing temperature differences. Rio Grande City in the west is only 153 feet higher than Port Isabel on the coast.

Records of monthly average temperatures do not tell the complete story. It is essential to know the high and low temperatures. Summer high temperatures can reach more than 100 degrees in all areas of the Valley. Winter temperatures are generally mild, but the region does experience occasional hard freezes. Extreme low temperatures reported by the United States Weather Bureau for Cameron County are 15 degrees (1901), 22 degrees (1905), 15–17 degrees (1989), 19–22 degrees (1962), and 21–23 degrees (1951, 1963). Extreme lows for Starr County are 10 degrees (1962), 15 degrees (1951, 1973, 1975, 1983), and 24 degrees (1985). The Valley was presented with a white Christmas in 2004, with a blanket of snow over much of the area. Such an event had not occurred in the Valley in more than 100 years. Low temperatures did not last long, and there was little damage to the native vegetation.

The Weather Bureau reports average rainfall from data collected during the years 1971 to 2000: Cameron County 28.08 inches, Willacy County 28.27 inches, western Hidalgo County 22.61 inches, and Starr County 20.95 inches. The months of May to October experience more than twice as much rainfall as the other six months. The average rainfall, of course, does not happen every year. In her study of Boscaje de la Palma (the sabal palm grove in Brownsville), Anna May Tarrence Davis reports only slightly more than 12 inches of rain in 1898 and 1917. She also reports 1886 rainfall of 60.06 inches, with 30.57 inches in September that year (the month with the highest incidence of hurricanes). She could find no recognizable cycles of wet and dry periods but observed a drop in rainfall roughly every two to four years. The significant differences in rainfall across the region, plus the differences in soils and to a lesser extent differences in temperature, are considered the most influential factors determining the presence of various plant associations.

With the plentiful sunshine and relatively low rainfall, most of our native plants tend to have small leaves or lose them entirely, as in the cactus family. Green stems can produce all the food the plants need. The abundance of sunshine is illustrated by the fact that even many of the cacti tend to grow in the partial shade of shrubs or trees.

Information for the census of plants of the Lower Rio Grande Valley was derived from the University of Texas herbarium; Flora of Texas Database (TEX-LL) at the University of Texas Plant Resources Center; *Atlas of the Vascular Plants of Texas* (Turner et al. 2003); *Plants of the Rio Grande Delta* (Richardson 1995); and field observations. Occasionally a species new to the area is published in one of the botanical journals, and that species is added to the listing. Some of the plants photographed have not previously been reported for this area.

Common names were derived from *Manual of the Vascular Plants of Texas* (Correll and Johnston 1970), *Flora of the Texas Coastal Bend* (Jones 1975), *Guide to the Grasses of the Lower Rio Grande Valley, Texas* (Lonard 1993), *Plants of the Rio Grande Delta* (Richardson 1995), and *A 2003 Updated List of the Vascular Plants of Texas* (Jones and Wipff 2003). Knowledgeable local people were also consulted for common names used in this area.

When a plant is given a name, a rigid set of rules must be followed. That name can be changed only in certain circumstances. In the past, this has provided stability in plant names, with only occasional changes. In the last several years, however, there have been many name changes, and several different classification systems have been proposed. This is frustrating to almost everyone and can cause a certain amount of confusion: whose classification system does one use?

In this book we have in general followed the classification in Jones and Wipff (2003). For the cactus family we have followed the classification in *Flora of North America,* volume 4. We have also followed more recent publications of name changes. In cases where the name of a species has been changed fairly recently, the older name is also listed as a synonym, for the convenience of those who may still be using the older name. We have not listed all the synonyms for every species.

Plant classification is based on seven basic categories. The largest category is kingdom, which includes all plants. Following kingdom are the successively smaller categories of division, class, order, family, genus, and species. These categories can be divided into smaller groups; for example, the species category is often subdivided into subspecies, and the subspecies into varieties. Although variety is technically the smallest category, it is often used synonymously with subspecies. We focus on the smaller categories of family, genus, species, subspecies, and variety. A family contains a group of related genera (plural of genus); a genus contains a group of related species.

There are different ways of arranging the plant families in books. Some authors arrange them in alphabetical order, and some arrange them according to similarities. Others ignore the family category and arrange the plants by flower color or alphabetically by scientific name. All the arrangements have their positive and negative aspects. We have chosen to arrange the plants in five categories, arbitrarily beginning with the more simple-appearing to the more complex-appearing plants: (1) **Algae** are seedless plants without specialized conducting tissue. (2) **Ferns** are seedless plants with specialized conducting tissue, reproducing by spores. (3) **Gymnosperms** are woody plants producing seeds but without flowers. (4) **Monocots** are flowering seed plants with one "seed leaf." Usually the flower parts are in threes, and the leaf veins are all parallel. (5) **Dicots** are flowering seed plants with two "seed leaves." Usually the flower parts are in fours or fives, and the leaf veins form a branching network. The bulk of plants covered here are dicots. Within each category, the families and their genera are arranged in alphabetical order. We believe that this arrangement is the most convenient for rapid identification of a species. It is not intended to establish a taxonomic system.

References useful for learning more about the plants of the Lower Rio Grande Valley are Correll and Johnston's *Manual of the Vascular Plants of Texas* (1970); Alfred Richardson's *Plants of the Rio Grande Delta* (1995) and *Wildflowers and Other Plants of Texas Beaches and Islands* (2002); Robert Lonard's *Guide to the Grasses of the Lower Rio Grande Valley, Texas* (1993); *Illustrated Flora of North Central Texas* (Diggs et al. 2006); *Atlas of the Vascular Plants of Texas* (Turner et al. 2003); *Field Guide to the Broad-Leaved Herbaceous Plants of South Texas* (Everitt et al. 1999); *Trees, Shrubs and Cacti of South Texas* (Everitt et al. 2002); and *Weeds in South Texas and Northern Mexico* (Everitt et al. 2007). Others are listed in the bibliography.

Some terms that are in the glossary may need more explanation. To a botanist, the terms *spine*, *thorn*, and *prickle* are not synonymous. A spine is derived from a leaf or a leaf part. For example, the spines in the legume family (Fabaceae) are derived from the leaf stipules, which are in pairs at the base of the petiole. Correspondingly, the spines are usually paired. A thorn is the sharp end of a short stem. Sometimes they bear leaves. Prickles are outgrowths of the stem epidermis or bark. They are present on *Rubus riograndis* (dewberry) and on several species of *Mimosa*.

A stolon is a creeping stem usually on the surface of the soil. It often produces new plants and roots as it grows. A rhizome is a creeping stem usually below the surface of the soil. It can also produce new plants with roots. A tuber is a thickened, short underground stem.

Further explanation of some other plant characteristics might be helpful. For example, there are many terms describing the leaf margins. We have limited them to the ones illustrated. In the same way, the number of terms describing the hairs on the leaves and stems has been greatly reduced. In most cases, the plants have simply been described as hairy, or minutely hairy. Of special note is the stellate, or star-shaped, hair. Each hair is highly branched, under magnification looking like an asterisk. In order to make the text as easy as possible to read, we have sacrificed the fine detail of specific terms.

INTRODUCTION

About Botanical Terms

Botanical terms can be difficult. Besides being foreign to us, they are often long and difficult to pronounce and spell. Many can be replaced by ordinary words or phrases, and this has been done when possible. However, sometimes the scientific term is the only one available. A glossary is included at the back of the book to define those we have used, and terms for the shapes and structure of leaves and flowers are shown in the following illustrations. All the drawings are from *Wildflowers and Other Plants of Texas Beaches and Islands* by Alfred Richardson (2002) and are reproduced courtesy of the University of Texas Press.

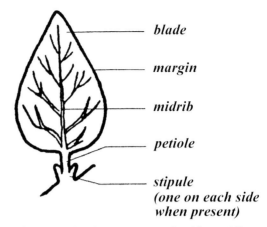

Leaf Parts. All plant drawings are reproduced from *Wildflowers and Other Plants of Texas Beaches and Islands* (Richardson 2002), courtesy of the University of Texas Press.

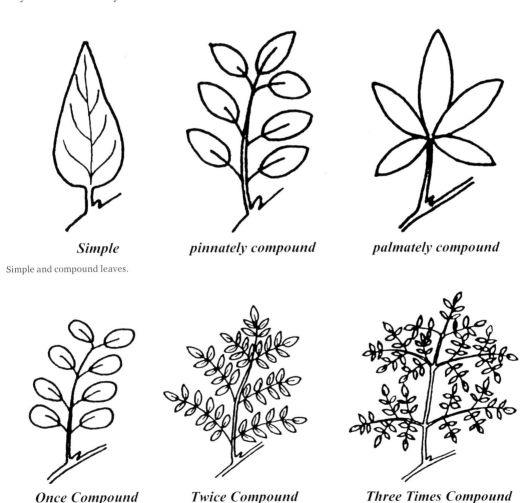

Simple and compound leaves.

Leaves can be one, two, or more times compound.

Alternate, opposite, or whorled leaves on the stem.

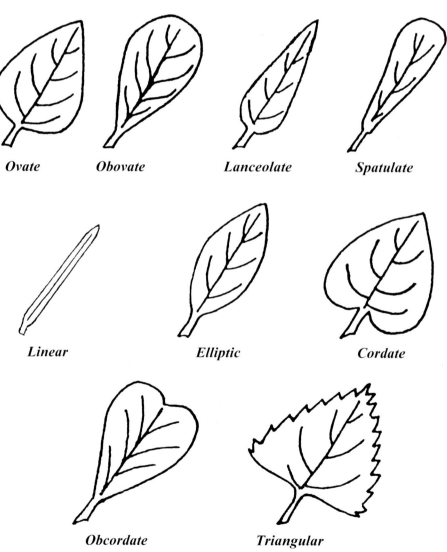

Some common leaf shapes.

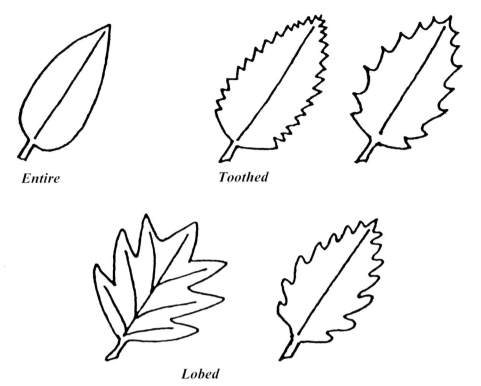

Leaf Margins. The edge (margin) of the leaf can be perfectly smooth (entire), or may have various kinds of teeth or indentations. Many botanical terms are used to distinguish fine differences in the leaf margins. In order to make things simpler, only a few terms are used in this book, each carrying a broad meaning useful in this context (if not as precise as full scientific terminology might be). Leaves with various kinds of cuts into the margin are simply called "toothed." Leaves with larger indentations into the margin are called "lobed."

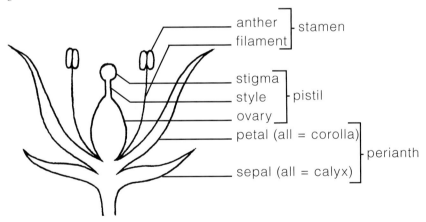

Parts of the flower.

Position of the ovary. An ovary is called superior if the other flower parts are joined to the receptacle below the ovary. It is called inferior if the other parts are joined to the ovary itself or above it.

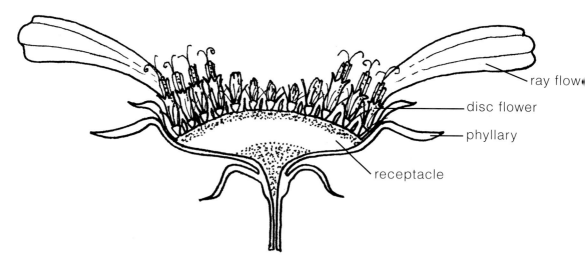

Asteraceae Family. The Asteraceae family (also called Compositae) has a unique flower structure that merits special attention. Using the common sunflower as an example, each of the yellow "petals" is a flower and is called a ray flower (or ray floret). Each of the tiny brown dots in the center is also a flower and is called a disk flower (or disk floret). The sunflower, then, is a composite of many flowers on a single receptacle; hence the name Compositae. The cluster of florets on one receptacle is called a "head." There is much variation of this pattern, and not all members of the family have both ray and disk florets. Each of the green parts at the base of a sunflower is called a phyllary. All the phyllaries together are called the involucre.

PLANT SPECIES ACCOUNTS

Algae

Seedless Plants without Specialized Conducting Systems

Stonewort, Brittlewort, Muskgrass
Family: Characeae
Scientific Name: *Chara* sp.
Habit: Small malodorous aquatic plants lacking specialized conducting tissue, growing on the bottoms of ponds.
Leaves: None.
Flowers: None. The tiny reproductive organs appear as tiny knots on the "stems."
Fruit: None. The product of fertilization, after a resting period, germinates into a juvenile plant without the formation of a seed.
Bloom Period: No flowers are produced. Reproductive time is not established.
Distribution: Hidalgo County.
Comments: The stoneworts are the only non-vascular plants we have included. They are included because of their similarity to the vascular plants. *Chara* is a member of the green algae but, with its stemlike growth, is not typical of that group. The common names are useful and descriptive. The outer parts are covered with calcium carbonate, giving the plant a rough texture, and it becomes malodorous when handled. Stoneworts are found in bodies of clear water, attached to the bottom.

Stonewort, Brittlewort, Muskgrass *Chara* sp.

Nitella, Stonewort
Family: Characeae
Scientific Name: *Nitella* sp.
Habit: Multicellular branching algae, lacking specialized conducting tissue, loosely attached to the bottoms of ponds.
Leaves: None.
Flowers: None. The tiny reproductive organs appear as knots on the "stems."
Fruit: None. The product of fertilization, after a resting period, germinates into a juvenile plant without the formation of a seed.
Bloom Period: No flowers are produced.
Distribution: Kenedy County, 4.8 miles north of the Willacy County border.
Comments: Like *Chara, Nitella* is a member of

Nitella, Stonewort *Nitella* sp.

the green algae but, with its stemlike growth, is not typical of that group. Unlike *Chara*, it is not malodorous when handled, and its "stems" are not covered with calcium carbonate. Although Nitella was found in Kenedy County, it was only a few miles north of Willacy County, which has similar habitats. It is highly probable that it exists in that county as well.

Ferns

Seedless Plants with Reproduction by Spores

Carolina Water Fern
Family: Azollaceae
Scientific Name: *Azolla caroliniana*
Habit: Tiny floating plants about ⅜" or less in diameter.
Leaves: Folded, notched at the fold, less than ¹⁄₁₆" long, with microscopic barbed hairs, red if in full sun.
Flowers: None
Fruit: None. Sporocarps on the lower leaf surface produce sporangia. Spores are produced within the sporangia.
Bloom Period: Flowers are not produced.
Distribution: Cameron and Hidalgo counties.
Comments: *Azolla* is an aquatic fern with roots, stems, and leaves. It is so small, however, that usually only the leaves are noticed. Hollows in the leaves are occupied by symbiotic blue green bacteria (cyanobacteria) that fix nitrogen, adding to the fertility of the habitat. Ferns reproduce by spores instead of flowers and seeds. Two species of *Azolla* are reported for Texas, *A. mexicana* and *A. caroliniana*. They can be distinguished only with a compound microscope.

Carolina Water Fern *Azolla caroliniana*

Big Foot Water Clover
Family: Marsileaceae
Scientific Name: *Marsilea macropoda*
Habit: Perennial ferns with creeping rhizomes.
Leaves: Compound, 4-foliate; leaflets up to 2" long.
Flowers: None.
Fruit: None. These ferns reproduce from spores produced in a sporocarp which is held above the ground.
Bloom Period: Flowers absent. Leaves are produced throughout the year, whenever there is ample moisture.
Distribution: Cameron and Hidalgo counties.
Comments: These ferns grow in ditches and wetlands. The leaves can be floating on the water surface or erect on moist soil. These

Big Foot Water Clover *Marsilea macropoda*

Pteridaceae

are different from the more familiar ferns. The compound leaf resembles a 4-leaf clover. The sporocarp has a thick coat which has to wear away in nature. If a mature sporocarp is notched and placed in water, it breaks open and a gelatinous-appearing material unrolls in the water. Embedded in the gelatinous material are larger female sporangia and smaller male sporangia. In the more familiar ferns, the sporangia are in clusters called sori, located on the lower surfaces of the leaves.

Bulb Lip Fern *Astrolepis sinuata*

Bulb Lip Fern
Family: Pteridaceae
Scientific Name: *Astrolepis sinuata* [*Notholaena sinuata*]
Habit: Xerophytic ferns with erect once-compound leaves 4½–50" tall. The leaves curl up during drought and unfurl after rains.
Leaves: Erect, once compound, up to 50" tall, usually less; upper surfaces of leaflets sparsely scaly or without scales; lower surfaces covered with silvery branched scales.
Flowers: None.
Fruit: None. Spores are produced on the lower surfaces of the leaflets.
Bloom Period: None.
Distribution: Starr County.
Comments: This fern was seen growing between sandstone rocks overlooking the Rio Grande. It has not previously been reported for our area. The species has been reported for Webb County, but it is more common in West Texas and New Mexico. It also occurs in Mexico and Central America.

Alabama Lip Fern *Cheilanthes alabamensis*

Alabama Lip Fern
Family: Pteridaceae
Scientific Name: *Cheilanthes alabamensis*
Habit: Ferns up to 20" tall, with creeping rhizomes.
Leaves: Blades once or twice compound, up to 14" long.
Flowers: None.
Fruit: None. Sporangia produced in clusters (sori) on the under surfaces.
Bloom Period: None.
Distribution: Cameron and Starr counties, rarely encountered.
Comments: Alabama lip ferns grow on rocky or sandy clay hillsides. They dry up during drought but quickly recover after rains. Ferns are non-flowering plants that reproduce by spores instead of seeds. The sporangia produce spores, and there is a mechanism that hurls the spores up to several feet from the parent plant. The spores germinate to produce a tiny green plant (gametophyte) that produces egg and sperm. Fertilization produces the fern plant (sporophyte), which grows up through the gametophyte plant to form another fern. The gametophyte soon dies.

Gymnosperms

Woody Plants Producing Seeds but without Flowers

Erect Joint Fir, Mormon Tea
Family: Ephedraceae
Scientific Name: *Ephedra antisiphylitica*
Habit: Erect shrubs up to 40" tall, appearing to be leafless.
Leaves: Scalelike, in pairs, about 1/8" long.
Flowers: None. Separate male and female cones are produced on different plants.
Fruit: Seeds are produced in the female cones, which are about 3/8" or slightly longer.
Bloom Period: Spring.
Distribution: Cameron, Hidalgo, and Starr counties.
Comments: This species was once thought to cure syphilis. The alkaloid ephedrine is extracted from another species of *Ephedra*. It or a synthetic form is used in some medications. *Ephedra* belongs to a non-flowering group of plants called gymnosperms. The stems are browsed by deer and rabbits.

Erect Joint Fir, Mormon Tea *Ephedra antisiphylitica*

Vine Joint Fir
Family: Ephedraceae
Scientific Name: *Ephedra pedunculata*
Habit: A trailing or vinelike shrub, climbing onto any convenient support and reaching a length of 22'. The stems appear to be leafless.
Leaves: Scalelike, in pairs, about 1/8" long.
Flowers: None. Separate male and female cones are produced on different plants.
Fruit: Seeds are produced in the female cones, which are about 3/8" or slightly longer. They turn red when mature.
Bloom Period: Midwinter to early spring.
Distribution: Hidalgo and Starr counties.
Comments: This species is easily distinguished from *E. antisiphylitica* by its vinelike habit. It was previously reported to grow only in counties farther west. We have found several populations in Hidalgo and Starr Counties. *Ephedra* belongs to a non-flowering group of plants called gymnosperms.

Vine Joint Fir *Ephedra pedunculata* male cones

Vine Joint Fir *E. pedunculata* female plant with red cones

Taxodiaceae

Montezuma Bald Cypress, Sabino

Family: Taxodiaceae

Scientific Name: *Taxodium distichum* var. *mexicanum* [*T. mucronatum*]

Habit: Deciduous trees up to 90' or taller.

Leaves: Needlelike, up to ½" long.

Flowers: None. This is a gymnosperm, a non-flowering seed plant. The reproductive parts are in tiny cones of separate sexes on the same tree.

Fruit: None. After fertilization, the globose female cones grow to about 1" in diameter and become woody. When the seeds are mature, the cones fall apart and release the seeds.

Bloom Period: Pollen is produced in the spring.

Distribution: Cameron, Hidalgo, and Starr counties.

Comments: Our cypresses are not as abundant as they once were. Since their water requirements are high, they grow on the banks of the Rio Grande and resacas in the area. Often they are planted on the banks of resacas, where they do well and become beautiful trees. The seeds are produced in one growing season, unlike the seeds of their relatives the pines, which may take two or more years to develop. The trees do not produce the "knees" typical of the cypresses often pictured in Louisiana and East Texas swamps. Until recently this species was known as *Taxodium mucronatum*. Cypress wood is prized for its resistance to insects and decay. It has been used to build water tanks. The most famous Montezuma cypress, called the "Tule Tree" in Oaxaca, Mexico, is one of the largest trees in the world. The tree pictured, growing in Abram, Hidalgo County, is thought to be 900 years old. The Spanish common name also means "soap."

Montezuma Bald Cypress, Sabino *Taxodium distichum* var. *mexicanum*

Montezuma Bald Cypress, Sabino *T. distichum* var. *mexicanum* branch with rounded cones

Monocots (Monocotyledoneae)

Flowering Seed Plants with One "Seed Leaf"

(Usually the flower parts are in threes and the leaf veins are all parallel)

Century Plant

Family: Agavaceae (Amaryllidaceae)
Scientific Name: *Agave americana*
Habit: Leaves large and succulent, crowded in a tight spiral; stems not visible.
Leaves: Stiff, gray or whitish, waxy, up to 5½' long, crowded in a tight spiral, hiding the stems; margins and leaf tips with sharp spines.
Flowers: Greenish, on inflorescences over 20' tall.
Fruit: Dehiscent, Splitting into 3 parts.
Bloom Period: Summer. The inflorescence appears after the plant is several years old, and then the plant dies.
Distribution: Cameron and Starr Counties.
Comments: Century plant is a popular ornamental. Contrary to common folklore, agaves do not live 100 years. The plants live several years before flowering, and then they die. They reproduce by seeds, bulblets, and young plants from the roots and base. This species is considered introduced but the senior author noted it as abundant near the coast more than 60 years ago. Possibly it was introduced by early Native Americans. Agaves have been important to the survival of ancient American cultures as a source of food, fiber, and drink. The leaves of this species are smooth, whereas those of *A. scabra* are rough. Different species of *Agave* have been used for fibers (leaves), to produce various alcoholic drinks (the stem or apical growth), and as food (the stem with leaves removed). Needle and thread are produced from some agaves by snapping the spine backward and peeling the fiber away with it.

Century Plant *Agave americana*

Century Plant *A. americana* inflorescence, from *Wildflowers and Other Plants of Texas Beaches and Islands* (Richardson 2002 Courtesy University of Texas Press)

Agavaceae

Thorn Crested Agave
Family: Agavaceae (Amaryllidaceae)
Scientific Name: *Agave lophantha*
Habit: Leaves green, in a tight spiral, hiding the stems.
Leaves: Green with a pale central stripe, about 16" long, in a tight spiral, hiding the stems; margins and tips with sharp spines.
Flowers: Greenish, on inflorescences 10–16' tall.
Fruit: Capsules splitting into 3 parts; seeds black.
Bloom Period: Summer.
Distribution: Starr County.
Comments: In Texas, this species is reported only from Starr and Kenedy counties. It is more widespread in Mexico. It reproduces prolifically by offsetting and forming large clumps or colonies.

Thorn Crested Agave *Agave lophantha*

Rough Leaved Agave
Family: Agavaceae (Amaryllidaceae)
Scientific Name: *Agave scabra*
Habit: Leaves large and succulent, crowded in a tight spiral; stems not visible.
Leaves: Gray or whitish, waxy, 4' or longer, crowded in a tight spiral, hiding the stems; margins and leaf tips with sharp spines; upper and lower surfaces rough to the touch.
Flowers: Yellow, on inflorescences 13–20' tall.
Fruit: Capsules splitting into 2 parts.
Bloom Period: Spring, summer. The inflorescence appears after the plant is several years old, and then the plant dies.
Distribution: Starr County.
Comments: Although not common in Texas, this plant is occasionally seen in Starr County growing on hilltops. Large patches of this plant in Starr County are associated with historic Indian camps. The leaves are rough, whereas those of *A. americana* are smooth. The specific epithet is derived from the Latin *scabrum* (rough), referring to the rough leaves. The agaves have been important to the survival of ancient cultures as a source of food, fiber, and drink. The sap of this and other Agaves is a skin irritant.

Rough Leaved Agave *Agave scabra*

Rough Leaved Agave *A. scabra* spent inflorescence

Runyon's Huaco

Family: Agavaceae (Amaryllidaceae)

Scientific Name: *Manfreda longiflora* [*Polianthes runyonii*]

Habit: Leaves long and narrow, succulent, resting on the ground in basal rosettes. Stems underground, swollen.

Leaves: Often with purple dashes, succulent, 8" or longer, resting on the ground in basal rosettes.

Flowers: Greenish, turning red with age, on inflorescences up to 20" tall; the stamens remain inside the floral tube.

Fruit: Capsules leathery, splitting into 3 parts. Seeds flattened.

Bloom Period: Summer.

Distribution: Hidalgo and Starr counties.

Comments: This species is only occasionally seen. It is distinguished from the other Manfredas by the stamens, which remain inside the corolla tube. Its bloom spike is shorter than those of most other Manfredas.

Runyon's Huaco *Manfreda longiflora*

Spotted Tuberose

Family: Agavaceae (Amaryllidaceae)

Scientific Name: *Manfreda maculosa* [*Polianthes maculosa*]

Habit: Leaves resting on the ground in basal rosettes. Stems underground, swollen.

Leaves: Green or brown-spotted, succulent, 8" or longer.

Flowers: Purplish or greenish, on inflorescences 24" or taller. The stamens extend outside the corolla tube ¾" or less.

Fruit: Capsules leathery, splitting into 3 parts. Seeds flattened.

Bloom Period: Spring, summer.

Distribution: Starr County.

Comments: This species is distinguished from *A. variegata* and *M. sileri* by the stamens, which extend outside the corolla tube ¾" or less. It is a host plant for the Manfreda Giant Skipper butterfly. The specific epithet is a Latin word meaning "spotted," referring to the leaf pattern.

Spotted Tuberose *Manfreda maculosa*

Spotted Tuberose *M. maculosa* inflorescence

Agavaceae

Siler's Tuberose

Family: Agavaceae (Amaryllidaceae)
Scientific Name: *Manfreda sileri*
Habit: Leaves resting on the ground in basal rosettes. Stems underground, swollen.
Leaves: Succulent, brown-spotted, up to 16" long.
Flowers: Inflorescences 6–8' or taller; flowers green, the stamens exserted 1½–2½."
Fruit: Capsules leathery, splitting into 3 parts. The seeds are flattened.
Bloom Period: Spring, summer.
Distribution: Cameron, Hidalgo, and Starr counties.
Comments: This species is recognized by its very tall inflorescences and by the anther filaments pointing mostly upward.

Siler's Tuberose *Manfreda sileri*

Siler's Tuberose
M. sileri inflorescence

Mottled Tuberose, Liverspot Lily

Family: Agavaceae (Amaryllidaceae)
Scientific Name: *Manfreda variegata*
Habit: Leaves resting on the ground in basal rosettes. Stems underground, swollen.
Leaves: Succulent, green or brown-spotted, up to 16" long.
Flowers: Green to brownish, on inflorescences up to 4' tall; stamens extend 1⅛–2½" outside the corolla tube, pointing outward.
Fruit: Capsules leathery, splitting into 3 parts. Seeds are black, flattened.
Bloom Period: Spring, summer, fall.
Distribution: Cameron and Hidalgo counties.
Comments: This species is distinguished from *M. maculosa* and *M. sileri* by the stamens, which extend outside the corolla tube 1 ⅛" or more, pointing outward (not upward). The specific epithet is a Latin word meaning irregularly colored, referring to the leaves.

Mottled Tuberose, Liverspot Lily *Manfreda variegata*

Mottled Tuberose, Liverspot Lily *M. variegata* inflorescence

Buckley's Yucca

Family: Agavaceae (Liliaceae)

Scientific Name: *Yucca constricta*

Habit: Stems absent or short and falling over. Leaves mostly linear, crowded in spirals around the stems, hiding the upper stems.

Leaves: Flexible, sharp-pointed, mostly linear, in dense spirals hiding the stems. Stringy white threads hang from the margins.

Flowers: Whitish, in dense clusters from the plant apices on long inflorescences up to 6' or taller.

Fruit: Capsules up to 2 ½" long, constricted about the middle.

Bloom Period: Spring, summer.

Distribution: Cameron, Hidalgo, and Starr counties.

Comments: This species is common in sandy soils and also occurs on sandstone in Starr County. The flexible leaves distinguish this species from *Y. treculeana,* which has rigid leaves.

Buckley's Yucca
Yucca constricta

Buckley's Yucca *Y. constricta* constricted fruit

Spanish Dagger, Palma Pita

Family: Agavaceae (Liliaceae)

Scientific Name: *Yucca treculeana*

Habit: Woody, branched and treelike, up to 25' tall. Leaves crowded in spirals, hiding the upper stems.

Leaves: Stiff and sharp pointed, U-shaped in cross section, up to 40" long, crowded in spirals.

Flowers: White, in dense clusters from the stem apices on inflorescences reaching well above the tallest leaves.

Fruit: Capsules up to 4" long, splitting into 3 parts.

Bloom Period: Spring.

Distribution: Cameron, Hidalgo, Willacy, and Starr counties.

Comments: Pollination of yuccas depends on an unusual relationship with a small moth. After she is fertilized, the female moth collects pollen from one flower and goes to another flower, where she lays her eggs on the flower's ovary and then rubs the pollen on the stigma of the same flower. The young larvae feed on the developing seeds, but there are plenty left over to produce mature seeds. In effect, the Spanish dagger pays a fee of a few seeds in order to be pollinated. Rhinoceros beetles feed inside the roots and stems, causing them

Spanish Dagger, Palma Pita *Yucca treculeana*

Spanish Dagger, Palma Pita
Y. treculeana
detached flower

Alismataceae

to collapse. The flowers are harvested and eaten as a delicacy, especially in Mexico. The tuberous roots are sold in markets in Mexico as a soap. Yuccas are also popular landscape plants. The rigid leaves of this species separate it from *Y. constricta,* which has flexible leaves.

Beaked Burhead

Family: Alismataceae
Scientific Name: *Echinodorus berteroi* [*E. rostratus*]
Habit: Growing near water or in recently dried places. Stems hidden by the leaves.
Leaves: Basal, about 8" tall, the blades ovate or almost rounded, up to 8" long. Submerged leave are linear with wavy margins. Floating leaves have very long petioles, with blades similar to those of the emergent leaves.
Flowers: Inflorescences up to 24" or taller. Flower petals 3, white, about ⅜" long.
Fruit: Achenes, crowded in roundish, burlike heads.
Bloom Period: Spring, summer.
Distribution: Cameron, Hidalgo, and Willacy counties.
Comments: Beaked burhead grows in or near water or in recently dried places. The ovate or almost rounded leaves separate this species from related species. The genus name is derived from the Greek, meaning spiny achenes. A dwarf-growing burhead, once called *Echinodorus rostratus* var. *lanceolatus,* is now treated as a synonym of *E. berteroi*.

Beaked Burhead *Echinodorus berteroi*

Beaked Burhead *E. berteroi* floating leaves of submerged plants

Beaked Burhead, a dwarf form of *E. berteroi* once called *Echinodorus rostratus* var. *lanceolatus*

Creeping Burhead

Family: Alismataceae
Scientific Name: *Echinodorus cordifolius*
Habit: Plants of wet or recently dried places. Leaves erect. Inflorescences running along the ground, sometimes producing small plants as well as flowers.
Leaves: Erect, the petiole about 8" long; blades heart shaped, about 8" long. Submerged leaves were not observed.
Flowers: Petals 3, white, about ¼" long, on inflorescences that grow downward and run along the ground.
Fruit: Achenes, crowded in roundish, burlike heads.
Bloom Period: Spring, summer.
Distribution: Cameron and Willacy counties.
Comments: These are plants of wet or recently dried places. Their roots are important food for Sandhill Cranes. The heart-shaped leaves and the inflorescences growing along the ground separate this species from related species. The specific epithet is derived from the Latin, meaning "having heart-shaped leaves."

Creeping Burhead *Echinodorus cordifolius*

Flecha De Agua, Arrowhead

Family: Alismataceae
Scientific Name: *Sagittaria longiloba*
Habit: Plants of wet places, up to 39" tall, usually less. Leaves more or less erect.
Leaves: Erect or leaning, arrowhead shaped, up to 10" long.
Flowers: On branching inflorescences; petals 3, white or rarely pink, up to ⅝" long.
Fruit: Achenes crowded on a roundish receptacle.
Bloom Period: Spring, summer, fall.
Distribution: Cameron, Hidalgo, and Willacy counties.
Comments: The arrowhead-shaped leaves separate this species from related species. The tubers are eaten by Sandhill Cranes. The genus name comes from Latin and translates as "arrowhead." The Spanish common name translates as "water arrow."

Flecha De Agua, Arrowhead *Sagittaria longiloba*

Alliaceae

Delta Arrowhead

Family: Alismataceae

Scientific Name: *Sagittaria platyphylla*

Habit: Aquatic perennials up to 3' tall, usually less, rooting in the mud, reproducing by tubers from long rhizomes, often forming large colonies.

Leaves: Some simply bladeless petioles (underwater), intergrading through narrow blades to ovate blades (floating or above water) which are up to 15 ⅞" long; petioles long, triangular.

Flowers: Unisexual, usually both sexes on the same plant; petals 3, white, about ⅜" long.

Fruit: Achenes crowded on a roundish receptacle about ½" tall.

Bloom Period: Spring, summer, and fall.

Distribution: Cameron and Starr counties, probably also Hidalgo County.

Comments: This versatile species can flourish in fresh water and also in brackish water. The tubers are said to be edible but must be cooked to eliminate an acrid taste.

Delta Arrowhead *Sagittaria platyphylla*

Delta Arrowhead *S. platyphylla* submerged leaves

Prairie Onion

Family: Alliaceae (Liliaceae)

Scientific Name: *Allium drummondii*

Habit: Leaves and inflorescence arising from a bulb.

Leaves: Usually 3 or more, more or less linear, up to 18" tall, having an odor of onion when crushed.

Flowers: On tall stalks up to 18" tall, with 10–25 flowers; flowers in umbels (pedicels all originating from the same point); pedicels of different lengths. Sepals and petals look alike; white, up to ⅜" long.

Fruit: 3-lobed, the seeds black.

Bloom Period: Spring.

Distribution: Willacy County

Comments: *Allium* species smell of onion or garlic when bruised. The inflorescence bracts of *A. drummondii* have one vein, whereas those of *A. runyonii* have 3 or more veins. Although this species and other *Alliums* are said to be edible, great care must be taken to identify them properly. Several quite similar species are toxic. *A. drummondii* was named in honor of Thomas Drummond, from Scotland, who made extensive collections of plants in the Edwards Plateau area of Texas in the 1830s.

Prairie Onion *Allium drummondii*

Runyon's Onion

Family: Alliaceae (Liliaceae)

Scientific Name: *Allium runyonii*

Habit: Leaves and inflorescence arising from a bulb.

Leaves: 3 or more, more or less linear, up to 18" tall, having odor of onion or garlic when crushed.

Flowers: On stalks up to 18" tall, with 10–25 flowers; flowers in umbels (pedicels all originating from the same point). Sepals and petals look alike, white with midrib pink to red or purple, about ¼" long.

Fruit: 3-lobed, the seeds black.

Bloom Period: Spring.

Distribution: Willacy and Starr counties.

Comments: This species was named for Robert Runyon, a Brownsville photographer, mayor, and self-taught botanist. It is endemic to Texas. The inflorescence bracts of this species have 3 or more veins, whereas those of *A. drummondii* have only one vein. *Allium* species smell of onion or garlic when bruised. Although *Allium* species are said to be edible, great care must be taken to identify them properly. Several similar species are considered toxic, and they sometimes grow together.

Runyon's Onion *Allium runyonii*

Crow Poison

Family: Alliaceae (Liliaceae)

Scientific Name: *Nothoscordum bivalve*

Habit: Leaves and inflorescence arising from a bulb.

Leaves: Linear, arising from the bulb, up to 20" or taller. Do not have odor of onion or garlic when crushed.

Flowers: On stalks up to 20" or taller, with 6–12 flowers; flowers in umbels (pedicels all originating from the same point). Sepals and petals white, looking alike.

Fruit: 3-lobed, the seeds black.

Bloom Period: All seasons. Frequent after rains.

Distribution: Cameron, Hidalgo, and Willacy counties.

Comments: This species is said to be poisonous. The leaves do not smell of onion or garlic when bruised, whereas *Allium* species do.

Crow Poison *Nothoscordum bivalve*

Amaryllidaceae

Cebolleta, Rain Lily

Family: Amaryllidaceae
Scientific Name: *Cooperia drummondii*
Habit: Plants from bulbs.
Leaves: Linear, up to 12" long.
Flowers: Often appearing after rains, on stalks up to 13" tall. Sepals and petals white, in threes, looking alike, grown together to form a tube 2 ⅜–4 ¾" long; lobes up to ¾" long.
Fruit: A capsule, splitting into 3 parts.
Bloom Period: Spring, summer, fall.
Distribution: Cameron, Hidalgo, Willacy, and Starr counties.
Comments: This is our most abundant rain lily (often flowering after rains). It is endemic to Texas. *C. drummondii* was named in honor of Thomas Drummond, from Scotland, who made extensive collections of Central Texas plants in the 1830s.

Cebolleta, Rain Lily *Cooperia drummondii*

Golden Zephyr Lily, Yellow Rain Lily

Family: Amaryllidaceae
Scientific Name: *Zephyranthes pulchella*
Habit: Plants from bulbs.
Leaves: Linear, up to 8" long.
Flowers: Often appearing after heavy rains, on stalks originating from bulbs; sepals and petals yellow, in threes, grown together to form a tube up to 2" long, lobes up to ⅝" long.
Fruit: A capsule, splitting into 3 parts.
Bloom Period: Summer, fall.
Distribution: Cameron and Hidalgo counties.
Comments: This plant is endemic to Texas. It flowers after rains, sometimes in huge colonies where water stands for some time. *Cooperia smallii,* reported for our area, is now thought to be a hybrid between *Zephyranthes pulchella* and *Cooperia drummondii.*

Golden Zephyr Lily, Yellow Rain Lily *Zephyranthes pulchella*

Lily of the Lomas, Lila De Los Llanos
Family: Anthericaceae (Liliaceae)
Scientific Name: *Echeandia chandleri* [*Anthericum chandleri*]
Habit: Herbs from fleshy roots. Stems with 1 or 2 bracts or none.
Leaves: Basal, linear, up to 18" long.
Flowers: On branching inflorescences up to 39" tall. Sepals and petals yellow, in threes, up to ½" or ¾" long, looking alike. Anthers separate, up to ⅛" long.
Fruit: A capsule splitting into 3 parts.
Bloom Period: Spring, summer, fall.
Distribution: Cameron County.
Comments: This species is typically found growing in the lomas (clay hills) near the coast. It is inconspicuous when not in bloom because of its grasslike appearance. *E. chandleri* has separate anthers, distinguishing it from *E. texensis,* in which they are joined. Lila de los llanos is an attractive native ornamental. The common name Lila de los llanos is a misnomer, as the plant is found on the lomas rather than on the plains (*llanos*).

Lily of the Lomas, Lila De Los Llanos *Echeandia chandleri*

Lily of the Lomas, Lila De Los Llanos *E. chandleri* inflorescence

Green Island Echeandia
Family: Anthericaceae (Liliaceae)
Scientific Name: *Echeandia texensis*
Habit: Herbs from fleshy roots.
Leaves: Basal leaves linear, up to 24" long; stem leaves shorter.
Flowers: On branching inflorescences up to 39" tall. Sepals and petals yellow, in threes, up to ½" or ¾" long, looking alike. Anthers joined at the tips, forming a cone, up to ¼" long.
Fruit: A capsule splitting into 3 parts.
Bloom Period: Spring, summer, fall.
Distribution: Cameron County.
Comments: This species is somewhat rare. It was first collected from Green Island in the Laguna Madre; it is now known to occur on the clay lomas with *E. chandleri.* It occurs in the LRGV and in Mexico. It is very similar to *E. chandleri,* which has separate anthers. The anthers of *E. texensis* are joined at the tips, forming a cone, easily distinguishing the two species. This is the most commonly cultivated *Echeandia.*

Green Island Echeandia *Echeandia texensis*

Arecaceae

Water Lettuce
Family: Araceae
Scientific Name: *Pistia stratiotes*
Habit: Floating herbaceous plants somewhat resembling an open head of lettuce.
Leaves: Clustered together to form a rounded "boat."
Flowers: Compared with the rest of the family, the inflorescence is highly reduced. It consists of one small female flower with a few male flowers above it. Sepals and petals are absent. These are surrounded by a modified leaf called a spathe, about ⅝" long.
Fruit: Very tiny.
Bloom Period: Spring.
Distribution: Cameron and Hidalgo counties.
Comments: This introduced aquatic plant is widespread in the tropics of Africa, Asia, and the Americas. It is found all along the Gulf Coast in fresh water. The plants are extremely invasive, blocking waterways. Although the plants resemble lettuce, they contain toxins and are not edible. In the photo, the inflorescence is located in the central part of the plant. It resembles a microscopic calla lily. More familiar members of this family are the cultivated philodendrons, elephant ears, and calla lilies. The white, attractive part of the flower of the calla lily is the spathe, which wraps around the rod-shaped inflorescence.

Water Lettuce *Pistia stratiotes*

Rio Grande Palmetto, Texas Sabal Palm *Sabal mexicana*

Rio Grande Palmetto, Texas Sabal Palm
Family: Arecaceae (Palmae)
Scientific Name: *Sabal mexicana* [*Sabal texana*]
Habit: Trees up to 50' tall (see p. viii).
Leaves: Petioles 40" or longer, not spiny, but the edges are very sharp; blades are palmate and pleated, up to 40" broad. The midvein curves back.
Flowers: Cream colored, tiny, and very fragrant. The inflorescences emerge from the leaf axils and are shorter than the leaves.
Fruit: Soft, black, about ¾" broad.
Bloom Period: Winter, spring, summer.
Distribution: Cameron and Hidalgo counties.
Comments: This palm was once thought to be limited to Cameron County in Texas, but populations have been found as far north as Brazoria County. In the past it was more widespread in Texas. Now it is best known in

Rio Grande Palmetto, Texas Sabal Palm *S. mexicana* fruit

the Sabal Palm Audubon Sanctuary, Brownsville. It is more abundant in Mexico. It is distinguished from the abundant introduced *Washingtonia* palms by the recurved midvein of the frond and its leaf petioles, which do not have spines.

Guapilla

Family: Bromeliaceae
Scientific Name: *Hechtia glomerata*
Habit: Spiny plants, almost stemless.
Leaves: In a tight spiral; blades more or less lanceolate, the margins purplish, especially where the spines emerge.
Flowers: Inflorescences up to 6' tall, arising from the center of the plant; flowers unisexual, male and female on separate plants; petals 3, scaly, up to ¼" long.
Fruit: Capsules up to ⅜" long.
Bloom Period: Spring, summer.
Distribution: Starr County.
Comments: This is a plant to be admired but not touched. It is beautiful but full of spines. It is our only terrestrial bromeliad, and can be confused with *Agave*. The clumps grow outward, forming a circular pattern.

Guapilla *Hechtia glomerata*

Guapilla *H. glomerata* flowers

Bailey's Ball Moss

Family: Bromeliaceae
Scientific Name: *Tillandsia baileyi*
Habit: Epiphytes growing in clumps.
Leaves: Gray, in a spiral, curled, up to 12" long, covered with water-absorbing scales.
Flowers: Subtended by pink or red bracts; petals 3, bluish, up to 1 ⅛" long.
Fruit: Capsules up to 1½" long, splitting into 3 parts to release hairy seeds.
Bloom Period: Spring.
Distribution: Cameron, Hidalgo, and Willacy counties.
Comments: This species is sometimes mistaken for an orchid, since it grows on trees, as do the better known orchids. In spite of the common name, *T. baileyi* is not a true moss (mosses do not produce flowers). It is larger and more robust than *T. recurvata,* and the flowers are showier. They are pollinated by hummingbirds. Bailey's ball moss has an affinity for Texas ebony trees, *Chloroleucon ebano.* This ball moss species is threatened by fox squirrels, which were introduced into our area. They eat the hearts of the plants, discarding the rest.

Bailey's Ball Moss *Tillandsia baileyi*

Ball Moss

Family: Bromeliaceae
Scientific Name: *Tillandsia recurvata*
Habit: Epiphytes growing in tight clumps resembling balls.

Ball Moss *Tillandsia recurvata*

Leaves: Gray, curled, up to 4" long.
Flowers: Subtended by greenish bracts; petals 3, bluish, about ⅜" long.
Fruit: Capsules up to 1 ⅛" long, splitting into 3 parts to release hairy seeds.
Bloom Period: Spring.
Distribution: Cameron and Hidalgo counties.
Comments: The growth habit gives this species its common name, although it is not a true moss. Although they grow on trees, these plants are not parasites. They use photosynthesis like other plants and can also grow on telephone wires.

Spanish Moss *Tillandsia usneoides*

Spanish Moss
Family: Bromeliaceae
Scientific Name: *Tillandsia usneoides*
Habit: Epiphytic plants loosely draped on tree branches; stems scaly and gray.
Leaves: Gray and scaly, up to 1 ⅛" long.
Flowers: Subtended by greenish bracts; petals yellow-green, up to ⅜" long.
Fruit: Capsules 1" long, splitting into 3 parts and releasing hairy seeds.
Bloom Period: Spring.
Distribution: Cameron and Hidalgo counties.
Comments: Common names can sometimes be confusing. This plant is not a moss (a non-flowering plant) but is a flowering plant. The flowers are rarely noticed. Spanish moss was used extensively for stuffing cushions until synthetic materials became available. A word of warning: various pests can be hiding in the Spanish moss. Many years ago, some relatives gathered bundles of it and used it to decorate their home during the Christmas season. Soon the walls were covered with ticks.

Spanish Moss *T. usneoides* flowers

Southern Coastal Roseling
Family: Commelinaceae
Scientific Name: *Callisia micrantha* [*Tradescantia micrantha*]
Habit: Perennial fleshy herbs with creeping brittle stems.
Leaves: Alternate; blades lanceolate or elliptic, up to 1 ⅜" long.
Flowers: Radial; petals 3, rose or pink, about ¼" long.
Fruit: Capsules about 1/16" tall.
Bloom Period: Spring, summer, fall.
Distribution: Cameron, Hidalgo, and Starr counties.

Southern Coastal Roseling *Callisia micrantha* flowers

Comments: This species makes an attractive potted plant. It is not recommended for the garden, because it is invasive and very difficult to remove. Wild populations disappear in dry weather, surviving by jointed, underground rhizomes. *Manual of the Vascular Plants of Texas* (Correll and Johnston 1970) lists this species as endemic to Texas, but we suspect that it is also present in Mexico. The Commelinaceae family is commonly referred to as the snotweed family because of the mucilaginous sap.

Spreading Dayflower

Family: Commelinaceae

Scientific Name: *Commelina diffusa*

Habit: Succulent annuals with erect stems when young, creeping when older.

Leaves: Alternate; blades broadly lanceolate, up to 4 ¼" long.

Flowers: Bilateral, emerging from a folded bract, the edges not grown together; petals 3, blue, with 2 larger up to 5/16" broad and one smaller pale one. Flowers open in early morning and usually wilt by noon.

Fruit: Capsules about 1/16" tall.

Bloom Period: Spring, summer, fall.

Distribution: Cameron, Hidalgo, and Starr counties.

Comments: This plant has a preference for moist soils. It is distinguished from *C. erecta* by the flower bracts, which are not grown together. The genus *Commelina* was described and named by C. Linnaeus. The two showy flower petals and the one insignificant petal reminded him of the three Commelin brothers: Two of them were respected botanists, but the third did not accomplish anything in botany. The common name is appropriate, since the flowers last less than a day.

Spreading Dayflower *Commelina diffusa*

Widow's Tears

Family: Commelinaceae

Scientific Name: *Commelina erecta* [*C. elegans*]

Habit: Succulent perennials with erect or leaning stems, sometimes falling over.

Leaves: Alternate; ovate to elliptic, up to 3 1/8" long.

Flowers: Bilateral, emerging from a folded bract, the edges grown together at the base; petals 3, blue, with 2 larger up to 7/8" broad and one smaller pale one. Flowers open in early morning and usually wilt by noon.

Fruit: Capsules about 1/16" tall.

Bloom Period: All seasons.

Distribution: Cameron, Hidalgo, Willacy, and Starr counties.

Comments: Plant classification is not totally objective. Two camps that may disagree are often referred to as the splitters and the lumpers. Some recognize only one species, *C. erecta*, with some variation; others recognize *C. elegans* as a separate species. The floral

Widow's Tears *Commelina erecta*

Widow's Tears *C. erecta*, sandland form

Cyperaceae

bracts of *C. erecta,* which are partially grown together, distinguish it from *C. diffusa.*
C. erecta is the species most commonly found growing in local flower beds. It is almost impossible to eliminate because of the tenacious fleshy roots.

Stemless Spiderwort
Family: Commelinaceae
Scientific Name: *Tradescantia subacaulis*
Habit: Succulent perennials with very short stems.
Leaves: Alternate, narrowly lanceolate, up to 7 ¼" long.
Flowers: Radial; petals 3, blue, up to 9/16" long.
Fruit: Capsules up to ¼" tall.
Bloom Period: Spring, early summer.
Distribution: Hidalgo and Willacy counties.
Comments: This species is endemic to the sandy soils of Texas. The specific epithet is from Latin, meaning "without much of a stem."– the common name reflects this characteristic. The tradescantias have 3 noticeable petals, while the commelinas have 2 noticeable ones plus one small, almost colorless one.

Stemless Spiderwort *Tradescantia subacaulis*

Cyperaceae: Sedge Family

The sedges somewhat resemble the grasses. They tend to grow in wet or moist places. The Lower Rio Grande Valley has numerous species in this family, and many of them are quite similar in general appearance. Often photographs of different species look the same. Because of this, we have included a relatively small number of sedge species.

Alkali Bulrush
Family: Cyperaceae
Scientific Name: *Bolboschoenus maritimus* [*Scirpus maritimus*]
Habit: Perennials up to 5' tall with 3-cornered stems.
Leaves: Blades up to ⅜" broad.
Flowers: Tiny, in several groups of spikelets up to 1" long and ⅜" broad
Fruit: Achenes flattened, 2-edged, up to 3/16" long.
Bloom Period: Spring, summer, fall.

Alkali Bulrush *Bolboschoenus maritimus* var. *macrostachyos*

Distribution: Cameron, Hidalgo, and Willacy counties.
Comments: This plant is usually found in wet places near the coast.

Jointed Flat Sedge

Family: Cyperaceae
Scientific Name: *Cyperus articulatus*
Habit: Perennials up to 56" tall; stems round in cross section, with partitions as in bamboo.
Leaves: Few, reduced to sheaths without extended blades.
Flowers: Tiny, in crowded groups of spikelets.
Fruit: Achenes 3-cornered, about 1/16" tall.
Bloom Period: Spring, summer, fall.
Distribution: Cameron, Hidalgo, and Willacy counties.
Comments: This species is abundant around freshwater wetlands.

Jointed Flat Sedge *Cyperus articulatus*

Finger Flat Sedge

Family: Cyperaceae
Scientific Name: *Cyperus digitatus*
Habit: Erect perennials 20–60" tall; stems 3-cornered
Leaves: W- or V-shaped, 16–40" long; undersurfaces whitish.
Flowers: Tiny, in crowded groups of spikelets; inflorescences subtended by 8–12 bracts.
Fruit: Achenes ellipsoid, less than 1/16" tall.
Bloom Period: Summer, winter.
Distribution: Cameron and Hidalgo counties.
Comments: Like most sedges, this species prefers moist or wet places. Major characteristics are the V- or W-shaped leaves with whitish undersurfaces.

Finger Flat Sedge *Cyperus digitatus* subsp. *digitatus*

Cyperaceae 33

Umbrella Plant

Family: Cyperaceae
Scientific Name: *Cyperus involucratus* [*C. alternifolius*]
Habit: Perennials up to 5' tall; stems 3-cornered.
Leaves: Blades on stems absent. Bracts subtending inflorescences up to 11" long.
Flowers: Tiny, in crowded groups of spikelets.
Fruit: Achenes brown, ellipsoid, less than $1/16$" tall.
Bloom Period: Summer, fall.
Distribution: Hidalgo and Willacy counties.
Comments: This species is usually found growing in wet soils. It is an introduced ornamental. When the flowering stems fall over into water, new plantlets grow from them.

Umbrella Plant *Cyperus involucratus*

Rice Field Flat Sedge

Family: Cyperaceae
Scientific Name: *Cyperus iria*
Habit: Erect annuals up to 24" tall.
Leaves: Crowded near the base, smaller up the stem.
Flowers: Tiny, in crowded groups of spikelets. Groups of spikelets in umbrella-like arrangement.
Fruit: Achenes brown, 3-cornered, about $1/16$" tall.
Bloom Period: Summer, fall.
Distribution: Cameron County
Comments: This species is widespread in southeast Asia, Africa, and Australia. It was introduced to the United States and is found in scattered localities. In Texas, it is found mostly in counties near the Gulf Coast. It has not previously been reported for our area. Because of its short stature, it would be attractive in a small water garden.

Rice Field Flat Sedge *Cyperus iria*

Large-Head Flat Sedge

Family: Cyperaceae
Scientific Name: *Cyperus macrocephalus* var. *eggarsii*
Habit: Erect perennials up to 36" tall.
Leaves: from the base, about 24" tall; bracts in clusters around the inflorescences, about 7" long.
Flowers: Tiny, in globose clusters.
Fruit: Achenes brownish, more or less 3-cornered.
Bloom Period: Spring, summer, fall.
Distribution: Cameron, Hidalgo, and Willacy counties.
Comments: This species has been considered by many to be a synonym of *C. odoratus*. the large, more or less globose clusters of spikelets call attention to this species. Like most sedges, this is a wetland plant.

Large-Head Flat Sedge *Cyperus macrocephalus*

Pond Flat Sedge

Family: Cyperaceae
Scientific Name: *Cyperus ochraceus*
Habit: Perennials up to 32" tall. Stems more or less 3-cornered.
Leaves: Blades up to 3/16" broad.
Flowers: Tiny, in flattened spikelets up to 5/16" long.
Fruit: Achenes cylindrical or somewhat 3-cornered, about 1/16" tall.
Bloom Period: Summer, fall.
Distribution: Cameron, Hidalgo, Willacy, and Starr counties.
Comments: This species is usually found growing in damp or wet soils.

Pond Flat Sedge *Cyperus ochraceus*

Cyperaceae

Fragrant Flat Sedge
Family Cyperaceae
Scientific Name: *Cyperus odoratus* subsp. *engelmannii*
Habit: Perennials up to 24" tall, occasionally as much as 36" tall; stems 3-cornered.
Leaves: Blades about 3/16" broad, V-shaped.
Flowers: Tiny, in small flattened spikelets. Large bracts surround the base of the inflorescence.
Fruit: Achenes 3-cornered, about 1/16" tall.
Bloom Period: Spring, summer, fall.
Distribution: Cameron and Hidalgo counties.
Comments: This species is fairly widespread in Texas. It prefers moist localities.

Fragrant Flat Sedge *Cyperus odoratus* subsp. *engelmannii*

Many Spiked Flat Sedge
Family: Cyperaceae
Scientific Name: *Cyperus polystachyos*
Habit: Perennials up to 17" or taller.
Leaves: Tiny, short.
Flowers: Tiny, inflorescences of loose clusters of spikelets about 3/8" long.
Fruit: Achenes flattened, 2-edged, about 1/16" tall.
Bloom Period: Spring, summer, fall.
Distribution: Cameron and Hidalgo counties.
Comments: The achenes are interesting in having a tiny fin along one side. Also, they are contained in a small basal cup.

Many Spiked Flat Sedge *Cyperus polystachyos*

Cylindric Sedge
Family: Cyperaceae
Scientific Name: *Cyperus retrorsus*
Habit: Perennials up to 34" tall. Stems 3-cornered.
Leaves: Up to 22" long, V- or M-shaped in cross section. Bracts flat or V-shaped.
Flowers: Tiny; spikelets crowded into globose or cylindrical inflorescences.
Fruit: Brown, ellipsoid to oblong, about 1/16" tall.
Bloom Period: Summer.
Distribution: Willacy County.
Comments: This species is usually found growing in damp or wet soils.

Cylindric Sedge *Cyperus retrorsus*

Southern Flat Sedge

Family: Cyperaceae
Scientific Name: *Cyperus thyrsiflorus* [*C. hermaphroditus, C. tenuis*]
Habit: Perennials up to 24" tall. Stems 3-cornered.
Leaves: Up to 16" long, V-shaped in cross section.
Flowers: Tiny; spikelets in loose cylindrical inflorescences.
Fruit: Achenes brown to reddish, narrowly oblong, 1/16" or slightly taller.
Bloom Period: Summer.
Distribution: Cameron, Hidalgo, and Willacy counties.
Comments: In our area, this sedge is usually found growing near the coast.

Southern Flat Sedge *Cyperus thyrsiflorus*

Jointed Spikerush

Family: Cyperaceae
Scientific Name: *Eleocharis interstincta*
Habit: Perennials up to 39" tall; stems round in cross section, with many visible partitions. The partitions are much closer together near the inflorescence.
Leaves: Blades absent. A few sheaths are present at the base of the stem.
Flowers: Inflorescences cone shaped, unbranched, at the stem apex; flowers tiny.
Fruit: Achenes less than 1/8" tall, 2-edged, slightly concave on both surfaces.
Bloom Period: Summer, fall.
Distribution: Cameron County.
Comments: The conelike inflorescence and the "jointed" stems call attention to this species.

Jointed Spikerush *Eleocharis interstincta*

Dwarf Spike Rush

Family: Cyperaceae
Scientific Name: *Eleocharis parvula*
Habit: Mostly annuals up to 4" tall.
Leaves: Sheaths around the stems, without blades.
Flowers: Tiny, in spikelets arranged in cone-shaped inflorescences.
Fruit: Achenes more or less pear shaped, less than 1/16" tall.
Bloom Period: Summer, fall.
Distribution: Cameron County.
Comments: This plant is usually found growing in wet or moist places.

Dwarf Spike Rush *Eleocharis parvula*

Cyperaceae

Square Stem Spike Rush
Family: Cyperaceae
Scientific Name: *Eleocharis quadrangulata*
Habit: Perennials up to 32" tall, with 4-sided stems.
Leaves: None.
Flowers: Tiny, in spear-shaped clusters at the ends of the stems.
Fruit: Achenes flattened, almost 1/8" tall.
Bloom Period: Spring, summer, fall.
Distribution: Cameron, Hidalgo, and Willacy counties.
Comments: This species is easily recognized by its lack of leaves and its square stems.

Square Stem Spike Rush *Eleocharis quadrangulata*

Coastal Marsh Fimbry
Family: Cyperaceae
Scientific Name: *Fimbristylis castanea*
Habit: Perennials up to 5' tall, sometimes taller.
Leaves: Blade length half the total height or more, about 1/8" broad.
Flowers: Tiny, in spikelets arranged in cone-shaped inflorescences.
Fruit: Achenes flattened, about 1/16" tall.
Bloom Period: Summer, fall.
Distribution: Cameron and Willacy counties.
Comments: This species is usually found growing in wet places.

Coastal Marsh Fimbry *Fimbristylis castanea*

Annual Western Umbrella Grass
Family: Cyperaceae
Scientific Name: *Fuirena simplex*
Habit: Perennials up to 20" tall.
Leaves: Blades along the stem, up to 8" long.
Flowers: Tiny; in spikelets arranged in cone-shaped structures; these in turn are crowded together.
Fruit: Achenes 3-cornered, about 1/16" tall.
Bloom Period: Summer, fall.
Distribution: Cameron County.
Comments: This plant is usually found growing in moist places. The common name for the genus is umbrella grass. Some species may produce growth reminiscent of an umbrella, although this species does not.

Annual Western Umbrella Grass *Fuirena simplex*

White Topped Umbrella Grass
Family: Cyperaceae
Scientific Name: *Rhynchospora colorata* [*Dichromena colorata*]
Habit: Perennials up to 22" tall.
Leaves: Blades about 3/16" broad. Bracts surrounding the inflorescences white on the basal half, green on the distal half.
Flowers: Tiny; crowded at the stem tips, surrounded by the bracts.
Fruit: Achenes flattened, less than 1/16" tall.
Bloom Period: Summer, fall.
Distribution: Cameron and Willacy counties.
Comments: This plant is usually found growing near the coast. It is a popular water garden ornamental and is not a true grass.

White Topped Umbrella Grass *Rhynchospora colorata*

Giant Bulrush, Tule
Family: Cyperaceae
Scientific Name: *Schoenoplectus californicus* [*Scirpus californicus*]
Habit: Perennials up to 6' or taller; stems more or less 3-cornered.
Leaves: Blades absent.
Flowers: Tiny; in spikelets arranged in cone-shaped structures that are crowded together. Flowering bract pointing above the inflorescence, resembling a continuation of the stem.
Fruit: Achenes flattened, about 1/16" tall.
Bloom Period: Spring, summer, fall.
Distribution: Cameron and Hidalgo counties.
Comments: This plant is usually found growing in wet or moist places. *Tule* is the Spanish word for "rush" or "reed."

Giant Bulrush, Tule *Schoenoplectus californicus*

Raynal's Upright Braided Sedge
Family: Cyperaceae
Scientific Name: *Schoenoplectus erectus* subsp. *raynalii*
Habit: Annuals with erect to horizontal or arching stems; stems up to 20" long, usually less.
Leaves: 3 or 4, nearly as long as the stems; cross section near the base is C-shaped.
Flowers: Tiny, in a 1–5 spikelets.
Fruit: Achenes nearly 1/8" tall, of two types. Some are unequally convex on both surfaces to vaguely 3-cornered; others are strongly 3-cornered with rounded surfaces.
Bloom Period: Spring, summer, fall.

Raynal's Upright Braided Sedge *Schoenoplectus erectus* subsp. *raynalii*

Distribution: Willacy County.
Comments: This species is found growing in wet places or recently dried areas. The spikelet can vary from one to five.

American Bulrush
Family: Cyperaceae
Scientific Name: *Schoenoplectus pungens* var. *longispicatus* [*Scirpus pungens* var. *longispicatus*]
Habit: Perennials up to 5' tall; stems 3-cornered. Flowering bracts pointing above the inflorescences, resembling a continuation of the stem.
Leaves: All basal, up to 30" long, narrow and V-shaped in cross section near the base.
Flowers: Tiny, in spikelets arranged in cone-shaped structures that are crowded together.
Fruit: Achenes flattened, about 1/8" tall.
Bloom Period: Spring, summer.
Distribution: Cameron, Hidalgo, and Willacy counties.
Comments: This species is usually found growing in wet or moist places.

American Bulrush *Schoenoplectus pungens* var. *longispicatus*

Great Braided Sedge
Family: Cyperaceae
Scientific Name: *Schoenoplectus tabernaemontani*
Habit: Perennials up to 9' or sometimes taller.
Leaves: One or two, reduced to basal bladeless sheaths.
Flowers: Tiny, crowded into small groups of spikelets.
Fruit: Achenes flattened, less than 1/8" tall.
Bloom Period: Spring and fall.
Distribution: Cameron, Hidalgo, and Starr counties.
Comments: This species is found in freshwater wetlands.

Great Braided Sedge *Schoenoplectus tabernaemontani*

Hydrilla

Family: Hydrocharitaceae
Scientific Name: *Hydrilla verticillata*
Habit: Plants growing submerged in the water, rooted in the mud, producing long stolons.
Leaves: Whorled, sometimes as few as two per node; blades oblong, minutely toothed, rough to touch, about ⅝" long.
Flowers: from the leaf axils, rising to the surface on long peduncles; sepals 3, petals 3 (occasionally absent), white, about 3/16" long.
Fruit: Ripening underwater, about 3/16" tall.
Bloom Period: Late summer, early fall.
Distribution: Cameron, Hidalgo, and Starr counties.
Comments: Hydrilla was introduced from the Old World for use in aquariums. It reproduces by fragments, buds, seeds, tubers, and stolons. It has become a pest in the Rio Grande and is federally listed as a noxious weed. The pollination process is interesting. The anthers pop open, throwing the pollen grains away from the flower. Some land by chance on nearby flowers. Pollen grains landing in the water will not function.

Hydrilla *Hydrilla verticillata*

Hydrilla *H. verticillata* flower

Purple Pleat Leaf

Family: Iridaceae
Scientific Name: *Alophia drummondii*
Habit: Bulb-forming perennials up to 24" tall.
Leaves: Narrow, pleated, up to 16" long.
Flowers: Bilateral, 1 ¾" or broader; petals purplish, some with yellow basal mottling.
Fruit: Capsules more or less cylindrical, up to 1" long, splitting into 3 parts.
Bloom Period: Spring.
Distribution: Hidalgo and Willacy counties.
Comments: The flower is very showy, resembling the cultivated irises, although smaller. Seeds planted in the spring often produce flowers in about three months. It would make a beautiful planting, substituting for the bearded iris, which will not flower here. It is a plant of sandy soils. This species was named in honor of Thomas Drummond, from Scotland, who made extensive collections of plants in Central Texas in the 1830s.

Purple Pleat Leaf *Alophia drummondii*

Lemnaceae 41

Blue-Eyed Grass
Family: Iridaceae
Scientific Name: *Sisyrinchium biforme*
Habit: Erect herbs up to 15" tall, with green stem bracts.
Leaves: Simple, linear, erect, and grasslike.
Flowers: Radial, sepals 3, petals 3, looking alike, blue, up to ⅜" long.
Fruit: Capsules about ¼" tall, splitting into 3 parts.
Bloom Period: Spring, summer.
Distribution: Cameron, Hidalgo, and Willacy counties.
Comments: The floral bracts are green; *S. langloisii* has bracts that are purplish basally. The common name can be misleading. *Sisyrinchium* is not a grass, although it may be thought to resemble grasses.

Blue-Eyed Grass *Sisyrinchium biforme*

Pale Blue-Eyed Grass
Family: Iridaceae
Scientific Name: *Sisyrinchium langloisii*
Habit: Erect herbs up to 14" tall, with stem bracts that are purplish basally.
Leaves: Simple, linear, erect, and grasslike.
Flowers: Radial; sepals 3, petals 3, looking alike, blue, up to ⅜" long.
Fruit: Capsules about 3/16" tall, splitting into 3 parts.
Bloom Period: Spring.
Distribution: Hidalgo and Willacy counties.
Comments: The purple markings on the stem bracts distinguish this species from *S. biforme*. As explained for *S. biforme*, *Sisyrinchium* is not a grass.

Pale Blue-Eyed Grass *Sisyrinchium langloisii*

Lesser Duckweed
Family: Lemnaceae
Scientific Name: *Lemna aequinoctialis*
Habit: Tiny plants growing on mud or floating in water, each plant with a single root ⅜–1⅛" long.
Leaves: Ovate, about ¼" long.
Flowers: Very small.
Fruit: Tiny.
Bloom Period: Spring, fall.
Distribution: Cameron, Hidalgo, and Willacy counties.
Comments: *Lemna obscura* also grows in our area. In order to distinguish them, a microscope and a key are needed.

Lesser Duckweed *Lemna aequinoctialis*

Lesser Duckweed *L. aequinoctialis* individual plants

Pointed Watermeal

Family: Lemnaceae
Scientific Name: *Wolffia brasiliensis*
Habit: Tiny plants, consisting of a leaf without roots.
Leaves: More or less egg shaped, up to 1/16" long, with a small point in the middle. They appear to be small green specks in the water.
Flowers: Tiny, but the plants rarely bloom.
Fruit: Minute.
Bloom Period: The plants rarely bloom.
Distribution: Hidalgo County.
Comments: The southernmost point in Texas previously reported for this species is Jim Wells County. It has probably been overlooked because of its small size. *Wolffia* species are the smallest known flowering plants. Like all the species in this family, the plants float on the surfaces of ponds and lakes.

Pointed Watermeal *Wolffia brasiliensis*

Common Watermeal

Family: Lemnaceae
Scientific Name: *Wolffia columbiana*
Habit: Very tiny plants consisting of a leaf without roots.
Leaves: Egg shaped or rounded, less than 1/16" long. They appear as tiny green specks in the water.
Flowers: Tiny, one per leaf, but the plants rarely bloom.
Fruit: Minute.
Bloom Period: The plants rarely bloom.
Distribution: Hidalgo County.
Comments: *Wolffia* has not previously been reported for the Lower Rio Grande Valley. In Texas, it has been reported for Jim Wells County and farther northward. It differs from pointed watermeal in lacking the point in the middle of the leaf. This tiny plant could be growing abundantly in other parts of the LRGV, overlooked because of its small size. *Wolffia* species are the smallest known flowering plants in the world.

Common Watermeal *Wolffia columbiana*

Common Water Nymph

Family: Najadaceae
Scientific Name: *Najas guadalupensis*
Habit: Small annuals, growing submerged in fresh water. Stems slender, brittle, branching, up to 24" long.
Leaves: Narrow, up to 3/4" long; arrangement varies.
Flowers: Tiny, solitary in the leaf axils; male and female flowers on separate plants.
Fruit: About 1/16" tall, purplish brown.
Bloom Period: Spring, summer, fall.
Distribution: Cameron and Hidalgo counties.

Common Water Nymph *Najas guadalupensis*

Orchidaceae

Comments: This is a small, inconspicuous plant found in shallow water of ponds, lakes, or ditches, occasionally on mud.

Spring Ladies' Tresses
Family: Orchidaceae
Scientific Name: *Spiranthes vernalis*
Habit: Perennial herbs 4–24" tall.
Leaves: Simple, mostly linear, about 5" long.
Flowers: In spiraled spikes up to 16" or taller; corollas bilateral, white, sepals 3, petals 3, looking alike, up to ⅜" long.
Fruit: Capsules roundish, about ¼" tall.
Bloom Period: Spring.
Distribution: Cameron County.
Comments: The plants are erratic in their blooming. Some seasons they are abundant; in other seasons they may be hard to find; or blooming may be in between those extremes. The plants are limited to the coastal sands. Another orchid, *Habenaria repens* (water spider orchid), has been reported for our area but has not been seen in recent times.

Spring Ladies' Tresses *Spiranthes vernalis*

Lawn Orchid
Family: Orchidaceae
Scientific Name: *Zeuxine strateumatica*
Habit: Erect or leaning herbs with succulent stems about 10" tall.
Leaves: Spirally arranged, erect, pressed against the stem; blades mostly narrowly lanceolate, up to 3" or sometimes longer.
Flowers: Small, about ¼" long, white with yellowish green lip.
Fruit: More or less egg shaped, about ¼" tall.
Bloom Period: Fall, winter.
Distribution: Cameron County.
Comments: This is a new orchid for our area. Native to Asia and some Pacific islands, it was found in Florida in 1936. It was suspected that seeds arrived in shipments of centipede grass imported from China. From there, it spread along the Gulf Coast. It was first reported in Texas in 1989, in Montgomery County, and was later found in Harris County. This is the first report for our area. It has been observed at two nurseries in the Harlingen area. There was an unconfirmed sighting in 2006 on the campus of the University of Texas at Brownsville/Texas Southmost College (Cameron County). This orchid is probably spreading in various nursery plants, evidently germinating and growing well from seed. It has been suggested that lawn orchid reproduces by self-pollination or by nonsexual means (Diggs et al. 2006).

Lawn Orchid *Zeuxine strateumatica*

Water Hyacinth

Family: Pontederiaceae
Scientific Name: *Eichhornia crassipes*
Habit: Floating herbs.
Leaves: Petioles inflated, functioning as floats; blades broadly elliptic, about 4" long.
Flowers: Bilateral, blue, about 2 ⅜" broad, in spikes resembling those of a hyacinth.
Fruit: Capsules many-seeded.
Bloom Period: Summer, fall.
Distribution: Cameron and Hidalgo counties, probably also Starr County.
Comments: This introduced species from Brazil is beautiful when in flower. It has been used in water gardens for many years and was probably introduced to our resacas from one of these. Because of its rank growth it has become a pest, clogging canals, the Rio Grande, and other waterways. Now it is illegal to buy, sell, or release this plant in the United States. Water hyacinth has been used to absorb excess nutrient from sewage treatment pond effluent. The dense root systems provide refuge and spawning sites for fish, amphibians, and aquatic invertebrates.

Water Hyacinth *Eichornia crassipes*

Water Stargrass

Family: Pontederiaceae
Scientific Name: *Heteranthera dubia*
  [*H. liebmannii*]
Habit: Submerged herbs, or rooting in the mud at the edges of bodies of water.
Leaves: Linear, up to 6" long.
Flowers: Reaching above the water surface; radial, yellow, about ¾" broad, with a long tube up to 4 ¾" long partially enclosed in a bract.
Fruit: Capsules many-seeded.
Bloom Period: Spring, summer.
Distribution: Cameron, Hidalgo, and Starr counties.
Comments: This is a beautiful water plant but can become a pest in our resacas.

Water Stargrass *Heteranthera dubia*

Pontederiaceae

Blue Mud Plantain, Mud Babies

Family: Pontederiaceae
Scientific Name: *Heteranthera limosa*
Habit: Annuals rooting in mud, forming rosettes.
Leaves: Blades lanceolate, about 2 ⅜" long.
Flowers: Radial, blue to white, about 1 ⅛" broad, with a long tube partially enclosed in a bract.
Fruit: Capsules many-seeded.
Bloom Period: Spring, summer, fall.
Distribution: Cameron, Hidalgo, Willacy, and Starr counties.
Comments: *H. limosa* grows abundantly in depressions or potholes in flooded cultivated fields. We have four species of *Heteranthera*. *H. limosa* and *H. dubia* have only one flower per inflorescence, and they are easily distinguished by flower color; *H. multiflora* (some leaves with petioles) and *H. mexicana* (leaves without petioles) have two or more white or pale to dark blue flowers per inflorescence. The last two named species are not common in the LRGV.

Blue Mud Plantain, Mud Babies *Heteranthera limosa*

Blue Mud Plantain, Mud Babies *H. limosa* white-flowering form

Mexican Mud Plantain

Family: Pontederiaceae
Scientific Name: *Heteranthera mexicana* [*Eurystemon mexicanum*]
Habit: Erect annuals up to 16" tall, partially submersed in water, rooting in mud; upper parts with small glandular hairs.
Leaves: Simple, alternate, sheathing the stems, broadly linear, up to 6" long.
Flowers: In terminal spikes; corollas bilateral, pale blue to dark blue; sepals and petals grown together, lobes 6, up to ½" long. Stamens 3, not alike.
Fruit: Capsules many-seeded.
Bloom Period: Summer.
Distribution: Cameron and Hidalgo counties.
Comments: Mexican mud plantain occurs in ephemeral ponds. This habit makes it difficult to find the plants and confirm distribution records (Poole et al. 2007). In effect, study of the plants in the field is limited to times of abundant rainfall. The flowers open in the morning and wilt in the afternoon. The population photographed by Bill Carr was relocated in Nueces County by Tom Patterson in 2008.

Mexican Mud Plantain *Heteranthera mexicana* (Courtesy Bill Carr)

Long Leaf Pondweed

Family: Potamogetonaceae
Scientific Name: *Potamogeton nodosus*
Habit: Herbs growing in shallow freshwater, rooting in the mud; rhizomes white with red spots.
Leaves: Alternate; the submerged blades lanceolate to elliptic, tapering gradually to the base, up to 8" long; floating leaves broader, usually up to 4½" long.
Flowers: Inflorescences with flowers in dense whorls; flowers radial, greenish to brown, about ⅛" broad.
Fruit: Obovate, about 3/16" tall.
Bloom Period: Spring and summer.
Distribution: Cameron, Hidalgo, and Starr counties.
Comments: In stable areas, the plants can establish large colonies that form floating mats with overlapping leaves. The seeds provide food for water birds. Four other species of *Potamogeton* are reported for our area: *P. diversifolius, P. latifolius, P. pulcher,* and *P. pusillus.* They are not easy to distinguish.

Long Leaf Pondweed *Potamogeton nodosus*

Cat Brier

Family: Smilacaceae
Scientific Name: *Smilax bona-nox*
Habit: Vines with tendrils, often woody basally, with 4 ridges on the stems. Lower stems with prickles; upper stems may not have them.
Leaves: Alternate, variable in shape, up to 4" long (occasionally more), with 3, sometimes more, major veins originating from the base.
Flowers: Radial, male and female flowers on separate plants in umbrella-like clusters; sepals and petals looking alike, greenish, about 3/16" long.
Fruit: Berries roundish, about 3/16" tall, black when mature.
Bloom Period: Spring, early summer.
Distribution: Cameron and Hidalgo counties.
Comments: This vine is abundant in Central Texas but less so in our area. It greatly resembles variable leaf snailseed, *Cocculus diversifolius,* which is abundant here. The prickly lower stems and the flowers in umbrella-like clusters distinguish cat brier. The fruit is eaten by various species of wildlife.

Cat Brier *Smilax bona-nox*

Typhaceae

Cattail, Tule

Family: Typhaceae
Scientific Name: *Typha domingensis*
Habit: Aquatic herbs.
Leaves: Tall and narrow, up to 8' tall.
Flowers: Minute, in separate male and female cylindrical, brown inflorescences; the male inflorescence is positioned above the female inflorescence and falls away early.
Fruit: Tiny, hidden by the cottony material of the female inflorescence.
Bloom Period: Summer.
Distribution: Cameron, Hidalgo, Willacy, and Starr counties.
Comments: The leaves are sometimes woven to make mats. The swollen roots are eaten by wildlife and sometimes by humans. The thick stands are important for wildlife habitat and in filtering toxins and excess nutrients from wetlands. Another species, *T. latifolia,* has not been officially reported for this area but may be present. Its male and female inflorescences are touching or almost so; those of *T. domingensis* are farther apart. *Tule* is the Spanish word for "rush" or "reed."

Cattail, Tule *Typha domingensis*

Cattail, Tule *T. domingensis* male inflorescence above the female inflorescence

Dicots (Dicotyledoneae)

Flowering Seed Plants with Two "Seed Leaves"

(Usually the flower parts are in fours or fives and the leaf veins form a branching network.)

Small Flowered Wrightwort
Family: Acanthaceae
Scientific Name: *Carlowrightia parviflora*
Habit: Creeping to erect subshrubs, up to 20" tall, forming colonies.
Leaves: Opposite. Petioles short or absent; blades lanceolate, up to 1 ⅝" long and ½" broad.
Flowers: Blue, bilateral, 4-lobed, with a narrow tube, ¼" or slightly longer. Anthers 2.
Fruit: a ⅛" capsule with 4 seeds.
Bloom Period: Fall, winter, spring.
Distribution: Cameron, Hidalgo, and Starr counties.
Comments: This species is similar to some Justicias, which have corollas ⅝" or longer. It is a host plant for the Crimson Patch butterfly. The specific epithet is derived from the Latin *parvus* and *flora* (small + flower). It grows at edges of brush or in shady areas.

Small Flowered Wrightwort *Carlowrightia parviflora*

Texas Wrightwort
Family: Acanthaceae
Scientific Name: *Carlowrightia texana*
Habit: Creeping perennials, sometimes up to 12" tall, minutely hairy.
Leaves: Opposite. Petioles very short or absent; blades grayish green, ovate to lanceolate, up to ¾" long and ½" broad.
Flowers: White with maroon veins, bilateral, 4-lobed, with a narrow tube and 2 anthers, about ¼" long.
Fruit: a capsule with 4 seeds.
Bloom Period: Spring, summer.
Distribution: Hidalgo and Starr counties.
Comments: This species is similar to *C. parviflora*, which has leaves three or more times longer than they are broad. *C. texana* has leaves twice as long as they are broad, or less. Also, the corolla lobes have maroon-colored veins. It grows in sunny, disturbed areas.

Texas Wrightwort *Carlowrightia texana*

Acanthaceae

Six Angle Fold Wing

Family: Acanthaceae
Scientific Name: *Dicliptera sexangularis* [*D. vahliana*]
Habit: Erect herbs up to 39" tall.
Leaves: Opposite. Petioles absent or up to ⅜" long; blades spatulate, up to 2" long and ¾" broad.
Flowers: Reddish to orange, bilateral, up to 1 ⅛" long, with 2 stamens.
Fruit: a capsule that opens suddenly, dispersing the seeds. Rows of curved, sharp stalks remain.
Bloom Period: Spring
Distribution: Cameron and Hidalgo counties.
Comments: This species is shade tolerant and opportunistic, forming large colonies in disturbed places. However, it is not usually abundant in undisturbed areas. It is a host plant for the Pale-banded Crescent, Texan Crescent, and Crimson Patch butterflies and is a nectar source for hummingbirds and Sulphur butterflies. The genus name is derived from the Greek *diclis* and *pteron* (2 + wing), referring to the winglike appearance of the open capsule.

Six Angle Fold Wing *Dicliptera sexangularis*

Crenate Leaf Snake Herb

Family: Acanthaceae
Scientific Name: *Dyschoriste crenulata*
Habit: Erect or leaning perennials forming low mounds 12" or less in height.
Leaves: Opposite, hairy, the blades oblanceolate, up to 9" long. Margins usually with minute teeth.
Flowers: Axillary. Corollas bilateral, bluish lavender, with 4 stamens, up to ¾" long.
Fruit: Dry, dehiscent, with rows of curved, sharp stalks. Seeds tan colored, more or less disk shaped, ⅛" broad or less.
Bloom Period: Spring, summer
Distribution: Cameron, Hidalgo, Willacy, and Starr counties.
Comments: This species is found at the edge of brush and in disturbed places. It is used as a foreground ornamental in native gardens. The flowers resemble those of *Ruellia,* but they are smaller. The family is characterized by a dry, dehiscent fruit, dispersing the seeds and leaving rows of curved, sharp stalks (where seeds were attached).

Crenate Leaf Snake Herb *Dyschoriste crenulata*

Wheatspike Scaly Stem

Family: Acanthaceae
Scientific Name: *Elytraria bromoides*
Habit: Perennials from a basal rosette, up to 4" tall.
Leaves: Mostly basal; blades spatulate, up to 4" long, tapering onto the petioles.
Flowers: Crowded in terminal inflorescences, with bracts and calyxes overlapping; corollas purple to white, 5-lobed, somewhat bilateral, about ¼" long, the lobes unequal, notched apically; stamens 2.
Fruit: Dry, dehiscent, dispersing the seeds and leaving rows of curved, sharp stalks. Seeds tan colored, more or less disk shaped, ⅛" broad or less.
Bloom Period: Summer.
Distribution: Cameron, Hidalgo, Willacy, and Starr counties.
Comments: This small species resembles sweet shaggy tuft, *Stenandrium dulce,* which has rounded corolla lobes and 4 stamens. Wheatspike scaly stem has notched corolla lobes and 2 stamens.

Wheatspike Scaly Stem *Elytraria bromoides*

Tube Tongue

Family: Acanthaceae
Scientific Name: *Justicia pilosella* [*Siphonoglossa pilosella, S. greggii*]
Habit: Woody herbs or shrubs; stems erect, up to 12" tall, or creeping.
Leaves: Opposite, the blades ovate, up to ¾" long.
Flowers: Axillary, usually without pedicels. Calyxes 4-lobed. Corollas usually lavender, bilateral, ⅝–1" long, with a narrow tube; stamens 2.
Fruit: Dry, dehiscent, dispersing the seeds and leaving rows of curved, sharp stalks. Seeds tan colored, more or less disk shaped, ⅛" broad or less.
Bloom Period: Spring, summer, fall.
Distribution: Cameron, Hidalgo, Willacy, and Starr counties.
Comments: Tube tongue is a host plant for the Tiny Checkerspot and Elada Checkerspot butterflies and also the Texan Crescent and Vesta Crescent butterflies. It makes an attractive ground cover.

Tube Tongue *Justicia pilosella*

Acanthaceae

Runyon's Water Willow

Family: Acanthaceae
Scientific Name: *Justicia runyonii*
Habit: Stems erect or falling over, woody when old.
Leaves: Opposite. Blades lanceolate to elliptic, up to 5 ½" long.
Fruit: Dry, somewhat club shaped, about ⅝" long, dehiscent, leaving rows of curved, sharp stalks (where seeds were attached).
Flowers: In axillary groups. Corollas purplish, bilateral, up to 1" long; stamens 2.
Bloom Period: Summer, fall.
Distribution: Cameron and Hidalgo counties.
Comments: This species was named in honor of Robert Runyon, a Brownsville photographer, mayor, and self-taught botanist. It prefers low, moist places but is not commonly found. It is an attractive plant, easily grown from cuttings, and does well in cultivation. It is a host plant for the Malachite butterfly. The genus is named for James Justice, a Scottish horticulturist of the 1700s.

Runyon's Water Willow *Justicia runyonii*

Turner's Tube Tongue

Family: Acanthaceae
Scientific Name: *Justicia turneri*
Habit: Herbs with woody bases, 8" or taller. Stem hairs in two rows on alternating sides of the stem.
Leaves: Opposite; blades lanceolate to ovate, up to 1" long and ¼" broad.
Flowers: Axillary, without pedicels; corollas white (sometimes purple), bilateral, up to ¾" long.
Fruit: Dry, somewhat club shaped, about ⅜" long, dehiscent, leaving rows of curved, sharp stalks (where seeds were attached).
Bloom Period: Spring, summer, fall.
Distribution: Cameron, Hidalgo, and Starr counties.
Comments: The alternating rows of hairs are an identifying characteristic of *J. turneri*. This plant blooms heavily in brush thickets after good rains. There has been some taxonomic confusion regarding this taxon in the past. It was misidentified as *Siphonoglossa greggii* and was so treated in the *Manual of the Vascular Plants of Texas* (Correll and Johnston 1970). In 1990 Richard Hilsenbeck declared *S. greggii* to be a synonym of *Justicia pilosella* and described this taxon as *Justicia turneri,* a new species. He named it in honor of Billie L. Turner, who has made great contributions to the study of plants in Texas.

Turner's Tube Tongue *Justicia turneri*

Turner's Tube Tongue *J. turneri*; two rows of hairs on alternating sides of the stem

Violet Petunia

Family: Acanthaceae
Scientific Name: *Ruellia nudiflora* var. *nudiflora*
Habit: Perennials with woody bases about 10" tall when young, growing up to 28" tall.
Leaves: Opposite; blades oblong to ovate, up to 4 ¾" long.
Flowers: In terminal inflorescences; corollas violet to pink, occasionally white, bilateral, up to 2 ⅛" long.
Fruit: Dry, dehiscent, leaving rows of curved, sharp stalks (where seeds were attached).
Bloom Period: All seasons, especially summer.
Distribution: Cameron and Willacy counties.
Comments: This variety of *R. nudiflora* is not commonly found here. It is distinguished from var. *runyonii* by its bigger flowers, up to 2 ⅛" long.

Violet Petunia *Ruellia nudiflora* var. *nudiflora* white form

Runyon's Violet Wild Petunia

Family: Acanthaceae
Scientific Name: *Ruellia nudiflora* var. *runyonii* [*Ruellia runyonii*]
Habit: Perennials, sometimes woody basally, up to 24" tall.
Leaves: Opposite; blades mostly ovate, up to 3 ½" long.
Flowers: Terminal and axillary; corollas bluish lavender or sometimes white, slightly bilateral, funnel shaped, up to 1 ⅝" long.
Fruit: Dry, dehiscent, leaving rows of curved, sharp stalks (where seeds were attached).
Bloom Period: Spring, summer.
Distribution: Cameron, Hidalgo, Willacy, and Starr counties.
Comments: This is our most common *Ruellia*. It is a host plant to the Pale-banded Crescent, Malachite, Texan Crescent, and White Peacock butterflies. The variety name was given in honor of Robert Runyon, a former Brownsville mayor, photographer, and self-taught botanist. The variety *nudiflora* is similar to var. *runyonii*. Its flowers are usually violet, but occasionally white, up to 2 ⅛" long. The fruits pop explosively when wet, scattering the seeds. When wet, the seeds have a sticky, gelatinous covering.

Runyon's Violet Wild Petunia *Ruellia nudiflora* var. *runyonii*

Acanthaceae

Western Wild Petunia

Family: Acanthaceae
Scientific Name: *Ruellia occidentalis*
Habit: Viscid perennials, sometimes woody basally, up to 28" tall.
Leaves: Opposite; hairy, glandular; blades ovate, up to 4" long.
Flowers: Self-pollinating flowers tiny; normal flowers with corollas bluish lavender, slightly bilateral, funnel shaped, up to 2 ¾" long and 1 ½" broad.
Fruit: Dry, dehiscent, leaving rows of curved, sharp stalks (where seeds were attached).
Bloom Period: Spring, summer, fall.
Distribution: Cameron, Hidalgo, and Starr counties.
Comments: The fruiting capsules of *Ruellias* all look more or less alike, making it difficult to distinguish species without using keys. As in most members of this family, the fruits (when water falls on them) pop and throw seeds several feet from the parent plant. The tackiness and the larger corolla, 2" or longer, are noticeable characteristics of this species.

Western Wild Petunia *Ruellia occidentalis*

Yucatán Wild Petunia

Family: Acanthaceae
Scientific Name: *Ruellia yucatana*
Habit: Perennials up to 14" tall.
Leaves: Opposite; blades more or less elliptic to spatulate, up to 5" long.
Flowers: In branching inflorescences; corollas bluish lavender, slightly bilateral, funnel shaped, about 1 ¼" long.
Fruit: Dry, dehiscent, leaving rows of curved, sharp stalks (where seeds were attached).
Bloom Period: Spring, summer, fall.
Distribution: Cameron, Hidalgo, Willacy, and Starr counties.
Comments: This is a South Texas plant, extending from Goliad County southward to Cameron County. The family is characterized by a dry, dehiscent fruit, the seeds falling and leaving rows of curved, sharp stalks (where seeds were attached). The stems are glandular and sometimes viscid (sticky), especially near the inflorescences.

Yucatán Wild Petunia *Ruellia yucatana*

Sweet Shaggy Tuft

Family: Acanthaceae

Scientific Name: *Stenandrium dulce*

Habit: Small perennials from a basal rosette; stems short or absent.

Leaves: Crowded basally; blades ovate, up to 1 ¾" long.

Flowers: Crowded, with bracts and sepals overlapping; corollas purplish or almost white, 5-lobed, slightly bilateral, ½" long or less; stamens 4.

Fruit: Dry, dehiscent, leaving rows of curved, sharp stalks (where seeds were attached).

Bloom Period: Spring, summer, fall.

Distribution: Cameron, Hidalgo, Willacy, and Starr counties.

Comments: This small species is a host plant for the Definite Patch butterfly. It is often found growing on bare patches beneath ebony trees. It resembles wheatspike scaly stem, *Elytraria bromoides,* which has notched corolla lobes and 2 stamens. *Stenandrium dulce* has rounded corolla lobes and 4 stamens.

Sweet Shaggy Tuft *Stenandrium dulce*

Texas Shrimp Plant

Family: Acanthaceae

Scientific Name: *Yeatesia platystegia* [*Tetramerium platystegium*]

Habit: Small woody plants up to 30" tall.

Leaves: Opposite; blades elliptic to lanceolate, sometimes linear, up to 2" long and ¼–½" broad, with hairs along the veins.

Flowers: Pale lavender, ¾–1" long, 2-lipped and 5-lobed, with a long narrow tube; enclosed in prominent overlapping green bracts (like the shrimp plant).

Fruit: Dry, dehiscent, leaving rows of curved, sharp stalks (where seeds were attached).

Bloom Period: Summer.

Distribution: Starr County.

Comments: This species is worthy of cultivation. It is a host plant for the Elf butterfly. Rarely encountered, it grows on dry, rocky hillsides under brush. The green floral bracts identify this species. Seedlings are often seen growing around the parent plant.

Texas Shrimp Plant *Yeatesia platystegia*

Aizoaceae

Snake Eyes

Family: Achatocarpaceae [Phytolaccaceae]
Scientific Name: *Phaulothamnus spinescens*
Habit: Thorny shrubs up to 8' tall, usually less.
Leaves: Petioles absent; blades thickened, spatulate, gray-green, up to 1 ⅜" long.
Flowers: Tiny, greenish, about ⅛" broad, sepals present; petals absent. Male and female flowers are on separate plants.
Fruit: Globose translucent "berries," the single seed visible.
Bloom Period: Summer, fall.
Distribution: Cameron, Hidalgo, Willacy, and Starr counties.
Comments: The translucent berries resembling snake eyes are a reliable guide to the identity of the species. If fruit are not present, this species resembles narrow leaf elbow bush, *Forestiera angustifolia,* which has similar growth form, leaf shape, and size. The thorns of *P. spinescens* can be rather soft, and the stems have a blackish tint. *F. angustifolia* has no spines.

Snake Eyes *Phaulothamnus spinescens*

Slender Sea Purslane

Family: Aizoaceae
Scientific Name: *Sesuvium maritimum*
Habit: Prostrate, succulent, forming mats up to 6' broad.
Leaves: Opposite, succulent; blades mostly spatulate, up to 1" long.
Flowers: From the leaf axils; calyxes pink to almost white, with green tips, about ⅛" long. Petals absent.
Fruit: Capsules roundish, about 3⁄16" broad.
Bloom Period: All seasons.
Distribution: Cameron County.
Comments: The few localities in Texas that have previously been reported for this species are farther north along the coast. We have found it on salt flats. The very small flowers are easily overlooked.

Slender Sea Purslane *Sesuvium maritimum*

Cenicilla

Family: Aizoaceae
Scientific Name: *Sesuvium portulacastrum*
Habit: Prostrate, stems rooting at the nodes. Plants partially turning purple in the winter.
Leaves: Succulent, oblong to elliptic, up to 2 ⅜" long.
Flowers: Star shaped. Petals absent; sepals partially united, purple, the lobes about ⅜" long.
Fruit: a capsule, splitting around the middle.
Bloom Period: All seasons.
Distribution: Cameron and Willacy counties.
Comments: The sepals form the colorful part of the flower. Petals are absent. This species forms bigger plants and leaves than winged sea purslane, *S. verrucosum,* and the stems root at the nodes. *S. portulacastrum* occurs on coastal sands and clay dunes. *Cenicilla* is a Spanish word for "powdery mildew." The connection with the common name is not clear, since the leaves are basically green.

Cenicilla *Sesuvium portulacastrum*

Winged Sea Purslane

Family: Aizoaceae
Scientific Name: *Sesuvium verrucosum* [*S. sessile*]
Habit: Erect, leaning, or prostrate plants with opposite, succulent leaves.
Leaves: Opposite, succulent, linear to oblanceolate, 1 ¼" or longer.
Flowers: Star shaped. Petals absent; sepals partially united, purple, the lobes ¼" or longer.
Fruit: A capsule, splitting around the middle.
Bloom Period: All seasons.
Distribution: Cameron, Hidalgo, Willacy, and Starr counties.
Comments: This species grows on saline and alkaline soils. The sepals form the colorful part of the plant. Petals are absent. The stems do not root at the nodes, separating this species from cenicilla, *S. portulacastrum.* The common name is probably a reference to the appendages on the outer surfaces of the sepals. Sesuviums are host plants for the Western Pigmy Blue butterfly.

Winged Sea Purslane *Sesuvium verrucosum*

Amaranthaceae

Horse Purslane

Family: Aizoaceae
Scientific Name: *Trianthema portulacastrum*
Habit: Plants prostrate, the stem ends pointing upward.
Leaves: Opposite, unequal, succulent, ovate to roundish, up to 1 ⅝" long.
Flowers: Star shaped. Petals absent; sepals partially united, purple or pinkish, the lobes about ⅛" long.
Fruit: A capsule, splitting around the middle.
Bloom Period: Summer, fall.
Distribution: Cameron, Hidalgo, and Willacy counties.
Comments: This species grows in disturbed soils. It is a host plant for the Western Pigmy Blue butterfly. The sepals form the colorful part of the plant. Petals are absent. *Trianthema* has leaf stipules; *Sesuvium* species do not.

Horse Purslane *Trianthema portulacastrum*

Chaff Flower

Family: Amaranthaceae
Scientific Name: *Achyranthes aspera*
Habit: Herbs with stems 3' or longer, leaning or falling over.
Leaves: Opposite, up to 8" long; blades ovate.
Flowers: In terminal and axillary spikes, the stalks white-hairy. Sepals scalelike, 4 or 5, subtended by scalelike bracts.
Fruit: A one-seeded utricle (membranous, slightly inflated), indehiscent.
Bloom Period: Summer, fall.
Distribution: Cameron and Hidalgo counties. It is more abundant in Mexico.
Comments: This species is not reported to occur elsewhere in Texas. Characteristics that help separate this species from other Amaranthaceae are the opposite leaves and the perfect flowers in axillary and terminal elongate inflorescences. The leaves and seeds have been used for food.

Chaff Flower *Achyranthes aspera*

Matt Chaff Flower

Family: Amaranthaceae
Scientific Name: *Alternanthera caracasana*
Habit: Perennial herbs from tuberous roots. Stems prostrate, sometimes hairy.
Leaves: Clustered; blades spatulate to roundish, up to ¾" long, narrowing onto the petioles.
Flowers: Scaly, in dense clusters in the leaf axils.
Fruit: a one-seeded utricle (membranous, slightly inflated), indehiscent.
Bloom Period: Spring, summer, fall.
Distribution: Cameron and Hidalgo counties.
Comments: This species is somewhat spiny but not extremely prickly to touch, as is the spiny leaf chaff flower, *A. pungens*.

Matt Chaff Flower *Alternanthera caracasana*

Smooth Chaff Flower

Family: Amaranthaceae
Scientific Name: *Alternanthera paronychioides*
Habit: Herbs; stems prostrate, rooting at the nodes.
Leaves: Opposite; blades spatulate to elliptic, hairy on the lower surfaces, up to 1" long.
Flowers: Scaly, in dense clusters in the leaf axils.
Fruit: A one-seeded utricle (membranous, slightly inflated), indehiscent.
Bloom Period: Summer, fall, winter.
Distribution: Cameron, Hidalgo, Willacy, and Starr counties.
Comments: Characteristics that help separate this species from other Amaranthaceae are the opposite leaves, the flowers in dense axillary clusters, and leaves that are hairless. Smooth chaff flower is found at edges of wet areas or bottoms of dried ponds.

Smooth Chaff Flower *Alternanthera paronychioides*

Amaranthaceae

Alligator Weed

Family: Amaranthaceae
Scientific Name: *Alternanthera philoxeroides*
   [*Achyranthes philoxeroides*]
Habit: Usually in or near wet places. Aquatic forms produce mats of floating vegetation. Terrestrial-growing plants have stems that are erect, leaning or falling over, often rooting at the nodes.
Leaves: Opposite, fleshy, linear to obovate, up to 4 ⅜" long.
Flowers: In axillary or terminal spikes. Flowers subtended by bracts, clustered in globose or cylindrical clusters; sepals white, 4 or more, scalelike; petals absent.
Fruit: A one-seeded utricle (membranous, slightly inflated), indehiscent.
Bloom Period: Spring, summer.
Distribution: Hidalgo County.
Comments: This species is an exotic pest, found growing in or near waterways. Characteristics which help separate this species from other Amaranthaceae are the opposite leaves, and the flower clusters in white globose or cylindrical clusters. The sepals are white, and the petals are absent.

Alligator Weed *Alternanthera philoxeroides*

Spiny Leaf Chaff Flower

Family: Amaranthaceae
Scientific Name: *Alternanthera pungens*
Habit: Prostrate-growing herbaceous plants with opposite leaves.
Leaves: Opposite, up to 2" long, usually less; blades roundish to broadly ovate.
Flowers: In axillary clusters. Flowers subtended by bracts. Sepals and bracts scalelike, about ¼" long, very sharp pointed; petals none.
Fruit: A one-seeded utricle (membranous, slightly inflated), indehiscent.
Bloom Period: Summer.
Distribution: Cameron and Hidalgo counties.
Comments: This is a common yard pest in some parts of the LRGV. It grows in disturbed soils. Characteristics that help separate this species from other Amaranthaceae are the prostrate growth, the fact that it is very prickly to touch, and the opposite leaves. The specific epithet is a Latin word meaning "sharp pointed."

Spiny Leaf Chaff Flower *Alternanthera pungens*

Sandhills Amaranth

Family: Amaranthaceae

Scientific Name: *Amaranthus arenicola*

Habit: Erect, whitish, hairless herbs up to 6' tall, usually much less.

Leaves: Alternate; blades ovate or oblong, up to 3 1/8" long.

Flowers: Tiny, male and female flowers on separate plants. Sepals and bracts scalelike; petals none.

Fruit: A one-seeded utricle (membranous, slightly inflated), dehiscent, the tops falling as lids.

Bloom Period: Summer, fall.

Distribution: Cameron, Hidalgo, and Willacy counties.

Comments: The specific epithet comes from a Latin word meaning "a dweller on sand." The name is appropriate, since this species usually grows on sandy soil.

Sandhills Amaranth *Amaranthus arenicola*

Gregg's Pigweed

Family: Amaranthaceae

Scientific Name: *Amaranthus greggii*

Habit: Erect or sprawling plants, sparsely hairy or hairless.

Leaves: Alternate, thick and fleshy, up to 1 3/8" long; blades rhombic or oblong.

Flowers: In terminal inflorescences, male and female flowers on separate plants. Sepals 5, scalelike, subtended by scalelike bracts. Petals absent.

Fruit: A one-seeded utricle (membranous, slightly inflated), indehiscent.

Bloom Period: All seasons.

Distribution: Cameron and Willacy counties.

Comments: This is a common beach amaranth. Characteristics that help separate this species from other Amaranthaceae are the thick and fleshy leaves and having male and female flowers on separate plants.

Gregg's Pigweed *Amaranthus greggii*

Amaranthaceae 61

Palmer's Pigweed, Quelite

Family: Amaranthaceae
Scientific Name: *Amaranthus palmeri*
Habit: Erect growing, to 3' or taller.
Leaves: Alternate, the petioles as long as or longer than the blades; blades rhombic to ovate, up to 3 ⅛" long.
Flowers: Male and female flowers on separate plants in slender terminal spikes, or in clusters along the stem; bracts twice as long as the sepals; sepals 5, scalelike; petals none.
Fruit: A one-seeded utricle (membranous, inflated), the top falling as a lid.
Bloom Period: All seasons.
Distribution: Cameron, Hidalgo, Willacy, and Starr counties.
Comments: We have many species of *Amaranthus,* and they can be difficult to identify without optic aid and using keys. Members of this genus are also called carelessweed or pigweed. Seeds and leaves of *Amaranthus* species have been important food items for ancient cultures in North and South America. Seed heads of *A. palmeri* are used for cage bird food. The plant is frequent in disturbed sites and along edges of cultivated fields.

Palmer's Pigweed, Quelite *Amaranthus palmeri*

Low Amaranth

Family: Amaranthaceae
Scientific Name: *Amaranthus polygonoides* [*A. berlandieri*]
Habit: Erect annuals up to 18" tall, usually less.
Leaves: Alternate; blades oblong to lanceolate, up to 1" long, usually with gray markings.
Flowers: Male and female flowers tiny, on the same plant, crowded in small clusters in the leaf axils.
Fruit: A one-seeded utricle (membranous, inflated), the top falling as a lid.
Bloom Period: All seasons.
Distribution: Cameron, Hidalgo, Willacy, and Starr counties.
Comments: This *Amaranthus* is usually smaller than the other species, and the gray markings on the leaves point to this species. However, the gray markings are not unique to *A. polygonoides.* Seeds and leaves of amaranths in general have been an important source of food for ancient cultures in the Americas. This species produces smaller leaves, but it produces an abundance of seeds.

Low Amaranth *Amaranthus polygonoides*

Spiny Pigweed

Family: Amaranthaceae

Scientific Name: *Amaranthus spinosus*

Habit: Erect plants up to 4' tall, with paired spines at many of the nodes.

Leaves: Alternate; blades ovate or lanceolate, up to 4" long.

Flowers: In terminal and axillary spikes or clusters, male and female flowers on the same plant. Sepals and bracts scalelike; petals absent.

Fruit: A one-seeded utricle (membranous, slightly inflated), the top sometimes falling as a lid.

Bloom Period: Summer, fall.

Distribution: Cameron County.

Comments: Characteristics that help separate this species from other Amaranthaceae are pairs of spines at many of the nodes and having male and female flowers on the same plant. Although many of the amaranths appear to be spiny, this is the only one in our area that actually has spines. The seeds and leaves of Amaranths have been important food items for ancient cultures in North America and South America. They produce a surprisingly large number of seeds.

Spiny Pigweed *Amaranthus spinosus*

Silverhead

Family: Amaranthaceae

Scientific Name: *Blutaparon vermiculare* [*Philoxerus vermicularis*]

Habit: Prostrate perennials, forming mats.

Leaves: Opposite, without petioles, thick and fleshy, linear to oblong.

Flowers: Perfect, in dense cylindrical or roundish white, ½"-long heads at the stem ends. Sepals and bracts scalelike; petals absent.

Fruit: A one-seeded utricle (membranous, slightly inflated), indehiscent.

Bloom Period: All seasons.

Distribution: Cameron and Willacy counties.

Comments: The plants form mats in sandy soils, and the white, silvery flower heads are held above the plants. These characteristics help separate this species from other Amaranthaceae.

Silverhead *Blutaparon vermiculare*

Amaranthaceae 63

Shiny Cock's Comb

Family: Amaranthaceae
Scientific Name: *Celosia nitida*
Habit: Perennial hairless herbs from a large, woody taproot, with weak stems that are sometimes vinelike. The young stems are reddish.
Leaves: Alternate, the blades ovate, up to 2 ¾" long.
Flowers: Perfect, tiny, with 3 stigmas, in terminal or axillary white or silvery spikes. Sepals and bracts scalelike; petals absent.
Fruit: A one-seeded utricle (membranous, slightly inflated), the top falling as a lid.
Bloom Period: Summer, fall.
Distribution: Cameron, Hidalgo, Starr counties.
Comments: This species grows in brush lands. Characteristics that help separate it from other Amaranthaceae are weak stems, the young ones reddish, and the perfect flowers in white or silvery spikes.

Shiny Cock's Comb *Celosia nitida*

Shiny Cock's Comb *C. nitida* flower

Palmer's Cock's Comb

Family: Amaranthaceae
Scientific Name: *Celosia palmeri*
Habit: Low branching perennials about 2' tall, somewhat woody, sometimes called shrubs.
Leaves: Alternate, the blades lanceolate to triangular, up to 2" long, often lobed basally. Older leaves sometimes turning reddish.
Flowers: Inflorescences short, roundish, from the branch ends; flowers perfect, scaly, enclosed in scaly bracts, with 2 stigmas; petals absent.
Fruit: A utricle (membranous, slightly inflated), with 3 or 4 seeds, the top falling as a lid.
Bloom Period: Spring, summer.
Distribution: Starr County.
Comments: For more than 50 years, there have been no reports of this species. It was rediscovered recently by Tom Patterson. Its 2 stigmas distinguish it from *C. nitida* which has 3 stigmas. The two species were seen growing together and may be hybridizing.

Palmer's Cock's Comb *Celosia palmeri*

Drummond's Snake Cotton

Family: Amaranthaceae
Scientific Name: *Froelichia drummondii*
Habit: Erect annuals up to 4' tall.
Leaves: Opposite; petioles present on lower part, very short or absent on upper part; blades lanceolate to ovate, up to 5 ½" long and 1 ½" broad, densely hairy on the lower surfaces.
Flowers: Perfect, congested in whitish clumps, concealing the flower stalks, very hairy.
Fruit: A one-seeded utricle (membranous, slightly inflated), indehiscent.
Bloom Period: Mostly spring, but sporadically throughout the year, depending on the rains.
Distribution: Hidalgo, Willacy, and Starr counties.
Comments: The genus was named for German physician and botanist Joseph Aloys von Froelich and for Thomas Drummond. This species is a larger version of slender snake cotton, *F. gracilis*. Its smallest leaves are at least ⅜" broad; those of *F. gracilis* are narrower. *F. drummondii* is fairly widespread in sandy soils in the southern part of Texas.

Drummond's Snake Cotton *Froelichia drummondii*

Slender Snake Cotton

Family: Amaranthaceae
Scientific Name: *Froelichia gracilis*
Habit: Hairy annuals. Stems branching, about 24" long, erect or falling over.
Leaves: Opposite, the upper ones without petioles; blades lanceolate.
Flowers: Perfect, in dense heads, the sepals and bracts scalelike; when in fruit, the sepals are white with woolly hairs.
Fruit: A one-seeded utricle (membranous, slightly inflated), indehiscent.
Bloom Period: All seasons.
Distribution: Cameron, Hidalgo, Willacy, and Starr counties.
Comments: This species is common in sandy soils. Characteristics that help separate this species from other Amaranthaceae are the hairiness of the plants; the stems often falling over; the opposite leaves; and the white-woolly fruiting sepals.

Slender Snake Cotton *Froelichia gracilis*

Amaranthaceae

Coast Globe Amaranth

Family: Amaranthaceae
Scientific Name: *Gomphrena nealleyi*
Habit: Hairy herbs. Stems weak, often falling over.
Leaves: Opposite, without petioles; blades spatulate or oblong, up to 1 ⅝" long.
Flowers: Perfect, in terminal or axillary heads. Sepals scalelike; bracts usually purplish, rarely white.
Fruit: A one-seeded utricle (membranous, slightly inflated), indehiscent.
Bloom Period: All seasons.
Distribution: Cameron, Hidalgo, and Willacy counties.
Comments: Characteristics that help separate this species from other Amaranthaceae are the short plants with flowers in purplish, or rarely white, heads.

Coast Globe Amaranth *Gomphrena nealleyi*

Coast Globe Amaranth *G. nealleyi* with white inflorescences

Woolly Cotton Flower

Family: Amaranthaceae
Scientific Name: *Gossypianthus lanuginosus* var. *lanuginosus* [*Guilleminea lanuginosa*]
Habit: Perennial herbs. Stems woolly to hairless, weak, often falling over, or prostrate, radiating outward from the base.
Leaves: Basal leaves in a tight cluster, up to 3 ½" long; upper stem leaves smaller, opposite, obovate, with soft straight hairs.
Flowers: Perfect, woolly, in axillary clusters. Sepals and bracts scalelike.
Fruit: A one-seeded utricle (membranous, slightly inflated), indehiscent.
Bloom Period: Spring, summer, fall.
Distribution: Cameron, Hidalgo, and Starr counties.
Comments: The genus name is derived from a Latin word meaning "cottony." The specific epithet means "woolly." We have two varieties in our area: var. *lanuginosus* has broader leaves and is hairier; var. *tenuiflorus* has narrower leaves and is less hairy.

Woolly Cotton Flower *Gossypianthus lanuginosus* var. *lanuginosus.*

Woolly Cotton Flower *G. lanuginosus* var. *tenuiflorus.*

Palmer's Bloodleaf

Family: Amaranthaceae
Scientific Name: *Iresine palmeri*
Habit: Weak-stemmed shrubs or twining vines.
Leaves: Opposite, with petioles; blades ovate or lanceolate, up to 2 ⅜" long.
Flowers: Male and female flowers on separate plants; inflorescences several from the leaf axils; densely woolly in fruit. Sepals scalelike; petals absent.
Fruit: A one-seeded utricle (membranous, slightly inflated), indehiscent.
Bloom Period: Spring, summer, fall.
Distribution: Cameron County.
Comments: Although this species does not have colorful leaves, some other species in the genus do have colorful leaves, hence the common name. In Texas, this species is reported only from the Brownsville and Olmito areas in Cameron County. The genus name is derived from Greek, and refers to the woolly nature of the fruit.

Palmer's Bloodleaf *Iresine palmeri*

Espanta Vaqueros

Family: Amaranthaceae
Scientific Name: *Tidestromia lanuginosa*
Habit: Mostly prostrate annuals forming large mounds.
Leaves: Opposite, grayish to whitish and shiny, covered with branched or star-shaped hairs; blades obovate, up to 1 ¼" long.
Flowers: Perfect, inconspicuous, in clusters in the leaf axils; sepals scalelike; petals absent.
Fruit: A one-seeded utricle (membranous, slightly inflated), indehiscent.
Bloom Period: Spring, summer, fall.
Distribution: Cameron, Hidalgo, Willacy, and Starr counties.
Comments: The plants form large whitish or silvery mounds or mats that glisten in the sunlight or moonlight. They are abundant in the sand dunes at the coast. The meaning of the Spanish common name is not clear. Translated roughly as "startle cowboys" (or "cowherds"), it might suggest a person being startled by the brightly gleaming plants in the moonlight.

Espanta Vaqueros *Tidestromia lanuginosa* var. *lanuginosa*

Anacardiaceae 67

Brazilian Pepper, Peruvian Pepper

Family: Anacardiaceae
Scientific Name: *Schinus terebinthifolius*
Habit: Dioecious trees (male and female flowers on separate plants) up to 10' or taller, with drooping branches.
Leaves: Alternate, compound, with about 7 ovate leaflets about 1 ¾" long. When crushed, the leaves have a turpentine smell.
Flowers: Small, white, in a dense inflorescence.
Fruit: Roundish drupes about 5/16" broad, reddish when mature, in tight clusters. When crushed, they have a tangy, pepperlike smell.
Bloom Period: Fall, winter.
Distribution: Cameron and Hidalgo counties.
Comments: Brazilian pepper is native to Brazil. It has been cultivated in warmer parts of the United States because of the showy drupes produced in winter. The seeds are quickly spread by birds, and the plants have become pests in Florida (infesting over one million acres), southern Texas, and southern California. They are especially fond of damp sites. Once established, and given the opportunity, the plants overtop and shade out other vegetation. The leaves also produce a chemical that inhibits germination and growth of other plants. Contact with the plant can cause dermatitis.

Brazilian Pepper, Peruvian Pepper *Schinus terebinthifolius*

Poison Ivy *Toxicodendron radicans* with flowers

Poison Ivy

Family: Anacardiaceae
Scientific Name: *Toxicodendron radicans*
Habit: Vines or sometimes shrubs.
Leaves: Alternate, compound with 3 leaflets; leaflets more or less ovate, up to 8" long, usually less, usually turning red in the fall or winter.
Flowers: Small, whitish, in axillary inflorescences.
Fruit: Whitish, roundish capsules up to 5/16" broad.
Bloom Period: Generally spring and summer; in mild seasons, also fall and winter.
Distribution: Cameron County.
Comments: Only one vigorous population is known, in Cameron County. It is suspected to have been introduced through nursery plants. This population is definitely out of the normal distribution of the species. It has not been reported farther south in Texas than San Patricio County but is common along waterways in

Poison Ivy *T. radicans* fruit

Mexico. The leaves are very toxic to the touch, causing severe dermatitis, and some especially sensitive individuals are affected by the evaporated materials from the plants. Fumes from burning plants are especially toxic.

Large Bishop's Weed

Family: Apiaceae (Umbelliferae)

Scientific Name: *Ammi majus*

Habit: Erect annuals with hollow stems up to 32" tall.

Leaves: Once or twice compound, up to 8" long and 5½" broad.

Flowers: In compound umbels (peduncles or pedicels all originating from the same point). Petals tiny, white, about 1/16" long.

Fruit: Dry, separating into 2 equal halves, about 1/16" tall.

Bloom Period: Spring, summer.

Distribution: Cameron, Hidalgo, and Starr counties.

Comments: This is a cultivated plant that has become naturalized. The tiny individual flowers occur in large, lacy-looking umbels. Children used to punch holes in a card or paper to produce a design or spell a name. Then they would pull away the smaller clusters of flowers, push the peduncles through the holes, and laboriously tie them on the reverse side, producing a lacy design. Large bishop's weed is a host plant for the Black Swallowtail butterfly.

Large Bishop's Weed *Ammi majus*

Butler's Sand Parsley

Family: Apiaceae (Umbelliferae)

Scientific Name: *Ammoselinum butleri*

Habit: Short-lived annuals. Stems slender, erect or leaning.

Leaves: Once or twice compound.

Flowers: In compound axillary umbels (peduncles or pedicels all originating from the same point). Peduncles very short or absent. Petals tiny, white, about 1/16" long.

Fruit: Dry, separating into 2 equal halves, about 3/16" tall, with corky longitudinal ribs.

Bloom period: Spring.

Distribution: Cameron and Hidalgo counties.

Comments: This species is very common in our area. It is a host plant for the Black Swallowtail butterfly. Although commonly seen, it has not been reported for our area, probably because of confusion with *A. popei* (not pictured). *A. popei* is distinguished from *A. butleri* by its mostly terminal inflorescences with noticeable peduncles.

Butler's Sand Parsley *Ammoselinum butleri*

Apiaceae

Rabbit Lettuce

Family: Apiaceae (Umbelliferae)
Scientific Name: *Bowlesia incana*
Habit: Prostrate to leaning annuals with star-shaped hairs, or sometimes hairless.
Leaves: Opposite; petioles up to 2 ¾" long; blades roundish, deeply lobed, up to 1 ¾" broad but usually less.
FLOWERS: Inflorescences from the leaf axils; flowers white, 6 or fewer, very tiny.
Fruit: Ellipsoid, about ⅛" broad.
Bloom Period: Spring, summer.
Distribution: Hidalgo County.
Comments: Most members of the Apiaceae are easily recognized, with their typical umbrella-shaped inflorescences. This species, with its few-flowered inflorescences with tiny flowers, is more difficult to recognize. Closer, careful observation is necessary. Rabbit lettuce is widespread in the United States, Mexico, Europe, and Asia. It generally occurs in moist places in the wild and in frequently watered localities such as lawns.

Rabbit Lettuce *Bowlesia incana*

Rattlesnake Weed

Family: Apiaceae (Umbelliferae)
Scientific Name: *Daucus pusillus*
Habit: Annuals up to 36" tall.
Leaves: Alternate, up to 4" long, several times compound, with narrow segments.
Flowers: White, in compound umbels (all the peduncles or pedicels originating from the same point); umbels subtended by several compound bracts up to ¾" long.
Fruit: Cylindrical, up to 3⁄16" long, with barbed spines.
Bloom Period: Spring.
Distribution: Cameron, Hidalgo, and Willacy counties.
Comments: The characteristic of the large compound bracts subtending the umbels is shared by large bishop's weed, *Ammi majus*. However, in that species the bracts are larger, and the fruit are not spiny. Carrots are also in the *Daucus* genus. Warning: If one member of a plant family is edible, it does not mean that other species are edible. Poison hemlock is also in this family.

Rattlesnake Weed *Daucus pusillus*

Hierba Del Sapo

Family: Apiaceae (Umbelliferae)
Scientific Name: *Eryngium nasturtiifolium*
Habit: Mostly prostrate, prickly plants.
Leaves: Lobed or compound, up to 4" long and 2" broad, the tips usually spiny.
Flowers: In spiny heads up to 1" tall; petals tiny, white.
Fruit: Dry, separating into 2 equal halves, about ⅛" tall, with many white scales.
Bloom Period: Winter, spring, summer.
Distribution: Cameron, Hidalgo, Willacy, and Starr counties.
Comments: Hierba del sapo is often found in low places that dry out slowly after rains. The common name, translated as "toad weed," comes from this habit. All parts of the plant are prickly, making this species easy to identify.

Hierba Del Sapo *Eryngium nasturtiifolium*

Hierba Del Sapo *E. nasturtiifolium* inflorescence

Texas Spread Wing

Family: Apiaceae (Umbelliferae)
Scientific Name: *Eurytaenia texana*
Habit: Annuals up to 39" tall.
Leaves: Alternate; blades compound, ovate in outline, up to 4" long; segments linear or various shapes.
Flowers: Umbels at the stem ends, compound; petals 4, white, about ⅛" long.
Fruit: Dry, winged, separating into 2 equal halves, up to ⁵⁄₁₆" tall.
Bloom Period: Spring, summer.
Distribution: Hidalgo and Starr counties.
Comments: This species is fairly widespread in scattered populations in Texas, except for West Texas.

Sombrerillo, Dollar Weed

Family: Apiaceae (Umbelliferae)
Scientific Name: *Hydrocotyle bonariensis*
Habit: Creeping perennials, rooting at the nodes. Large colonies look like a mass of small, green, flat umbrellas.
Leaves: Alternate; petioles attached to the center of the blade (peltate); blades roundish, up to 2" or broader, with rounded teeth.
Flowers: In compound umbels (peduncles or pedicels all attached at the same point). Petals white to yellow, tiny.
Fruit: Dry, separating into 2 halves about ¹⁄₁₆" tall.

Texas Spread Wing *Eurytaenia texana*

Apiaceae

Bloom Period: Spring, summer.
Distribution: Cameron, Hidalgo, and Willacy counties.
Comments: This species is easily recognized by the prostrate stems and the leaf blades with the petioles attached at the center. The plants occur in moist places. The Spanish word *sombrerillo* translates as "little hat," referring to the appearance of the leaves.

Sombrerillo, Dollar Weed *Hydrocotyle bonariensis*

Prairie Dog Sunshade

Family: Apiaceae (Umbelliferae)
Scientific Name: *Limnosciadium pumilum*
Habit: Annuals. Stems low and spreading out, up to 16" tall; stems hollow.
Leaves: Alternate, variable in shape, the margins entire to compound with long pinnae.
Flowers: In compound umbels (peduncles or pedicels all attached at the same point). Flowers tiny, white; calyx lobes prominent.
Fruit: Dry, separating into 2 oval to roundish halves about $\frac{1}{8}$" tall, ribbed, with corky wings.
Bloom Period: Spring, summer.
Distribution: Cameron and Willacy counties.
Comments: Characteristics that help identify this species are the corky ribbed fruit without bristles.

Prairie Dog Sunshade *Limnosciadium pumilum*

Hedge Parsley

Family: Apiaceae (Umbelliferae)
Scientific Name: *Torilis arvensis*
Habit: Erect hairy annuals with hollow stems up to 24" tall.
Leaves: Alternate, twice compound, the divisions linear or lanceolate.
Flowers: Tiny, white, on long peduncles.
Fruit: $\frac{1}{8}$" or slightly taller, densely covered with barbed bristles.
Bloom Period: Spring, summer.
Distribution: Hidalgo and Willacy counties.
Comments: A characteristic that helps identify this species is the fruit with its barbed spines. It is an introduction from the Old World.

Hedge Parsley *Torilis arvensis*

Knotted Hedge Parsley

Family: Apiaceae (Umbelliferae)
Scientific Name: *Torilis nodosa*
Habit: Hairy annuals with hollow stems up to 24" tall.
Leaves: Alternate, twice compound, the divisions sometimes threadlike.
Flowers: Tiny, white, in umbels or tight clusters originating opposite the leaves. Peduncles short or absent.
Fruit: ⅛" or taller, covered with barbed bristles.
Bloom Period: Spring.
Distribution: Cameron County.
Comments: Important characteristics are inflorescences originating opposite the leaves, with peduncles short or absent; and fruit covered with barbed bristles. This is an introduction from the Old World.

Knotted Hedge Parsley *Torilis nodosa*

Woolly Rocktrumpet, Flor De San Juan

Family: Apocynaceae
Scientific Name: *Telosiphonia lanuginosa* [*Macrosiphonia lanuginosa, M. macrosiphon*]
Habit: Low shrubs or woody herbs with milky sap, up to 12" or taller. Younger parts hairy.
Leaves: Opposite, simple, obovate to elliptic, up to 1 ¼" long.
Flowers: Solitary, regular; petals white, united, forming a tube up to 4" long and 5 lobes up to ¾" long.
Fruit: Separating into 2 dry, long and curving pods that open on one side (sometimes one aborts); seeds with long white hairs.
Bloom Period: Spring, summer, fall.
Distribution: Hidalgo and Starr counties.
Comments: The tufts of hairs on the seeds act like parachutes to carry them long distances on windy days. The flowers smell like gardenias and are open from evening to morning. They are pollinated by hawk moths.

Woolly Rocktrumpet, Flor De San Juan *Telosiphonia lanuginosa*

Woolly Rocktrumpet, Flor De San Juan *T. lanuginosa* fruit

Asclepiadaceae

Swan Flower

Family: Aristolochiaceae
Scientific Name: *Aristolochia erecta* [*A. longiflora*]
Habit: Low perennials, often prostrate, with fleshy roots, usually growing in sandy soils.
Leaves: Mostly linear, up to 5" or longer.
Flowers: Reddish purple, S-shaped, up to 4" long.
Fruit: a capsule about ¾" long, with triangular, black, flattened seeds.
Bloom Period: Spring, summer, fall.
Distribution: Hidalgo, Willacy, and Starr counties.
Comments: The leaves of this species are inconspicuous, resembling grass. The swan flower is a food source for the larvae of the Pipevine Swallowtail butterfly, and individual plants can frequently be located by following foraging Pipevine Swallowtail butterflies. The appearance of the flowers attracts flies, which are believed necessary for the plant to be pollinated. We have another species, *A. pentandra* (not pictured), which is a twining vine. Both species have the unusual purple S-shaped flowers.

Swan Flower *Aristolochia erecta*

Swan Flower *A. erecta* fruit with Pipevine Swallowtail larva

Veintiunilla

Family: Asclepiadaceae
Scientific Name: *Asclepias curassavica*
Habit: Perennial herbs up to 28" tall.
Leaves: Opposite; blades ovate, up to 4 ⅜" long.
Flowers: Clustered together, bright orange-red with yellow centers (some plants with all-yellow flowers); corolla lobes up to ⅜" long.
Fruit: Separating into 2 dry pods (follicles) up to 3 ⅛" long. Seed dry, ¼" tall with attached silk.
Bloom Period: Spring, summer.
Distribution: Cameron and Hidalgo counties.
Comments: This attractive milkweed is introduced from the American tropics. Generally, it is an escaped cultivar in our area. It is planted to attract Monarch and Queen butterflies. Like the other milkweeds, it bleeds a thick, milky sap when damaged. The genus is named for Asklepios, the Greek god of medicine, because some species are used for their medicinal properties.

Veintiunilla *Asclepias curassavica*

Emory's Milkweed

Family: Asclepiadaceae
Scientific Name: *Asclepias emoryi*
Habit: Perennial herbs up to 16" tall.
Leaves: Opposite; blades narrowly ovate, up to 4 ¾" long.
Flowers: In umbels from the leaf axils; petals basally grown together, the lobes greenish, up to ¼" or slightly longer.
Fruit: Separating into 2 dry pods (follicles) up to 3 ⅛" long (often one of them aborting), opening along one side and releasing seeds with long white hairs.
Bloom Period: Spring, summer, fall.
Distribution: Hidalgo and Starr counties.
Comments: The milkweeds are sometimes so similar that keys and optic aid are necessary in order to distinguish them. The hairs on the seeds act like parachutes to carry them long distances on a windy day. The milkweeds are toxic but have been eaten by Native Americans after careful boiling to remove the poisons.

Emory's Milkweed *Asclepias emoryi*

Slim Milkweed

Family: Asclepiadaceae
Scientific Name: *Asclepias linearis*
Habit: Erect perennials up to 24" tall, bleeding milky sap when damaged.
Leaves: Opposite, the blades linear, up to 3" or longer.
Flowers: In umbels from the upper leaf axils; petals basally grown together, the lobes greenish, ⅛" or longer.
Fruit: Separating into 2 dry pods up to 4" long (often one of them aborting), opening along one side and releasing seeds with long white hairs.
Bloom Period: Spring, summer, fall.
Distribution: Cameron, Hidalgo, and Willacy counties.
Comments: This species is endemic to Texas. The narrow leaves and greenish flowers are important characteristics for recognizing this species. Milkweeds are well known as food plants for Monarch butterfly larvae, which absorb the milkweed poisons without being harmed. These poisons make both larvae and adults unpalatable to predators.

Slim Milkweed *Asclepias linearis*

Asclepiadaceae

Savannah Milkweed

Family: Asclepiadceae
Scientific Name: *Asclepias obovata*
Habit: Plants up to 20" tall, bleeding milky sap when damaged.
Leaves: Opposite, the blades variable in shape and size, the lower ones broadly ovate to oblong, the upper ones becoming smaller.
Flowers: Inflorescences terminal and from the uppermost leaf axils; petals basally grown together, the lobes greenish, up to ⅜" long. Five hoods that are not noticeably dilated ring the center of each flower.
Fruit: Separating into 2 dry pods 4 ¾" or longer (often one of them aborting), opening along one side and releasing seeds with long white hairs.
Bloom Period: Spring, summer, fall.
Distribution: Willacy County.
Comments: This is a new record for the LRGV. The hairs on the seeds act like parachutes to carry them long distances on a windy day. The poisons milkweeds produce discourage foraging mammals. Native Americans have used milkweeds as a food source after carefully boiling them to remove the toxic compounds.

Savannah Milkweed *Asclepias obovata*

Prairie Milkweed

Family: Asclepiadaceae
Scientific Name: *Asclepias oenotheroides*
Habit: Erect plants up to 18" tall, bleeding milky sap when damaged.
Leaves: Opposite, the blades ovate, up to 3" or longer.
Flowers: In umbels, from the upper leaf axils; petals basally grown together, the lobes greenish, ½" or longer.
Fruit: Separating into 2 dry pods up to 3 ½" long (often one of them aborting), opening along one side and releasing seeds with long white hairs.
Bloom Period: Spring, summer, fall.
Distribution: Cameron, Hidalgo, Willacy, and Starr counties.
Comments: This is one of our most common species of milkweed. It is a host plant to Queen butterflies. The fruit is called a schizocarp (*schizo* means "divide") because it divides into two parts, each seeming to be a separate fruit.

Prairie Milkweed *Asclepias oenotheroides*

Prairie Milkweed *A. oenotheroides* plant in fruit

Prostrate Milkweed

Family: Asclepiadceae
Scientific Name: *Asclepias prostrata*
Habit: Plants prostrate, bleeding milky sap when damaged.
Leaves: Opposite, triangular to lanceolate, up to 1 ⅜" long.
Flowers: from the leaf axils; petals basally grown together, the lobes greenish, occasionally with pink, up to ⅝" or longer.
Fruit: Separating into 2 dry pods about 2 ⅛" long (often one of the aborting), opening along one side and releasing seeds with long white hairs.
Bloom Period: Summer.
Distribution: Hidalgo and Starr counties.
Comments: The prostrate habit distinguishes this species from other *Asclepias* species. It is fairly uncommon in our area, usually growing in sandy soils.

Prostrate Milkweed *Asclepias prostrata*

Prostrate Milkweed *A. prostrata* young plant

Rubber Vine, Purple Allamanda

Family: Asclepiadaceae
Scientific Name: *Cryptostegia grandiflora*
Habit: Woody shrubs or vines with milky juice.
Leaves: Opposite, leathery, elliptic to roundish, with a whitish midrib, up to 4" long.
Flowers: Bell shaped, 5-lobed, 2–3½" broad, white inside, reddish purple outside, fading to pink. They greatly resemble the cultivated *Mandevilla*.
Fruit: Paired dry pods sharply angled, up to 4" long.
Bloom Period: Variable, throughout the year.
Distribution: Cameron, Hidalgo, and Starr counties.
Comments: These vines were introduced from tropical Africa as ornamentals. In Starr County they are vigorously growing in trees along the Rio Grande at Roma and are cultivated in Rio Grande City. In Weslaco, Hidalgo County, the vines are being cultivated. They are beautiful in flower, but they are very invasive and aggressive, overtopping the trees they climb on and eventually killing them. They are also poisonous. The vines reproduce vegetatively as well as by seed. The seeds, as in other members of the family, have feathery "parachutes" that carry them long distances with even a moderate wind. In India the vines are grown

Rubber Vine, Purple Allamanda *Cryptostegia grandiflora* flowers and fruit

Asclepiadaceae

for their milky juice, which is used to produce rubber.

Narrow Leaf Swallow Wort

Family: Asclepiadaceae
Scientific Name: *Cynanchum angustifolium*
Habit: Perennial twining vines; stems 4' or longer.
Leaves: Alternate, somewhat succulent; blades linear, up to 3 1/8" long and 3/8" broad.
Flowers: Cream in color or purplish, about 5/16" broad, in umbrella-like clusters.
Fruit: Separating into 2 dry pods up to 2 3/4" long (often one of the aborting), opening along one side and releasing seeds with long white hairs.
Bloom Period: Spring, summer, fall.
Distribution: Western Hidalgo County and Willacy County.
Comments: This plant is uncommon in the LRGV. As with other members of the family, a broken stem or leaf bleeds milky sap. The genus name comes from the Greek *kyon* (dog) and *ancho* (strangle) because most species are poisonous.

Narrow Leaf Swallow Wort *Cynanchum angustifolium*

Thicket Threadvine, Aphid Vine

Family: Asclepiadaceae
Scientific Name: *Cynanchum barbigerum*
Habit: Twining vines, bleeding milky sap when damaged.
Leaves: Opposite, the blades linear to elliptic, up to 1 1/8" long.
Flowers: Tiny, whitish or cream in color, in umbrella-like clusters.
Fruit: Separating into 2 dry pods about 2" long (often one of them aborting), opening along one side and releasing seeds with long white hairs.
Bloom Period: Spring, summer, fall.
Distribution: Cameron, Hidalgo, Willacy, and Starr counties.
Comments: This species is a host plant for Queen butterflies. The hairs on the seeds act like parachutes to carry them long distances on a windy day. An important characteristic is the hairiness of the upper sides of the petals.

Thicket Threadvine, Aphid Vine *Cynanchum barbigerum* with seeds being released

Talayote

Family: Asclepiadaceae
Scientific Name: *Cynanchum racemosum* var. *unifarium* [*C. unifarium*]
Habit: Twining vines bleeding milky sap when damaged.
Leaves: Opposite, the blades broadly ovate to lanceolate, up to 3½" long and 2 ¾" broad.
Flowers: In umbrella-like clusters; petals basally grown together, the lobes whitish or cream in color, about ⅛" long.
Fruit: Separating into 2 dry pods up to 4" long (often one of them aborting), opening along one side and releasing seeds with long white hairs.
Bloom Period: Spring, summer, fall.
Distribution: Cameron, Hidalgo, and Starr counties.
Comments: The broader leaves segregate this species from climbing milkweed, *Funastrum cynanchoides*. The hairs on the seeds act like parachutes to carry them long distances on a windy day.

Talayote *Cynanchum racemosum*

White Twinevine, Wavy Twinevine

Family: Asclepiadaceae
Scientific Name: *Funastrum clausum* [*Sarcostemma clausum*]
Habit: Twining, herbaceous perennial vines, bleeding milky sap when damaged, often producing runners along the ground 20' or longer.
Leaves: Opposite, the blades oblong to linear, up to 4" long, and about 1" or broader.
Flowers: About ⅜" broad, fragrant, whitish, in umbrella-like clusters; petals with very short hairs on the inner surfaces.
Fruit: Separating into 2 dry pods 4" or longer (often one of them aborting), opening along one side and releasing seeds with long white hairs.
Bloom Period: Spring, summer, fall.
Distribution: Cameron, Hidalgo, and Starr counties.
Comments: This species produces long runners along the ground, 20' or longer. It grows in low areas near water. The hairy petals and the narrow leaves are distinctive for this species. Also, the peduncles are usually as broad as the stem below them.

White Twinevine, Wavy Twinevine *Funastrum clausum*

Asclepiadaceae

Climbing Milkweed

Family: Asclepiadaceae
Scientific Name: *Funastrum cynanchoides* [*Sarcostemma cynanchoides*]
Habit: Twining vines, bleeding milky sap when damaged.
Leaves: Opposite, the blades heart shaped or arrowhead shaped, up to 1 ¾" long.
Flowers: In umbrella-like clusters; petals basally grown together, the lobes greenish to purple, up to ¼" long.
Fruit: Separating into 2 dry pods up to 2 ⅜" long (often one of them aborting), opening along one side and releasing seeds with long white hairs.
Bloom Period: Summer, fall.
Distribution: Cameron, Hidalgo, and Willacy counties.
Comments: The heart-shaped or arrowhead-shaped leaves are a major characteristic of this species. Climbing milkweed grows aggressively, covering fences and slower-growing vegetation. It is used as a nectar source for insects and is a host plant for the Queen, Monarch, and Soldier butterflies.

Climbing Milkweed *Funastrum cynanchoides* fruit

Climbing Milkweed *F. cynanchoides*

Rio Grande Plains Milkvine

Family: Asclepiadaceae
Scientific Name: *Matelea brevicoronata*
Habit: Hairy, prostrate perennials.
Leaves: Opposite, more or less ovate, rounded to shallowly heart shaped basally, up to 1 ⅝" long.
Flowers: In small clusters; corollas greenish, the lobes about 3/16" long.
Fruit: Pods (follicles) ovoid, knobby, about 2 ¾" long, opening along one side and releasing seeds with long white hairs.
Bloom Period: Spring, summer, fall.
Distribution: Hidalgo and Starr counties.
Comments: The milky sap smells like burnt rubber. This species is endemic to Texas. It can be separated from mesquite plains milkvine, *M. parviflora,* by its larger flowers. The two species also resemble prostrate milkweed, *Asclepias prostrata,* which has much larger flowers than either *Matelea* species.

Rio Grande Plains Milkvine *Matelea brevicoronata*

Mesquite Plains Milkvine

Family: Asclepiadaceae
Scientific Name: *Matelea parviflora*
Habit: Stems prostrate, occasionally leaning and falling over, bleeding milky sap when damaged.
Leaves: Opposite, the blades hairy and mostly ovate, patterned with white venation, up to 2" long.
Flowers: Very tiny, greenish to yellow, in umbrella-like clusters, on long peduncles.
Fruit: Separating into 2 dry knobby pods up to 3 ½" long (often one of them aborting), opening along one side and releasing seeds with long white hairs.
Bloom Period: Spring, summer.
Distribution: Hidalgo, Willacy, and Starr counties.
Comments: This species is endemic to Texas, growing in sandy soils. The milky sap smells like burnt rubber. Queen butterfly larvae were observed feeding on the leaves. The prostrate stems and tiny flowers separate it from other mateleas. Rio Grande plains milkvine, *M. brevicoronata* also has a prostrate growth habit, but it has larger flowers. A similar species, prostrate milkweed, *Asclepias prostrata*, has much larger flowers than either prostrate growing *Matelea*.

Mesquite Plains Milkvine *Matelea parviflora*

Mesquite Plains Milkvine *M. parviflora* knobby fruit

Pearl Net Leaf Milkvine

Family: Asclepiadaceae
Scientific Name: *Matelea reticulata*
Habit: Twining vines with hairy stems, bleeding milky sap when damaged.
Leaves: Opposite, the blades heart shaped, up to 4 ¾" long.
Flowers: In umbrella-like clusters, greenish or cream in color, basally grown together with a "pearl" dot in the center, the lobes about ⅛" long, with branching markings.
Fruit: Separating into 2 knobby dry pods (often one of them aborting), opening along one side and releasing seeds with long white hairs.
Bloom Period: Spring, summer, and fall.
Distribution: Cameron and Hidalgo counties.
Comments: The hairs on the seeds act like parachutes to carry them long distances on a windy day.

Pearl Net Leaf Milkvine *Matelea reticulata*

Asteraceae

Arrow Leaf Milkvine

Family: Asclepiadaceae
Scientific Name: *Matelea sagittifolia*
Habit: Woody, twining vines, bleeding milky sap when damaged.
Leaves: Opposite, arrowhead shaped, up to ¾" long.
Flowers: One or two from the leaf axils; petals greenish, basally grown together, the lobes twisted, up to ¾" long.
Fruit: Separating into 2 dry pods (often one of them aborting), opening along one side and releasing seeds with long white hairs.
Bloom Period: Spring.
Distribution: Hidalgo and Starr counties.
Comments: This species is endemic to Texas. With its small leaves and stems, and its greenish flowers, *M. sagittifolia* is inconspicuous when growing within brush thickets.

Arrow Leaf Milkvine *Matelea sagittifolia* in flower

Arrow Leaf Milkvine *M. sagittifolia* fruit

Asteraceae: Sunflower Family

The Asteraceae family (also called Compositae or sunflower family) has a unique flower arrangement that merits special attention. Using the common sunflower as an example, each of the yellow "petals" is a flower and is called a ray flower (or ray floret). Each of the tiny brown dots in the center is also a flower and is called a disk flower (or disk floret). The sunflower, then, is a composite of many flowers on a common receptacle; hence the name Compositae. The cluster of florets on one receptacle is called a "head." There is much variation of this pattern, and not all members of the family have both ray and disk florets. Each of the green parts at the base of a sunflower is called a phyllary. All the phyllaries together are called the involucre. See figure on page 8 in the introduction.

The sunflower "seed" is technically a fruit, called an achene. It has a pair of tiny points at the top. Other members of the family have a plumelike top, as in a dandelion. The top part of the fruit, in both cases mentioned, is called the pappus.

Peonia

Family: Asteraceae (Compositae)
Scientific Name: *Acourtia runcinata* [*Perezia runcinata*]
Habit: Perennial herbs producing a basal rosette arising from fleshy tubers.
Leaves: Basal. Blades obovate, up to 6 ¾" long, the margins lobed and sharp toothed.
Flowers: Many small purple to pinkish flowers clustered on a single receptacle. Individual flowers all 2-lipped.
Fruit: Achenes about ³⁄₁₆" tall; pappus of bristles about ½" or longer.
Bloom Period: Spring, summer, fall.
Distribution: Cameron, Hidalgo, and Starr counties.
Comments: The 2-lipped florets occur in only five of our species within the Asteraceae. Their purple color, plus the spiny leaves and lack of stems, define this species. The specific epithet is derived from Latin, meaning "saw toothed." *Peonia* is the Spanish word for peony.

Peonia *Acourtia runcinata*

Pink Desert Peony

Family: Asteraceae (Compositae)
Scientific Name: *Acourtia wrightii* [*Perezia wrightii*]
Habit: Perennials up to 39" or taller.
Leaves: Alternate; blades obovate, up to 3 ⅜" long, the margins with sharp teeth.
Flowers: Many small pink to whitish flowers clustered on a single receptacle. Individual flowers all 2-lipped.
Fruit: Achenes about ¹⁄₁₆" tall; pappus bristles about ⁵⁄₁₆" long.
Bloom Period: Summer, fall.
Distribution: Cameron County.
Comments: This species is rarely encountered in the LRGV. The 2-lipped flowers are unusual for the family. Four other species in our area share this characteristic. This species is recognized by its pink flowers, sharp-toothed leaf margins, and presence of stems.

Pink Desert Peony *Acourtia wrightii*

Asteraceae

Huisache Daisy

Family: Asteraceae (Compositae)
Scientific Name: *Amblyolepis setigera*
Habit: Slender annuals up to 20" tall.
Leaves: Alternate, without petioles, the upper ones sometimes clasping the stem. Blades oblanceolate to ovate, up to 3 ¼" long; margins entire.
Flowers: In heads (many flowers clustered on a single receptacle) borne singly. Ray flowers yellow, up to ¾" long; disk flowers yellow.
Fruit: Achenes hairy, about 3⁄16" long; pappus usually of 5 tiny scales.
Bloom Period: Spring.
Distribution: Hidalgo, Willacy, and Starr counties.
Comments: Huisache daisy would make a beautiful landscape plant and is common in wildflower seed mixes. It contains toxic materials that interfere with blood clotting.

Huisache Daisy *Amblyopsis setigera*

Rio Grande Ragweed

Family: Asteraceae (Compositae)
Scientific Name: *Ambrosia cheiranthifolia*
Habit: Erect perennial, scented and resinous herbs spreading by rhizomes.
Leaves: Opposite below, alternate above. Petioles absent. Blades gray, up to 2 ⅛" long, the margins usually entire, occasionally toothed.
Flowers: Male and female flowers without corollas, in separate heads (clustered on one receptacle). Male heads hanging downward, about ⅛" broad, only the anthers emerging from the enclosing bracts; female heads hanging downward, about ⅛" tall, enclosed in spiny bracts (involucre).
Fruit: Achenes very small, without pappus.
Bloom Period: Summer, fall.
Distribution: Cameron County. Rare, only known here from historic records. The plant in the photograph was growing in Nueces County.
Comments: This plant is on the federal and state lists of endangered species. The tiny flowers in pendulous heads define this group of Asteraceae. The gray, mostly entire leaves identify this species. The pollen is well known to be an allergen.

Rio Grande Ragweed *Ambrosia cheiranthifolia*

Field Ragweed

Family: Asteraceae (Compositae)
Scientific Name: *Ambrosia confertiflora*
Habit: Scented and resinous perennials 3' or taller.
Leaves: Alternate. Blades lobed and relobed, up to 6" long.
Flowers: Male and female flowers without corollas, in separate heads (clustered on one receptacle). Male heads hanging downward, about ⅛" broad, only the anthers emerging from the enclosing bracts; female heads hanging downward, about ⅛" tall, the enclosing bracts (involucre) with scattered spines.
Fruit: Achenes small, without pappus.
Bloom Period: Summer, fall.
Distribution: Cameron, Hidalgo, and Starr counties.
Comments: The tiny flowers in pendulous heads define this group of Asteraceae. The scattered spines on the female involucres identify this species. The pollen is well known to be an allergen, and the plants are very invasive. The seeds, however, were a food source for ancient Native American cultures.

Field Ragweed *Ambrosia confertiflora*

Western Ragweed

Family: Asteraceae (Compositae)
Scientific Name: *Ambrosia psilostachya*
Habit: Scented and resinous perennials up to 24" or taller.
Leaves: Opposite below, alternate above. Blades 3 ½" or longer, usually lobed only one time.
Flowers: Male and female flowers without corollas, in separate heads (clustered on single receptacles). Male heads hanging downward, about ⅛" broad, only the anthers emerging from the enclosing bracts (involucre); female heads hanging downward, about ⅛" tall, the enclosing bracts (involucre) with a single row of spines around the top and a central beak.
Fruit: Achenes small, without pappus.
Bloom Period: Summer, fall.
Distribution: Cameron, Hidalgo, Willacy, and Starr counties.
Comments: The tiny flowers in pendulous heads define this group of Asteraceae. The female involucres with a single row of spines around the top identify this species. The pollen is well known to be an allergen, and the plants are very invasive. The seeds are edible.

Western Ragweed *Ambrosia psilostachya*

Asteraceae

Texas Giant Ragweed

Family: Asteraceae (Compositae)
Scientific Name: *Ambrosia trifida* var. *texana*
Habit: Tall annuals up to 10' or taller.
Leaves: Mostly opposite; blades rough, deeply divided 3 or 5 times, roundish to lanceolate in outline, up to 8" long; lobes on lower leaves often redivided.
Flowers: Male and female flowers tiny, without corollas, in separate heads. Male heads in clusters, hanging downward, about ⅛" broad; female heads in clusters, containing one flower.
Fruit: Fruiting heads with spines and ridges; achenes about 3/16" tall.
Bloom Period: Summer and fall.
Distribution: Cameron and Hidalgo counties.
Comments: This species is recognized by its tiny flowers in pendulous heads, combined with the large plant size and the deeply divided leaves. The pollen causes hay fever. Fortunately for us, the plants are not abundant in our area. They usually grow in seasonally moist areas.

Texas Giant Ragweed *Ambrosia trifida* var. *texana*

Texas Giant Ragweed
A. trifida inflorescence

Prairie Broomweed

Family: Asteraceae (Compositae)
Scientific Name: *Amphyachyris dracunculoides* [*Xanthocephalum dracunculoides*]
Habit: Highly branched sticky annuals up to 3' or taller.
Leaves: Alternate, resinous and sticky, very small and linear, about ¾" long.
Flowers: Heads about ⅝" broad, 1 per peduncle, but numerous. Ray and disk flowers yellow.
Fruit: Achenes flattened, tiny, with microscopic pappus.
Bloom Period: Fall.
Distribution: Cameron and Willacy counties.
Comments: The highly branched, sticky plants with insignificant leaves call attention to this species. The plants have been bundled together and used as brooms. Dense populations indicate disturbed areas.

Prairie Broomweed *Amphyachyris dracunculoides*

Lazy Daisy
Family: Asteraceae (Compositae)
Scientific Name: *Aphanostephus ramosissimus* var. *ramosissimus*
Habit: Annuals with weak stems that fall over.
Leaves: Alternate, ovate, up to 2" or longer, the margins lobed or toothed.
Flowers: Opening late morning or early afternoon. Heads up to ¾" broad. Ray flowers white to bluish; disk flowers yellow.
Fruit: Achenes tiny, pappus of tiny scales; hairs, when present, are straight.
Bloom Period: Spring.
Distribution: Cameron, Hidalgo, and Starr counties
Comments: *A. skirrhobasis* is similar to *A. ramosissimus* but has larger heads and slightly larger achenes. The hairs on its achenes, when present, are curved, whereas those on *A. ramosissimus* are straight. The genus name is derived from Greek, meaning "inconspicuous," referring to the pappus.

Lazy Daisy *Aphanostephus ramosissimus* var. *ramosissimus*

Kidder's Lazy Daisy
Family: Asteraceae (Compositae)
Scientific Name: *Aphanostephus skirrhobasis* var. *kidderi*
Habit: Annuals with weak stems that fall over.
Leaves: Alternate, oblong, up to 2 ¾" long, the margins toothed or lobed.
Flowers: Heads up to 1" broad. Ray flowers white to bluish, disk flowers yellow.
Fruit: Achenes tiny, with pappus of 5–10 tiny scales that taper to a long point.
Bloom Period: Spring, summer.
Distribution: Hidalgo and Starr counties.
Comments: The daisylike flowers are slightly larger than those of the lazy daisy, *A. ramosissimus*. Pappus scales taper to a long point, whereas in the coastal lazy daisy, var. *thalassius*, they are unequal and smaller. Var. *kidderi* is mostly seen inland, away from the coast.

Kidder's Lazy Daisy *Aphanostephus skirrhobasis* var. *kidderi*

Coastal Lazy Daisy
Family: Asteraceae (Compositae)
Scientific Name: *Aphanostephus skirrhobasis* var. *thalassius*
Habit: Annuals with weak stems that fall over.
Leaves: Alternate, oblong, up to 2 ¾" long, the margins toothed or lobed.

Coastal Lazy Daisy *Aphanostephus skirrhobasis* var. *thalassius*

Asteraceae 87

Flowers: Heads up to 1" broad. Ray flowers white to bluish, disk flowers yellow.
Fruit: Achenes tiny, with pappus of tiny scales.
Bloom Period: Spring, summer.
Distribution: Cameron, Willacy, and Starr counties.
Comments: Varieties of the same species usually look very much alike and are not easily distinguished. The pappus of var. *thalassius* usually has 5 unequal scales, whereas Kidder's lazy daisy, var. *kidderi* has 5–10 scales that taper to a long point. Var. *thalassius* is mostly seen along the coast. The hairs on the involucres of *A. skirrhobasis* var. *skirrhobasis* (Arkansas lazy daisy, not illustrated) are stiff and rough, whereas those on var. *thalassius* are soft. The pappi (plural for pappus) of these two varieties are similar, segregating them from var. *kidderi*.

White Sage *Artemisia ludoviciana* subsp. *mexicana*

White Sage
Family: Asteraceae (Compositae)
Scientific Name: *Artemisia ludoviciana* subsp. *mexicana*
Habit: Erect perennials, slightly woody basally, up to 3' or taller.
Leaves: Alternate, more or less linear or sometimes dissected, grayish on lower surfaces.
Flowers: Ray flowers white, one or few, usually infertile, often overlooked; disk flowers white.
Fruit: Achenes less than 1/16" long, without pappus.
Bloom Period: Fall.
Distribution: Cameron and Willacy counties.
Comments: At first glance, this species may suggest a dwarfed species of *Baccharis*. The *Baccharis* species are in bloom at the same time. A population was observed one mile north of the Willacy County border, making it reasonable to assume it is also present in Willacy County.

Consumption Weed *Baccharis halimifolia*

Consumption Weed
Family: Asteraceae (Compositae)
Scientific Name: *Baccharis halimifolia*
Habit: Shrubs with few branches, up to 6' or taller.
Leaves: Alternate, elliptic to ovate; margins toothed, often with irregular lobes.
Flowers: Male and female heads on separate plants; ray flowers absent; disk flowers whitish.
Fruit: Achenes about 1/16" tall, with pappus bristles about 3/8" long.
Bloom Period: Fall
Distribution: Cameron County.
Comments: This species is distinguished from the other *Baccharis* species by its irregular-shaped leaves. It has not been previously reported from the LRGV. It is more common in East Texas, where it is considered an aggressive pest. It occurs on disturbed sites such as fallow farmlands.

Roosevelt Weed, Depression Weed

Family: Asteraceae (Compositae)
Scientific Name: *Baccharis neglecta*
Habit: Shrubs with few branches, up to 9' or taller.
Leaves: Alternate, linear or narrowly elliptic, up to 3 1/8" or longer, with tiny glands.
Flowers: Male and female heads on separate plants. Ray flowers absent; disk flowers whitish.
Fruit: Achenes about 1/16" tall, with pappus bristles 3/8" long.
Bloom Period: Fall.
Distribution: Cameron, Hidalgo, and Starr counties.
Comments: The large shrubs resemble a young willow tree, with which it can be confused when not in bloom. The weight of the blooms causes the limbs to sag. The common name Roosevelt weed or Depression weed comes from the fact that during the Depression of the 1930s, fields were left unplanted. *Baccharis* began to grow in these fields. The pollen is carried by the wind and is allergenic. Until recently *B. neglecta* has been considered a synonym of *B. salicina.* They have again been split into two separate species in the *Flora of North America,* vol. 22 (2000). *B. neglecta* remains in our area, and *B. salicina* occurs farther west in Texas.

Roosevelt Weed, Depression Weed *Baccharis neglecta*

Seepwillow

Family: Asteraceae (Compositae)
Scientific Name: *Baccharis salicifolia*
Habit: Shrubs up to 11' or taller.
Leaves: Alternate; blades lanceolate, up to 5 1/8" long and 3/4" broad, with tiny glands; margins toothed.
Flowers: Male and female heads on separate plants; ray flowers absent; disk flowers whitish.
Fruit: Achenes about 1/16" tall, with pappus bristles about 3/16" long.
Bloom Period: Fall.
Distribution: Cameron, Hidalgo, and Starr counties.
Comments: The specific epithet refers to the similarity of the leaves to those of willow (genus *Salix*). The broad, lanceolate leaves distinguish this species from the other tall-growing *Baccharis* species.

Seepwillow *Baccharis salicifolia*

Asteraceae

Texas Baccharis
Family: Asteraceae (Compositae)
Scientific Name: *Baccharis texana*
Habit: Shrubs up to 36" tall.
Leaves: Alternate, linear, up to 1 ¼" long.
Flowers: Male and female heads on separate plants. Ray flowers absent; disk flowers whitish.
Fruit: Achenes about 3⁄16" tall; pappus bristles up to ½" long.
Bloom Period: Fall.
Distribution: Cameron and Willacy counties.
Comments: This species is much shorter than the other *Baccharis* species. It is often found growing in saline areas at the edges of lomas near the coast.

Texas Baccharis *Baccharis texana*

Whitened Leaf Bahia
Family: Asteraceae (Compositae)
Scientific Name: *Bahia absinthifolia* var. *dealbata*
Habit: Perennials up to 16" tall.
Leaves: Gray-hairy, mostly opposite; blades up to 2" long, the margins entire, lobed, or toothed. Minute resin globules occur in tiny pits along the leaf.
Flowers: Heads up to 1 ¼" broad. Ray flowers yellow, disk flowers yellow.
Fruit: Achenes about 3⁄16" tall, the pappus of tiny scales.
Bloom Period: Spring, fall.
Distribution: Cameron and Starr counties.
Comments: This would be an attractive ornamental in xeric gardens. The grayish plants have a resinous odor.

Whitened Leaf Bahia *Bahia absinthifolia* var. *dealbata*

Fragrant Beggar Ticks
Family: Asteraceae (Compositae)
Scientific Name: *Bidens pilosa* [*B. odorata*]
Habit: Annual or sometimes perennial herbs up to 3' or taller.
Leaves: Opposite; blades dissected into 3 or more leaflets, up to ⅞" long.
Flowers: Heads up to ¾" broad. Ray flowers white, disk flowers yellow.
Fruit: Achenes cylindrical, variable in height, up to 5⁄16" tall with pappus of 2–4 barbed awns about 1⁄16" long.
Bloom Period: Spring, fall.
Distribution: Cameron and Hidalgo counties.
Comments: The white ray flowers plus the dissected leaves without petioles help identify this species. It is considered by many to be an introduced species. The genus is named from the Latin *bis* (twice) and *dens* (tooth) for the barbed achenes.

Fragrant Beggar Ticks *Bidens pilosa*

Sea Ox Eye

Family: Asteraceae (Compositae)
Scientific Name: *Borrichia frutescens*
Habit: Woody plants up to 32" tall.
Leaves: Opposite, without petioles. Blades thickened, oblanceolate, up to 2 ¼" long.
Flowers: Heads up to 1 ¼" broad. Ray flowers yellow, disk flowers yellow.
Fruit: Achenes 3- or 4-sided, almost ⅛" tall, with pappus of tiny scales.
Bloom Period: All seasons.
Distribution: Cameron, Hidalgo, Willacy, and Starr counties
Comments: This is a plant of saline soils. It would make an excellent landscape plant, especially in areas with saline content that would inhibit growth of other more popular plants. An easy method of identification is to place your thumb gently against the disk. You can feel the points of many sharp-pointed scales.

Sea Ox Eye *Borrichia frutescens*

Straggler Daisy, Prostrate Lawn Weed

Family: Asteraceae (Compositae)
Scientific Name: *Calyptocarpus vialis*
Habit: Sprawling plants.
Leaves: Opposite, blades ovate, up to 1 ¾" long, usually less, rough-hairy to touch.
Flowers: Heads ¼" or slightly broader; ray and disk flowers yellow.
Fruit: Achenes flattened, about ⅛" tall, with pappus of 2 awns.
Bloom Period: All seasons.
Distribution: Cameron, Hidalgo, Willacy, and Starr counties.
Comments: This is a common volunteer in gardens and lawns, besides occurring in various disturbed places.

Straggler Daisy, Prostrate Lawn Weed *Calyptocarpus vialis*

Common Least Daisy

Family: Asteraceae (Compositae)
Scientific Name: *Chaetopappa asteroides*
Habit: Annuals with weak and leaning, hairy stems.
Leaves: Basal leaves spatulate, up to 1" long, soon drying up; stem leaves alternate, smaller.
Flowers: Heads borne singly, about ⅜" broad; ray florets white turning lilac; disk flowers yellow.
Fruit: Achenes flattened, tiny, with pappus of 5 minute scales.

Common Least Daisy *Chaetopappa asteroides*

Bloom Period: Spring, summer.
Distribution: Hidalgo and Starr counties.
Comments: The erect or leaning stems of this species separate it from *Aphanostephus* and some *Erigeron* species, which have stems that are weak and fall over.

Mexican Devil Weed, Spiny Aster

Family: Asteraceae (Compositae)
Scientific Name: *Chlorocantha spinosa* [*Aster spinosus, Erigeron ortegae*]
Habit: Erect herbs or shrubs up to 6' or taller, with spines up to ⅜" long.
Leaves: Few, mostly absent.
Flowers: Heads up to ⅝" broad; ray flowers white, sometimes turning bluish; disk flowers yellow.
Fruit: Achenes very tiny; pappus bristles about ¼" long.
Bloom Period: All seasons, but mostly fall.
Distribution: Cameron, Hidalgo, Willacy, and Starr counties.
Comments: This species grows in low, disturbed, seasonally wet areas. The spines, combined with the small daisylike heads, make this species easy to recognize.

Mexican Devil Weed, Spiny Aster *Chlorocantha spinosa*

Crucita, Blue Mistflower

Family: Asteraceae (Compositae)
Scientific Name: *Chromolaena odorata* [*Eupatorium odoratum*]
Habit: Stems sometimes woody, sprawling, up to 6' or longer.
Leaves: Opposite; blades ovate or usually triangular, the sides longer than the base, up to 2¾" long.
Flowers: Heads ⅜" or slightly taller; ray flowers absent; disk flowers pinkish or bluish.
Fruit: Achenes ⅛" or slightly taller; pappus bristles about ¼" long.
Bloom Period: All seasons, but mostly in the fall.
Distribution: Cameron, Hidalgo, Willacy, and Starr counties.
Comments: This is a popular butterfly garden plant. It is a host plant to the Rounded Metalmark butterfly. The leaves give off a characteristic aroma when crushed. This species was for many years known as *Eupatorium*. Its cylindrical involucre separates it from similar species. Species of *Chromolaena, Conoclinium, Fleischmannia,* and *Tamaulipa* were all included in the genus *Eupatorium* until fairly recently. The Spanish common name is a diminutive of *cruz,* which translates as "cross," referring to the branching habit, which forms little crosses.

Crucita, Blue Mistflower *Chromolaena odorata*

Texas Thistle

Family: Asteraceae (Compositae)
Scientific Name: *Cirsium texanum*
Habit: Annuals up to 5' tall.
Leaves: Alternate, obovate, up to 18" long, the margins lobed and very spiny.
Flowers: Heads borne solitary, 2 ¾" broad, about 1 ½" tall; ray flowers absent; disk flowers pinkish or white; involucre very spiny.
Fruit: Achenes 3/16" tall; pappus bristles about the same length.
Bloom Period: Spring and summer.
Distribution: Cameron, Hidalgo, Willacy, and Starr counties.
Comments: This species is easily recognized by its attractive, large, purple or pink flowers on plants that are very spiny. It is an excellent butterfly nectar plant, and its seeds attract many birds. Thistles are not limited to Texas. In Scotland, the story is told that thistles saved Scotland from a sneak attack during a war with the Danes. Moving barefoot (to make less noise) at night, the Danes suddenly encountered a large patch of thistles, and their yelps of pain warned the Scots.

Texas Thistle *Cirsium texanum*

Texas Thistle *C. texanum* white-flowering form

Clappia, Fleshy Clapdaisy

Family: Asteraceae (Compositae)
Scientific Name: *Clappia suaedifolia*
Habit: Perennials, woody basally, up to 18" tall.
Leaves: Mostly alternate, the lowest ones opposite; blades linear, up to 2 ⅛" long.
Flowers: Heads solitary, up to 1 ⅝" broad; ray flowers yellow; disk flowers yellow.
Fruit: Achenes about ⅛" tall; pappus bristles about 3/16" long.
Bloom Period: Spring and fall.
Distribution: Cameron, Hidalgo, and Starr counties.
Comments: This species is found more commonly near the coast and in saline soils farther inland. The specific epithet comes from the linear leaves, which resemble those of *Suaeda*, another coastal plant.

Clappia, Fleshy Clapdaisy *Clappia suaedifolia*

Asteraceae

Betony Leaf Mistflower, Padre Island Mistflower

Family: Asteraceae (Compositae)

Scientific Name: *Conoclinium betonicifolium* [*Eupatorium betonicifolium*]

Habit: Perennials, usually slightly woody; stems up to 40" long, weak and falling over, rooting where they touch the ground.

Leaves: Opposite; blades succulent, oblong, up to 1 ¾" long, the margins toothed.

Flowers: Heads up to ⅜" tall; ray flowers absent; disk flowers bluish; involucres obconic.

Fruit: Achenes 1/16" tall; pappus bristles up to ⅛" long.

Bloom Period: Spring, summer, and fall.

Distribution: Cameron, Hidalgo, and Willacy counties, usually near the coast.

Comments: This is an important butterfly nectar plant. It is a host plant to the Rounded Metalmark butterfly. The succulent leaves, with the blade length more than twice the width, separate this species from similar species. Species of *Chromolaena, Conoclinium, Fleischmannia,* and *Tamaulipa* were until fairly recently included in the single genus *Eupatorium*.

Betony Leaf Mistflower, Padre Island Mistflower *Conoclinium betonicifolium*

Horse Weed

Family: Asteraceae (Compositae)

Scientific Name: *Conyza canadensis* var. *glabrata*

Habit: Annuals up to 40" tall.

Leaves: Alternate, mostly linear, up to 1 ¾", sometimes longer.

Flowers: Heads about 3/16" tall; ray flowers about 1/16" long, cream colored; disk flowers cream colored.

Fruit: Achenes flattened, very tiny, pappus bristles 1/16" or slightly longer.

Bloom Period: Spring, summer, and fall.

Distribution: Cameron, Hidalgo, and Starr counties.

Comments: Horse weed is commonly found on disturbed soils. The flowers are extremely short-lived. The leaves are toxic to livestock.

Horse Weed *Conyza canadensis* var. *glabrata*

Golden Tick Seed

Scientific Name: *Coreopsis basalis*

Habit: Annuals up to 16" tall, with hairy stems.

Leaves: Opposite; blades up to 4" long, the bigger ones compound with leaflets that can be linear, ovate, or obovate.

Flowers: Heads up to 2 ½" broad; ray flowers yellow, brownish red basally; disk flowers dark reddish color on the ends.

Fruit: Achenes flattened, about 1/16" tall; pappus of 2 tiny awns.

Bloom Period: Spring, summer.

Distribution: Willacy County

Comments: The hairy stems along with the reddish color on the disk flowers separate this species from other *Coreopsis*. The name of the genus comes from the Greek *koris* (bug). The achenes resemble a tick or bug.

Golden Tick Seed *Coreopsis basalis*

Rio Grande Tick Seed

Family: Asteraceae (Compositae)

Scientific Name: *Coreopsis nuecensoides*

Habit: Annuals up to 24" or taller, with hairy stems.

Leaves: Opposite; blades up to 6" long, the bigger ones compound with ovate leaflets.

Flowers: Heads up to 1 ⅛" broad; ray flowers yellow, brownish basally; disk flowers yellow.

Fruit: Achenes flattened, about 3/16" tall, winged; pappus of 2 tiny awns.

Bloom Period: Spring.

Distribution: Hidalgo and Willacy counties.

Comments: This species is endemic to Texas. The yellow disk flowers separate this species from the similar *C. basalis*.

Rio Grande Tick Seed *Coreopsis nuecensoides*

Golden Wave

Family: Asteraceae (Compositae)

Scientific Name: *Coreopsis tinctoria*

Habit: Annuals up to 20" or taller, the stems hairless.

Leaves: Opposite; up to 4" long, one or two times compound, the leaflets linear or sometimes lanceolate.

Flowers: Heads up to 1 ⅛" broad; ray flowers yellow, usually with reddish basally; disk flowers yellow.

Fruit: Achenes up to 3/16" tall, usually smaller; pappus of 2 tiny awns.

Golden Wave *Coreopsis tinctoria*

Bloom Period: Spring, summer, fall.
Distribution: Cameron, Hidalgo, Willacy, and Starr counties.
Comments: This is our most common *Coreopsis*. It is frequently found in ditches and edges of wet areas. It is very robust and continues to bloom even during the hottest, driest summer. Its hairless stems distinguish it from the other species of *Coreopsis*.

Scratch Daisy
Family: Asteraceae (Compositae)
Scientific Name: *Croptilon rigidifolium* [*C. divaricatum*]
Habit: Hairy annuals; stems erect or leaning, up to 16" tall.
Leaves: Alternate; blades oblanceolate, up to 1 ½" or longer.
Flowers: Heads scattered, about ¾" broad; ray flowers yellow; disk flowers yellow.
Fruit: Achenes cylindrical, about ⅛" tall, pappus bristles about ³⁄₁₆" long.
Bloom Period: Spring, summer, and fall.
Distribution: Cameron, Hidalgo, and Willacy counties.
Comments: Scratch daisy is found most often in sandy soils. The stiff outward-pointing leaves are an especially noticeable characteristic.

Scratch Daisy *Croptilon rigidifolium*

Rabbit Tobacco
Family: Asteraceae (Compositae)
Scientific Name: *Diaperia candida* [*Evax candida, Calymmandra candida*]
Habit: Low, hairy, gray annuals up to 6" tall.
Leaves: Alternate; blades without petioles; mostly spatulate, up to ⅝" long.
Flowers: Heads in the leaf axils, partially hidden by the woolly hairs.
Fruit: Achenes tiny; pappus absent.
Bloom Period: Spring
Distribution: Hidalgo and Starr counties.
Comments: The two *Diaperia* species are quite similar. *D. candida* bears flowering heads in the leaf axils; many stem evax, *D. verna*, bears them at the stem tips. The flowers are difficult to see because of the woolly hairs.

Rabbit Tobacco *Diaperia candida*

Many Stem Evax

Family: Asteraceae (Compositae)
Scientific Name: *Diaperia verna* [*Evax verna, E. multicaulis*]
Habit: Low-growing, gray-woolly annuals up to 6" tall.
Leaves: Alternate, without petioles; blades spatulate, up to ⅝" long.
Flowers: Heads at the branch tips, small, partially hidden by woolly hairs; ray flowers absent; disk flowers minute.
Fruit: Achenes tiny; pappus absent.
Bloom Period: Spring.
Distribution: Cameron, Willacy, and Starr counties.
Comments: This species often grows in sandy soils. The flowers are obscured by the woolly hairs, making it difficult to see them. This species is distinguished from rabbit tobacco, *D. candida,* by its habit of producing the flowering heads from the stem tips; *D. candida* produces them in the leaf axils.

Many Stem Evax *Diaperia verna*

Yerba De Tago

Family: Asteraceae (Compositae)
Scientific Name: *Eclipta prostrata* [*E. alba*]
Habit: Annuals with weak stems that can be erect, leaning, or falling over, forming roots where they are in contact with the ground.
Leaves: Opposite; blades elliptic, up to 3½" long, with toothed margins.
Flowers: Heads solitary, 3⁄16" broad; ray flowers tiny, whitish; disk flowers white.
Fruit: Achenes flattened, about 1⁄16" tall, with pappus of 2 tiny points.
Bloom Period: Spring, summer, fall.
Distribution: Cameron, Hidalgo, and Starr counties.
Comments: The small size of the ray flowers sometimes causes them to be overlooked. This species usually grows in moist places. The meaning of the Spanish common name is obscure. The same is true of some English common names.

Yerba De Tago *Eclipta prostrata*

Tasselflower

Family: Asteraceae (Compositae)
Scientific Name: *Emilia fosbergii*
Habit: Annuals up to 18" tall.

Tasselflower *Emilia fosbergii*

Asteraceae

Leaves: Alternate, without petioles; blades clasping the stem, roundish to ovate or oblong, up to 3" long, the margins toothed.
Flowers: In heads; corollas red or reddish purple, all radially symmetrical, up to 7/16" long.
Fruit: Achenes narrow, about 3/16" long with rows of white hairs; pappus bristles about 1/4" long.
Distribution: Cameron and Hidalgo counties.
Comments: Both tasselflower, *E. fosbergii,* and lilac tasselflower, *E. sonchifolia,* are introduced, probably through nursery stock. *E. fosbergii* was first seen in the LRGV by the junior author in 1983. This is the first report for *E. sonchifolia* in Texas.

Lilac Tasselflower, *E. sonchifolia* differs from *E. fosbergii* in its lavender or mauve flowers. It was recently found growing around a nursery in Weslaco, and on the University of Texas/Texas Southmost College campus in Brownsville.

Engelmann's Daisy

Family: Asteraceae (Compositae)
Scientific Name: *Engelmannia peristenia* [*E. pinnatifida*]
Habit: Hairy perennials up to 20" tall.
Leaves: Alternate; stem leaves oblong, up to 12" long, the margins deeply lobed.
Flowers: Heads about 1 1/8" broad; ray flowers yellow; disk flowers yellow.
Fruit: Achenes flattened, with pappus of a few scales that fall away.
Bloom Period: Spring, summer.
Distribution: Hidalgo County.
Comments: This species is recognized by its large, deeply lobed leaves and the tiny pappus scales that fall away. It grows commonly in sandy soils. It is popular in wildflower seed mixes. The genus is named for the German-born Georg Engelmann, an American physician and botanist.

Engelmann's Daisy *Engelmannia peristenia*

Prostrate Fleabane

Family: Asteraceae (Compositae)
Scientific Name: *Erigeron procumbens*
Habit: Prostrate perennials, rooting at the stem nodes.
Leaves: Alternate, spatulate, up to 3 1/8" long; margins lobed or toothed.
Flowers: Heads up to 3/4" broad; ray flowers narrow, white or bluish; disk flowers yellow.
Fruit: Achenes flattened, tiny; pappus bristles about 3/16" long.
Bloom Period: Spring, summer.

Prostrate Fleabane *Erigeron procumbens*

Distribution: Cameron, Hidalgo, and Willacy counties.
Comments: This species is usually found near the coast in sandy soils.

Variously Colored Fleabane
Family: Asteraceae (Compositae)
Scientific Name: *Erigeron versicolor* [*E. geiseri*]
Habit: Annuals up to 12" tall, somewhat hairy.
Leaves: Alternate, without petioles; oblong to oblanceolate, stem leaves up to 1 ½" long.
Flowers: Heads up to ⅝" broad; ray flowers narrow, white to bluish or purple; disk flowers yellow.
Fruit: Achenes flattened, tiny; pappus of bristles and tiny scales.
Bloom Period: Spring.
Distribution: Hidalgo and Starr counties.
Comments: This is a new record for the LRGV. This species is more abundant farther north and west in Texas. It is found on sandstone bluffs overlooking the Rio Grande and on caliche hillsides.

Variously Colored Fleabane *Erigeron versicolor*

Brown's Flaveria
Family: Asteraceae (Compositae)
Scientific Name: *Flaveria brownii* [*F. oppositifolia*]
Habit: Perennials (or long-lived annuals) up to 28" tall.
Leaves: Opposite; petioles absent; blades linear, up to 4 ¾" long; margins entire or sometimes slightly toothed.
Flowers: Heads crowded, ⅛" broad or less; ray flowers yellow, tiny, few; disk flowers yellow.
Fruit: Achenes tiny, 10-ribbed; pappus absent.
Bloom Period: Summer, fall.
Distribution: Cameron and Willacy counties.
Comments: *Flaveria* is seen mostly near the coast.

Brown's Flaveria *Flaveria brownii*

White Mistflower
Family: Asteraceae (Compositae)
Scientific Name: *Fleischmannia incarnata* [*Eupatorium incarnatum*]
Habit: Weak-stemmed and sprawling or climbing perennials.
Leaves: Opposite; blades triangular, up to 1" long, the margins toothed.
Flowers: Heads about ¼" tall; involucres obconic; ray flowers absent; disk flowers purple, pinkish, or whitish.
Fruit: Achenes about ¹⁄₁₆" tall; pappus bristles about ⅛" long.

White Mistflower *Fleischmannia incarnata*

Asteraceae

Bloom Period: Mostly fall and winter, but some blooms in spring and summer.
Distribution: Cameron, Hidalgo, Willacy, and Starr counties.
Comments: This plant can be trained on a trellis. It attracts many butterflies as a nectar source. The crushed leaves are malodorous. Species of *Chromolaena, Conoclinium, Fleischmannia,* and *Tamaulipa* were until fairly recently included in the single genus *Eupatorium*.

Three Lobed Florestina

Family: Asteraceae (Compositae)
Scientific Name: *Florestina tripteris*
Habit: Annuals up to 24" tall.
Leaves: Opposite below, alternate higher on the stem; compound with 3 leaflets, occasionally simple.
Flowers: Heads up to 5/16" tall; ray flowers absent; disk flowers white.
Fruit: Achenes about 1/8" tall; pappus of tiny scales.
Bloom Period: Spring, summer, fall.
Distribution: Cameron, Hidalgo, Willacy, and Starr counties.
Comments: *Florestina* is easily recognized by the compound leaves and the small heads of white disk flowers. It is easily grown from seed.

Three Lobed Florestina *Florestina tripteris*

Prairie Gaillardia

Family: Asteraceae (Compositae)
Scientific Name: *Gaillardia aestivalis*
Habit: Perennials up to 28" tall.
Leaves: Alternate; petioles mostly absent; blades oblanceolate, up to 3" long.
Flowers: Heads up to 1 5/8" broad; ray flowers yellow or red; disk flowers reddish or yellow.
Fruit: Achenes hairy, about 1/16" tall; pappus of scales about 3/16" long.
Bloom Period: Spring, summer, fall.
Distribution: Willacy County.
Comments: In our area, this species is found mostly in sandy areas near the coast.

Prairie Gaillardia *Gaillardia aestivalis*

Southern Indian Blanket, Fire Wheel

Family: Asteraceae (Compositae)
Scientific Name: *Gaillardia pulchella* var. *australis*
Habit: Annuals up to 24" tall.
Leaves: Basal and on the stems; stem leaves alternate; petioles absent or short; blades lanceolate, sometimes lobed, up to 5 ⅛" long.
Flowers: Heads 1 ¾" broad or more. Ray flowers red, tipped with yellow, but variable from mostly red to mostly yellow; disk flowers brown.
Fruit: Achenes hairy basally, about 1/16" tall; pappus scales about 3/16" long.
Bloom Period: Spring, summer.
Distribution: Cameron, Hidalgo, Willacy, and Starr counties.
Comments: This is one of our showier wildflowers, an especially dependable bloomer in the coastal sands. It is commonly sold in wildflower seed mixes.

Southern Indian Blanket, Fire Wheel *Gaillardia pulchella* var. *australis*

Pincushion Daisy

Family: Asteraceae (Compositae)
Scientific Name: *Gaillardia suavis*
Habit: Perennials up to 24" tall.
Leaves: All basal; blades obovate, toothed, or lobed, up to 6" long, the margins extending down the petiole.
Flowers: Heads up to 1 ⅜" broad; ray flowers yellow, orange, or red, often absent or falling early; disk flowers reddish.
Fruit: Achenes hairy, about ⅛" tall; pappus of scales about ¼" long.
Bloom Period: Spring, summer.
Distribution: Cameron, Hidalgo, and Starr counties.
Comments: This species can be confusing if a population of rayless heads is encountered. It shares this characteristic with ambiguous green thread, *Thelesperma ambiguum*. When the inner phyllaries (involucral bracts) are examined, those of *G. suavis* are entirely free; those of *T. ambiguum* are partially joined together.

Pincushion Daisy *Gaillardia suavis*

Pennsylvania Cudweed

Family: Asteraceae (Compositae)
Scientific Name: *Gamochaeta pensilvanica* [*Gnaphalium pensilvanicum*]
Habit: Annuals up to 16" tall; stems and leaves gray-woolly.

Pennsylvania Cudweed *Gamochaeta pensilvanica*

Asteraceae

Leaves: Alternate; petioles absent; blades oblanceolate, up to 3" long.
Flowers: Heads in the leaf axils, about 3/16" tall, mostly obscured by the woolly hairs; ray flowers absent.
Fruit: Achenes very tiny; pappus bristles about 1/8" long.
Bloom Period: Spring.
Distribution: Cameron, Hidalgo, and Willacy counties.
Comments: The gray woolly covering calls attention to this species. It also tends to obscure the characteristics of the family.

Chomonque

Family: Asteraceae (Compositae)
Scientific Name: *Gochnatia hypoleuca*
Habit: Shrubs 6' or taller.
Leaves: Alternate; petioles absent; blades elliptic, up to 2" long, with downward-turned margins, white-hairy on the lower surfaces.
Flowers: Heads up to 3/8" tall; flowers whitish or pinkish, all 2-lipped.
Fruit: Achenes about 1/8" tall; pappus bristles about 3/16" long.
Bloom Period: Fall, winter, spring.
Distribution: Hidalgo and Starr counties.
Comments: This plant grows on dry gravelly hills and in upland pockets of sand. It is used in xeriscaping and is an important butterfly nectar source. The whitish lower surfaces of the leaves are a noticeable characteristic of the species. The 2-lipped flowers are unusual for this family. In our area, there are only four other species of Asteraceae with this characteristic.

Chomonque *Gochnatia hypoleuca*

Little Head Gumweed

Family: Asteraceae (Compositae)
Scientific Name: *Grindelia microcephala* [*G. microcephala* var. *adenodonta*, *G. adenodonta*]
Habit: Branching, somewhat sticky annuals.
Leaves: Alternate, without petioles; blades more or less oblong, up to 1 1/4" long, the margins toothed.
Flowers: Heads up to 1 3/8" tall; ray flowers yellow, about 3/8" or longer; disk flowers yellow.
Fruit: Achenes about 1/8" tall; pappus of 2 or 3 bristles.
Bloom Period: Mostly spring and summer.

Little Head Gumweed *Grindelia microcephala*

Distribution: Cameron County; probably also Willacy County.
Comments: The plants in our area do not grow as tall as the plants found farther north. They somewhat resemble *Heterotheca,* but the achenes in that genus have a noticeable pappus of bristles.

Plains Gumweed

Family: Asteraceae (Compositae)
Scientific Name: *Grindelia oolepis*
Habit: Perennials up to 24" tall, sometimes leaning.
Leaves: Alternate; petioles absent; blades oblanceolate, up to 2 ½" long.
Flowers: Heads about ⅝" broad; ray flowers absent; disk flowers yellow.
Fruit: Achenes about ⅛" tall; pappus awns about 3/16" long
Bloom Period: Summer, fall.
Distribution: Cameron County.
Comments: This rare species is endemic to Texas. It is an endemic of coastal blackland prairies in the Corpus Christi and Brownsville areas, one of the scarcest habitats in Texas. Almost all of the blackland prairies were put under the plow in the early part of the last century. The habitat for this plant has become marginal along highway and railroad rights-of-way, in cemeteries, and on a few large prairie remnants. In the Brownsville area, this habitat is further disappearing with increased development. The genus is named for David Hieronymus Grinder, a German professor of botany.

Plains Gumweed *Grindelia oolepis*

Round Head Broomweed

Family: Asteraceae (Compositae)
Scientific Name: *Gutierrezia sphaerocephala* [*Xanthocephalum sphaerocephalum*]
Habit: Annuals up to 24" tall, highly branched, usually resinous and sticky.
Leaves: Alternate, narrow, about ⅛" long.
Flowers: Heads up to ½" broad; ray and disk flowers yellow.
Fruit: Achenes less than ⅛" long; pappus scales less than ⅛" long.
Bloom Period: Summer, fall.
Distribution: Hidalgo and Starr counties.
Comments: *G. sphaerocephala* grows in overgrazed or disturbed sites. It is very closely related to Texas sticky snakeweed, *G. texana*. Distinguishing the two species is difficult, requiring use of keys and optic aid. Broomweed is aptly named, since it used to be tied to sticks for use as brooms.

Round Head Broomweed *Gutierrezia sphaerocephala*

Asteraceae

Texas Sticky Snakeweed

Family: Asteraceae (Compositae)
Scientific Name: *Gutierrezia texana* var. *glutinosa* [*G. glutinosa*]
Habit: Annuals up to 32" tall, highly branched and usually resinous and sticky.
Leaves: Alternate, narrow.
Flowers: Heads up to ½" broad; ray and disk flowers yellow.
Fruit: Achenes about 1/16" tall; pappus scales about 1/16" tall, forming a jagged crown.
Bloom Period: All seasons.
Distribution: Cameron, Hidalgo, and Starr counties.
Comments: *G. texana* grows in overgrazed or disturbed sites. It is very similar to round head broomweed, *G. sphaerocephala*. Keys and optic aid are necessary to distinguish the two species. The resin content makes it useful as kindling for fires.

Texas Sticky Snakeweed *Gutierrezia texana* var. *glutinosa*

Gumhead, Tatalencho

Family: Asteraceae (Compositae)
Scientific Name: *Gymnosperma glutinosum*
Habit: Shrubs up to 6' or taller.
Leaves: Alternate; petioles absent; blades narrow, up to 2" long.
Flowers: Heads up to 1/8" broad; ray flowers yellow, few, with very short rays; disk flowers yellow, few.
Fruit: Achenes about 1/16" tall; pappus very small or absent.
Bloom Period: Spring, summer, fall.
Distribution: Cameron, Willacy, and Starr counties; probably also Hidalgo County.
Comments: Gumhead is reminiscent of *Gutierrezia*, but the plants are much bigger. They grow in brushy areas. The name of the genus comes from the Greek words for "naked" and "seed," referring to the tiny or absent pappus on the achene.

Gumhead, Tatalencho *Gymnosperma glutinosum*

Bitterweed

Family: Asteraceae (Compositae)
Scientific Name: *Helenium amarum*
Habit: Annuals up to 20" or taller, bad-smelling when crushed.
Leaves: Alternate; petioles short or absent; blades mostly linear and threadlike.
Flowers: Heads up to 1" broad; ray flowers yellow; disk flowers yellow, on a rounded receptacle.
Fruit: Achenes very tiny, with tiny pappus scales.
Bloom Period: Summer, fall.
Distribution: Hidalgo County.
Comments: The rounded receptacle is a characteristic of the genus. When eaten by cattle, the plants cause the milk to taste bitter. The genus was named for Helen of Troy.

Bitterweed *Helenium amarum*

Slim Leaf Sneezeweed

Family: Asteraceae (Compositae)
Scientific Name: *Helenium linifolium*
Habit: Annuals with winged stems up to 24" tall.
Leaves: Alternate; middle and upper blades mostly linear, up to 1 ½" long
Flowers: Heads about 1" broad; ray flowers yellow, yellow with reddish brown base, or yellow apically, the rest reddish brown.
Fruit: Achenes tiny; pappus scales about ⅛" long.
Bloom Period: Spring, summer.
Distribution: Cameron, Hidalgo, and Willacy counties.
Comments: This species is endemic to Texas. The rounded receptacle is a characteristic of the genus. The winged stems, combined with the narrow leaves, separate this plant from our other *Helenium* species.

Slim Leaf Sneezeweed *Helenium linifolium*

Small Headed Sneezeweed

Family: Asteraceae (Compositae)
Scientific Name: *Helenium microcephalum*
Habit: Annuals up to 39" tall, the stems winged.
Leaves: Alternate; basal leaves dying early; upper leaves lanceolate, up to 2 ¾" long.
Flowers: Heads up to ⅜" broad; ray flowers yellow; disk flowers reddish brown, yellow basally.
Fruit: Achenes tiny; pappus scales tiny.
Bloom Period: Spring, summer.
Distribution: Cameron, Hidalgo, Willacy, and Starr counties.
Comments: This species is commonly found growing in the beds of dried ponds. The rounded receptacle and the small heads distinguish it. The plants are highly toxic to livestock.

Small Headed Sneezeweed *Helenium microcephalum*

Common Sunflower, Mirasol

Family: Asteraceae (Compositae)
Scientific Name: *Helianthus annuus*
Habit: Annuals up to 8' or taller, with stiff hairs.
Leaves: Opposite on the lower stems, alternate on the upper ones; blades ovate to triangular, hairy, up to 6" broad, sometimes bigger.
Flowers: Heads up to 4" broad; ray flowers yellow; disk flowers purplish brown. Receptacles with chaffy scales among the flowers.
Fruit: Achenes enclosed in the chaffy scales, flattened, up to ¼" long with pappus of 2 tiny awns.
Bloom Period: Spring, summer, fall.
Distribution: Cameron, Hidalgo, Willacy, and Starr counties.
Comments: This species is commonly seen on roadsides and at the edges of fields. To distinguish it from Runyon's premature sunflower, *H. praecox,* the phyllaries must be examined. Those on the common sunflower, *H. annuus,* are abruptly pointed; those on *H. praecox* taper gradually to a point. The immature heads have been boiled and eaten, possibly tasting something like the artichoke, which is in the same family. Sunflower seeds are a popular food for humans and for wildlife. As might be expected, *mirasol* translates as "sunflower."

Common Sunflower, Mirasol *Helianthus annuus*

Silverleaf Sunflower

Family: Asteraceae (Compositae)

Scientific Name: *Helianthus argophyllus*

Habit: Annuals up to 13' tall, all over white-hairy, appearing silver-white from a distance.

Leaves: Alternate; blades wooly ovate, occasionally toothed, weakly heart shaped at the base, up to 8" long.

Flowers: Heads up to 4" broad; ray flowers yellow; disk flowers brown; receptacles with chaffy scales among the flowers.

Fruit: Achenes about 3/16" tall, usually hairy toward the apex, with 2 awns that fall away.

Bloom Period: Late summer and fall.

Distribution: Hidalgo and Willacy counties.

Comments: This species has not previously been reported to grow as far south as Hidalgo and Willacy counties, although it is abundant in the sandy areas in Kenedy County and farther north. Small populations were seen in northern Hidalgo County and in northern Willacy County in June of 2005. The Willacy County plants were growing together with Runyon's premature sunflower, *H. praecox,* and some plants appeared to be hybrids between the two species. Hybridization sometimes occurs in this genus. The seeds are eaten by Lesser Goldfinches.

Silverleaf Sunflower *Helianthus argophyllus*

Silverleaf Sunflower *H. argophyllus* flower head and leaf

Fringed Sunflower, Blue Weed Sunflower

Family: Asteraceae (Compositae)

Scientific Name: *Helianthus ciliaris*

Habit: Branching perennials up to 28" tall, often less. The plants spread from underground rhizomes.

Leaves: Opposite, blue-green, linear to lanceolate, occasionally lobed.

Flowers: Heads up to 1 1/4" broad; ray flowers yellow; disk flowers brownish red; receptacles with chaffy scales among the flowers.

Fruit: Achenes about 1/8" tall; pappus of 2–3 scales or none.

Bloom Period: Summer, fall.

Distribution: Cameron and Hidalgo counties.

Comments: This species is neither as widespread in the LRGV as the other *Helianthus* species nor as noticeable. The flower heads are unlike the typical heads of the other species. It often grows in dry, alkaline soils but also grows near canals and other sources of water. The blue-green color and the fringed (ciliate-hairy) phyllaries are good markers for this species. The common names reflect these characteristics.

Fringed Sunflower, Blue Weed Sunflower *Helianthus ciliaris*

Asteraceae

Runyon's Premature Sunflower

Family: Asteraceae (Compositae)
Scientific Name: *Helianthus praecox* subsp. *runyonii*
Habit: Annuals about 24" tall, rough to the touch.
Leaves: Mostly alternate; blades ovate to triangular, up to 2 1/8" long.
Flowers: Heads up to 2 3/8" broad; ray flowers orange-yellow; disk flowers purplish brown; receptacles with chaffy scales among the flowers.
Fruit: Achenes about 3/16" tall; pappus of 2 awns that fall away.
Bloom Period: Spring, summer, fall.
Distribution: Hidalgo, Willacy, and Starr counties.
Comments: This sunflower is endemic to Texas, growing in sandy soils. This species resembles short plants of common sunflower, *H. annuus*. The phyllaries on *H. praecox* taper gradually to a point; those of *H. annuus* are abruptly pointed.

Runyon's Premature Sunflower *Helianthus praecox* subsp. *runyonii*

Gray Golden Aster

Family: Asteraceae (Compositae)
Scientific Name: *Heterotheca canescens*
Habit: Densely hairy perennials up to 20" tall.
Leaves: Alternate, linear to oblanceolate, about 1 1/4" long.
Flowers: Heads about 3/4" or less broad; ray flowers yellow; disk flowers yellow.
Fruit: Achenes about 1/8" tall; pappus bristles about 5/16" long.
Bloom Period: Summer and fall.
Distribution: Brooks County.
Comments: The flowering heads resemble those of camphor weed, *H. subaxillaris,* but the smaller growth habit and the different leaves make it easy to distinguish the two species. Although this species has not been reported for the LRGV, its presence in sandy soils of Brooks County, just north of Hidalgo County, suggests that it might be encountered in the sandy soils in our area.

Gray Golden Aster *Heterotheca canescens*

Camphor Weed

Family: Asteraceae (Compositae)
Scientific Name: *Heterotheca subaxillaris* [*H. latifolia*]
Habit: Large annuals up to 40" tall, sometimes taller, with a camphor scent when bruised.
Leaves: Alternate, the blades somewhat rough, sometimes clasping the stem, oblong to obovate, up to 2 ⅜" long, the margins smooth or with small teeth.
Flowers: Heads about ⅝" broad; ray flowers yellow; disk flowers yellow.
Fruit: Achenes up to ⅛" tall; pappus bristles up to ¼" long.
Bloom Period: Spring, summer, fall.
Distribution: Cameron, Hidalgo, Willacy, and Starr counties.
Comments: The camphor smell is common to this species and also to the camphor daisy, *Rayjacksonia phyllocephala*. However, occasionally both species are without a noticeable odor. The leaf margins of camphor daisy usually have sharp pointed teeth; those of camphor weed do not.

Camphor Weed *Heterotheca subaxillaris*

Woolly White

Family: Asteraceae (Compositae)
Scientific Name: *Hymenopappus artemisiifolius* var. *riograndensis*
Habit: Biennials up to 39" tall.
Leaves: Green on upper surfaces, whitish below; in basal rosettes and alternate on the stem; blades ovate, up to 6 ¾" long; margins lobed and lobed again.
Flowers: Heads up to ⅝" tall; ray flowers absent; disk flowers whitish to yellowish red.
Fruit: Achenes about 3/16" tall; pappus scales about 1/16" long.
Bloom Period: Spring.
Distribution: Cameron, Hidalgo, and Willacy counties.
Comments: This species is endemic to Texas, growing in sandy soils. The white color on the lower surfaces of the leaves is a notable characteristic.

Woolly White *Hymenopappus artemisiifolius* var. *riograndensis*

Poison Bitterweed

Family: Asteraceae (Compositae)
Scientific Name: *Hymenoxys odorata*
Habit: Bushy annuals up to 18" tall, sometimes taller; dotted with resinlike material, aromatic when crushed.
Leaves: Alternate, divided into threadlike segments.
Flowers: Heads about 1 1/8" broad; ray flowers yellow, about 3/8" long; disk flowers yellow.
Fruit: Achenes 4-angled, about 3/16" tall; pappus scales pointed.
Bloom Period: Spring, summer, fall.
Distribution: Starr County.
Comments: The noticeably lobed ray flowers and the threadlike segments of the leaves call attention to this species. It would make an attractive ornamental. Although unpalatable, the plants are sometimes eaten by starving sheep. A toxic agent can accumulate in the animal's body over a long period and eventually cause death.

Poison Bitterweed *Hymenoxys odorata*

Rio Grande Pearlhead

Family: Asteraceae (Compositae)
Scientific Name: *Isocarpha oppositifolia* var. *achyranthes*
Habit: Perennials forming compact mounds up to 24", occasionally taller.
Leaves: Opposite; blades narrowly ovate, up to 1 3/8" long.
Flowers: Heads about 1/4" tall; ray flowers absent; disk flowers white.
Fruit: Achenes about 1/16" tall; pappus absent.
Bloom Period: Summer, fall.
Distribution: Cameron and Hidalgo counties.
Comments: This is a Mexican species, found in Texas in Cameron County and recently also found in eastern Hidalgo County. It is a good butterfly nectar plant and grows well in cultivation.

Rio Grande Pearlhead *Isocarpha oppositifolia* var. *achyranthes*

Common Jimmyweed *Isocoma coronopifolia*

Common Jimmyweed

Family: Asteraceae (Compositae)
Scientific Name: *Isocoma coronopifolia*
Habit: Sticky, woody perennials 36" or taller.
Leaves: Alternate, gray, about 1 1/2" long, divided into linear segments.
Flowers: Heads 1/2" long or less; ray flowers absent; disk flowers yellow.
Fruit: Achenes about 1/16" long or slightly more; more or less 4-sided, hairy; pappus of bristles.
Bloom Period: Fall.
Distribution: Starr County.
Comments: This species is readily separated from *I. drummondii* by its grayish, divided, and linear leaves.

Goldenweed

Family: Asteraceae (Compositae)
Scientific Name: *Isocoma drummondii*
Habit: Sticky woody perennials up to 32" or taller.
Leaves: Alternate; blades linear to obovate, up to 2 ¾" long, the margins entire.
Flowers: Heads up to ½" tall; ray flowers absent; disk flowers yellow.
Fruit: Achenes hairy, almost ⅛" tall; pappus bristles about ¼" long.
Bloom Period: Spring, summer, mostly fall.
Distribution: Cameron and Willacy counties.
Comments: This species is an attractive native ornamental. It is poisonous to livestock. It has green, undivided leaves whereas common jimmyweed, *I. coronopifolia,* has grayish, divided leaves. *I. drummondii* was named in honor of Thomas Drummond, from Scotland, who worked on Texas plants in the 1830s.

Goldenweed *Isocoma drummondii*

Narrow Leaf Sumpweed

Family: Asteraceae (Compositae)
Scientific Name: *Iva angustifolia*
Habit: Annuals up to 39" or taller.
Leaves: Opposite on lower part, alternate above; blades lanceolate to linear, up to 1 ⅝" long, the margins entire.
Flowers: Heads small, hanging downward, with noticeable pointed united bracts. Flowers tiny, barely protruding from the bracts; male and female in the same head.
Fruit: Achenes less than ⅛" tall; pappus absent.
Bloom Period: Summer, fall.
Distribution: Cameron and Willacy counties.
Comments: This species grows in sandy soils. The pollen is allergenic. The small pendulous heads are reminiscent of *Ambrosia,* but the heads are usually larger. The crushed leaves smell like bleach.

Narrow Leaf Sumpweed *Iva angustifolia*

Tailed Seacoast Sumpweed

Family: Asteraceae (Compositae)
Scientific Name: *Iva annua* var. *caudata*
Habit: Annuals up to 5' or taller.
Leaves: Opposite on lower part, alternate above; blades ovate, up to 2 ¾" long, with toothed margins.
Flowers: Heads small, hanging downward, the bracts not united. Flowers tiny, barely protrud-

Tailed Seacoast Sumpweed *Iva annua* var. *caudata*

Asteraceae

ing from the bracts; male and female flowers in the same head.
Fruit: Achenes up to 3/16" tall; pappus absent.
Bloom Period: Summer, fall.
Distribution: Cameron, Hidalgo, and Willacy counties.
Comments: The toothed leaf margins of this species distinguish it from narrow leaf sumpweed, *I. angustifolia*. The crushed leaves smell like bleach.

Tailed Seacoast Sumpweed *I. annua* flower heads

Shorthorn Jefea
Family: Asteraceae (Compositae)
Scientific Name: *Jefea brevifolia* [*Zexmenia brevifolia*]
Habit: Upright shrubs up to 39" tall.
Leaves: Mostly opposite; blades ovate, grayish, up to 1 1/4" long; margins entire or slightly toothed.
Flowers: Heads up to 1 1/4" broad; ray flowers and disk flowers yellow to orange.
Fruit: Achenes about 1/4" tall, winged (sometimes only slightly); pappus of 2 or 3 tiny awns.
Bloom Period: Spring, summer, fall.
Distribution: Hidalgo and Starr counties.
Comments: The small gray ovate leaves are rough-hairy to the touch. These characteristics are useful in identifying this species. The name of the genus was taken from the Spanish *jefe* (chief, boss) in honor of Dr. Billie L. Turner.

Shorthorn Jefea *Jefea brevifolia*

Coulter's Laennecia
Family: Asteraceae (Compositae)
Scientific Name: *Laennecia coulteri* [*Conyza coulteri*]
Habit: Annuals up to 39" tall.
Leaves: Alternate, spatulate or oblong, up to 3 1/2" long; margins coarsely toothed, sometimes lobed.
Flowers: Heads 3/16" tall; ray flowers and disk flowers cream colored, very similar.
Fruit: Achenes very tiny; pappus bristles about 1/8" long.
Bloom Period: Spring, summer.
Distribution: Cameron, Hidalgo, and Starr counties.
Comments: The flowers are extremely short-lived, and the pappus and achenes appear very quickly. The leaves are toxic to livestock.

Coulter's Laennecia *Laennecia coulteri*

Wild Lettuce

Family: Asteraceae (Compositae)
Scientific Name: *Launaea intybacea*
   [*Lactuca intybacea*]
Habit: Erect annuals up to 5' tall, bleeding milky sap; stems hollow.
Leaves: Alternate, up to 12" long, deeply lobed, sometimes lobed again; midribs sometimes lined with spines.
Flowers: Heads about ⅜" tall; flowers yellow to bluish, bilateral, 5-lobed; disk flowers absent.
Fruit: Achenes ⅛" long; pappus bristles about ¼" long.
Bloom Period: All seasons.
Distribution: Starr County.
Comments: This is an introduced species. It occurs in disturbed areas. In our area, we have a very similar species, western wild lettuce, *Lactuca ludoviciana*. In *Launaea intybacea*, the involucres surrounding the heads are less than ⅛" broad; in *Lactuca ludoviciana*, they are usually ³⁄₁₆" or more. These species are related to the commercial lettuce, *Lactuca sativa*.

Wild Lettuce *Launaea intybacea*

Elegant Gayfeather

Family: Asteraceae (Compositae)
Scientific Name: *Liatris elegans*
Habit: Erect plants with hairy stems, growing from underground stems that resemble bulbs.
Leaves: Blades sessile, linear to lanceolate, up to 4" long.
Flowers: Inflorescences cylinder-like, resembling the inflorescence of a hyacinth. Heads about ¾" broad; disk flowers usually 5, white to purplish; ray flowers none; outer phyllaries white to pinkish, spreading outward with the tips bent back
Fruit: Achenes about ¼" tall; pappus plumose, ⅜" or longer.
Bloom Period: Summer, fall.
Distribution: Southern Kenedy County and farther north.
Comments: Although this species has not been reported for our area, it has been found so close to Willacy County that it is likely to occur there. It occurs in deep sandy soils. The spreading phyllaries are noticeable and resemble the petals of a single flower.

Elegant Gayfeather *Liatris elegans*

Asteraceae

Plains Black Foot Daisy, Hoary Black Foot

Family: Asteraceae (Compositae)
Scientific Name: *Melampodium cinereum*
Habit: Perennial herbs, sometimes slightly woody, up to 8" tall. The weak stems grow along the ground.
Leaves: Opposite, up to 2 1/8" long; blades linear to oblong, the margins entire or sometimes divided.
Flowers: Heads up to 7/8" broad, usually less; ray flowers white; disk flowers yellow.
Fruit: Achenes very tiny, obconic, and spiny; pappus about twice the length of the achene, a single fringed scale about 1/16" long, wrapped almost completely around the top of the achene.
Bloom Period: Most of the year, especially spring and summer.
Distribution: Cameron, Hidalgo, and Starr counties.
Comments: The achenes are distinctive for this species but difficult to see without optic aid.

Plains Black Foot Daisy, Hoary Black Foot *Melampodium cinereum*

Climbing Hemp Weed

Family: Asteraceae (Compositae)
Scientific Name: *Mikania scandens*
Habit: Perennial twining vines.
Leaves: Opposite; blades triangular to heart shaped, up to 3 1/8" long.
Flowers: Heads about 1/4" tall; ray flowers absent; disk flowers white.
Fruit: Achenes about 1/16" tall; pappus bristles about 3/16" long.
Bloom Period: Summer, fall.
Distribution: Cameron and Hidalgo counties.
Comments: This is one of the few vines in the Asteraceae that occur in our area. It is usually found growing around wet places and is an important butterfly nectar plant.

Climbing Hemp Weed *Mikania scandens*

Palmer's Goldenweed, False Broomweed

Family: Asteraceae (Compositae)
Scientific Name: *Neonesomia palmeri* [*Xylothamia palmeri, Ericameria austrotexana*]
Habit: Dark green shrubs resembling Broomweed, up to 5' tall.
Leaves: Alternate; blades very narrow, up to 1 ¼" long.
Flowers: Heads up to ⅜" broad; ray flowers few, tiny, yellow; disk flowers pale yellow.
Fruit: Achenes pyramid shaped, hairy, about 1/16" tall; pappus bristles up to ⅛" long.
Bloom Period: Spring, summer, fall.
Distribution: Cameron, Hidalgo, Willacy, and Starr counties.
Comments: The small size of the flowering heads makes this plant easily overlooked, although it may be quite abundant in a given area.

Palmer's Goldenweed, False Broomweed *Neonesomia palmeri*

Palmer's Goldenweed, False Broomweed *N. palmeri* flowering heads

Butterweed

Family: Asteraceae (Compositae)
Scientific Name: *Packera tampicana* [*Senecio tampicanus, S. imparipinnatus*]
Habit: Annuals up to 16" or taller.
Leaves: Alternate; blades rectangular in outline, up to 2" long, the margins lobed.
Flowers: Heads about ⅝" broad; ray flowers and disk flowers yellow.
Fruit: Achenes about 1/16" tall; pappus bristles about 3/16" long.
Bloom Period: Spring.
Distribution: Cameron, Hidalgo, and Starr counties.
Comments: This is one of the first wildflowers to appear in the spring. It resembles our two species of *Senecio*. It is easily distinguished from them by its broad, lobed leaves. The genus is named for John G. Packer, a Canadian botanist.

Butterweed *Packera tampicana*

Showy Palafoxia

Family: Asteraceae (Compositae)
Scientific Name: *Palafoxia hookeriana*
Habit: Erect, sticky annuals up to 4' or taller.
Leaves: Alternate, lanceolate, about 2 ¾" long, the margins entire.
Flowers: Heads sticky, about 1 ¾" broad; ray flowers and disk flowers purple or pink.

Showy Palafoxia *Palafoxia hookeriana*

Asteraceae

Fruit: Achenes 4-sided, up to ⅜" long; pappus of scales.
Bloom Period: Fall.
Distribution: Starr County.
Comments: This species is endemic to Texas, growing in sandy soils. It is much larger and showier than Texas doubtful palafoxia, *P. texana*.

Texas Doubtful Palafoxia

Family: Asteraceae (Compositae)
Scientific Name: *Palafoxia texana* var. *ambigua*
Habit: Annuals up to 39" or taller.
Leaves: Alternate; blades lanceolate, up to 3 ⅛" long.
Flowers: Heads up to 15" tall; ray flowers absent; disk flowers purple or pinkish, rarely white.
Fruit: Achenes about ¼" tall; pappus of scales.
Bloom Period: Spring, summer, fall.
Distribution: Cameron, Hidalgo, and Willacy counties.
Comments: There are no ray flowers, and the heads are much smaller than those of showy palafoxia, *P. hookeriana*. This is an attractive addition to a butterfly garden. The seeds are eaten by Lesser Goldfinches.

Texas Doubtful Palafoxia *Palafoxia texana* var. *ambigua*

Texas Doubtful Palafoxia *P. texana* var. *ambigua* white-flowering form

Lyreleaf Parthenium

Family: Asteraceae (Compositae)
Scientific Name: *Parthenium confertum*
Habit: Herbs up to 24" tall, with a strong smell.
Leaves: Alternate, about 6" or longer, the margins lobed.
Flowers: Heads about ¼" broad; ray flowers 5, very small, white; disk flowers white, infertile.
Fruit: Achenes flattened, less than ⅛" tall; ray flowers remain attached; pappus of 2 awns.
Bloom Period: Spring, summer, fall.
Distribution: Cameron, Hidalgo, and Starr counties.
Comments: The flowering heads resemble miniature heads of cauliflower.

Lyreleaf Parthenium *Parthenium confertum*

False Ragweed

Family: Asteraceae (Compositae)
Scientific Name: *Parthenium hysterophorus*
Habit: Herbs up to 39" tall, with a strong smell.
Leaves: Alternate; blades up to 8" long, usually lobed and lobed again.
Flowers: Heads about 3/16" broad; ray flowers 5, very small, white; disk flowers white, infertile.
Fruit: Achenes flattened, less than 1/8" tall; ray flowers remain attached; pappus of 2 awns.
Bloom Period: Spring, summer, fall.
Distribution: Cameron, Hidalgo, and Willacy counties.
Comments: The flowering heads resemble miniature heads of cauliflower. This species is distinguished from lyreleaf parthenium, *P. confertum,* by its twice-lobed leaves.

False Ragweed *Parthenium hysterophorus*

Mariola

Family: Asteraceae (Compositae)
Scientific Name: *Parthenium incanum*
Habit: Shrubs up to 40" tall; branches cottony with hairs.
Leaves: Alternate, about 2 3/8" long; margins lobed.
Flowers: Heads about 1/4" broad; ray flowers 5, small, white; disk flowers white, infertile.
Fruit: Achenes flattened, less than 1/8" tall; pappus of 3 awns.
Bloom Period: Summer, fall.
Distribution: Starr County.
Comments: This species is infrequently encountered, growing on caliche hills and in saline soils. It is one of several plants involved in unfruitful experiments to replace commercial rubber, especially in World War II. It is distinguished from the other Partheniums by its woodiness and the cottony appearance of the branches. Also, the achenes have 3 awns rather than 2.

Mariola *Parthenium incanum*

Limoncillo

Family: Asteraceae (Compositae)
Scientific Name: *Pectis angustifolia* var. *tenella*
Habit: Annuals up to 4" tall, strongly scented, with reddish oil glands.
Leaves: Opposite; blades linear, mostly threadlike, up to 1 5/8" long.
Flowers: Heads about 3/8" broad; ray and disk flowers yellow.

Limoncillo *Pectis angustifolia* var. *tenella*

Fruit: Achenes ⅛" tall; pappus of 5 short awns.
Bloom Period: Spring, fall.
Distribution: Hidalgo and Starr counties.
Comments: Limoncillo is often found growing on rocky, gravelly hills. The strong scent and the linear leaves are important characteristics of this species. The Spanish common name translates as "small lemon." It probably originated from the small size and aromatic characteristics of the plant.

Short Ray Rock Daisy

Family: Asteraceae (Compositae)
Scientific Name: *Perityle microglossa* var. *microglossa*
Habit: Annuals up to 22" tall.
Leaves: Opposite at the base, alternate above; blades ovate in outline, about 2" long, the margins toothed or lobed.
Flowers: Heads up to ¼" broad; ray flowers white; disk flowers yellow.
Fruit: Achenes about ¹⁄₁₆" tall; pappus of minute scales and 2 bristles less than ¹⁄₁₆" long.
Bloom Period: Spring.
Distribution: Cameron and Willacy counties.
Comments: In Texas, this variety of *Perityle* occurs only in the LRGV. It has been introduced to our area. It also grows in northern Mexico. The opposite basal leaves, the small heads with white rays and yellow disks, and the unusual pappus help separate this species from similar ones.

Short Ray Rock Daisy *Perityle microglossa*

Basket Flower

Family: Asteraceae (Compositae)
Scientific Name: *Plectocephalus americanus* [*Centaurea americana*]
Habit: Annuals up to 5' tall.
Leaves: Alternate; blades lanceolate, up to 8" long.
Flowers: Heads up to 4" broad, usually less; bracts of involucre highly branched, giving a basketlike appearance; flowers all disk type, pink or occasionally white.
Fruit: Achenes flattened, about ³⁄₁₆" tall; pappus of bristles about ¼" long.
Bloom Period: Spring, summer.
Distribution: Cameron and Hidalgo counties, probably more widespread.

Basket Flower *Plectocephalus americanus*

Comments: This species is an important butterfly nectar plant. It resembles the Texas Thistle, *Cirsium texanum,* but it does not have the spiny leaves and involucres typical of thistles.

Cure-for-All

Family: Asteraceae (Compositae)
Scientific Name: *Pluchea carolinensis*
Habit: Woody perennials up to 5' or taller, with herbal aroma. Stems hairy.
Leaves: Alternate; blades dull gray-green, hairy, especially on the lower surfaces; narrowly ovate, 5½" or longer.
Flowers: Heads about ⅜" broad and ¼" tall, clustered together; ray flowers absent; disk flowers pinkish to white. The flowers take a long time to mature but are open for only a short time.
Fruit: Achenes very tiny; pappus bristles 1/16" or longer, persistent on the plant.
Bloom Period: Spring.
Distribution: Cameron and Hidalgo counties. Rarely seen.
Comments: This species is known to be native in tropical America and occurs in the southeastern United States. It is an introduced pest on South Pacific islands. It was not known to be in our area until 2005, when Richard Lehman reported a population growing near Santa Maria (Cameron County). Earlier unidentified populations were observed growing in Starr County in the late 1980s. Populations are increasing in the LRGV and occur continuously southward into Mexico.

Cure-for-All *Pluchea carolinensis*

Purple Marsh Fleabane

Family: Asteraceae (Compositae)
Scientific Name: *Pluchea odorata*
[*P. purpurascens*]
Habit: Scented annuals up to 5' tall, usually less; young stems with tiny hairs.
Leaves: Alternate; blades ovate, up to 6" long; margins entire to toothed.
Flowers: Heads ¼" tall; ray flowers absent; disk flowers pinkish.
Fruit: Achenes very tiny; pappus up to ⅛" long.
Bloom Period: Summer, fall.
Distribution: Cameron, Hidalgo, Willacy, and Starr counties.
Comments: This species is usually found growing in wet places, or low spots that retain water for a time. It is an excellent butterfly nectar plant. The leaves emit a noticeable odor when bruised.

Purple Marsh Fleabane *Pluchea odorata*

Asteraceae

South Texas False Cudweed

Family: Asteraceae (Compositae)
Scientific Name: *Pseudognaphalium austrotexanum*
Habit: Annual white-hairy herbs up to 28" tall.
Leaves: Congested; petioles absent; blades linear to lanceolate, up to 2" long; upper surfaces green and glandular; lower surfaces white-hairy; margins slightly curled under.
Flowers: Heads in tight clusters, up to 3/16" tall; phyllaries of involucre white-hairy.
Fruit: Achenes tiny, less than 1/16"; pappus of bristles that fall away early.
Bloom Period: Spring, summer, fall.
Distribution: Willacy County.
Comments: Our two species are similar. This species has leaves that are green on the upper surfaces; those of cotton batting false cudweed, *P. stramineum,* are gray-woolly on both surfaces, although less so on the upper surfaces.

South Texas False Cudweed *Pseudognaphalium austrotexanum*

Cotton Batting False Cudweed

Family: Asteraceae (Compositae)
Scientific Name: *Pseudognaphalium stramineum* [*Gnaphalium chilense*]
Habit: Woolly annuals or short-lived perennials 10" or taller.
Leaves: Alternate; blades mostly lanceolate, about 2" long.
Flowers: Heads 1/4" tall; ray flowers absent; disk flowers white, appearing yellow in bud.
Fruit: Achenes minute; pappus minute, together less than 1/16".
Bloom Period: Spring, summer, fall.
Distribution: Cameron and Hidalgo counties.
Comments: The leaves of this species are woolly on both surfaces, less so on the upper surface. South Texas false cudweed, *P. austrotexanum,* has leaves that are green on the upper surfaces.

Cotton Batting False Cudweed *Pseudognaphalium stramineum*

Paper Flower

Family: Asteraceae (Compositae)
Scientific Name: *Psilostrophe gnaphalioides*
Habit: Herbs up to 20" tall.
Leaves: Alternate, with long hairs; blades mostly oblanceolate, the lower ones up to 3 1/8" long; upper leaves smaller.
Flowers: Heads about 1/4" tall; involucres woolly; ray flowers yellow.
Fruit: Achenes tiny, hairy; pappus of hairy scales.
Bloom Period: Spring, summer, fall.
Distribution: Starr County.
Comments: This attractive plant also grows farther west in Texas, extending to the east no farther than Starr County. The petals become papery with age. It is poisonous to sheep.

Paper Flower *Psilostrophe gnaphalioides*

Blackroot

Family: Asteraceae (Compositae)
Scientific Name: *Pterocaulon virgatum*
Habit: Erect perennials up to 40" tall. Stems with narrow "wings" from leaf tissue.
Leaves: Alternate, without petioles; blades linear, up to 4 3/4" long, the lower surfaces white with woolly hairs.
Flowers: Heads about 3/8" tall; ray flowers absent; disk flowers tiny, purple.
Fruit: Achenes less than 1/16" tall; pappus bristles about 5/16" long.
Bloom Period: Summer, fall.
Distribution: Cameron, Hidalgo, and Willacy counties.
Comments: The white-woolly lower leaf surfaces and the winged stems are good indicators for this species. The name of the genus comes from two Greek words meaning "wings" and "stem."

Blackroot *Pterocaulon virgatum*

False Dandelion

Family: Asteraceae (Compositae)
Scientific Name: *Pyrrhopappus pauciflorus*
Habit: Annuals with milky sap, up to 20" or taller.
Leaves: Mostly crowded at the base; blades usually lobed, up to 6" long.
Flowers: Heads up to 1 3/8" broad; ray flowers yellow, 5-lobed; disk flowers absent.
Fruit: Achenes about 3/16" tall, with a small beak at the top; pappus bristles up to 1/4" long.
Bloom Period: Spring.

False Dandelion *Pyrrhopappus pauciflorus*

Distribution: Cameron, Hidalgo, Willacy, and Starr counties.
Comments: The flowering heads appear on a branching inflorescence; the dandelion, *Taraxacum,* produces a single head on the inflorescence.

False Dandelion *P. pauciflorus* fruiting head

Mexican Hat
Family: Asteraceae (Compositae)
Scientific Name: *Ratibida columnifera* [*R. columnaris*]
Habit: Herbs up to 30" tall.
Leaves: Opposite on lower stems, alternate above; blades once or twice lobed, up to 5 ½" long.
Flowers: Heads solitary, up to 2" broad; ray flowers variable, yellow to yellow with a reddish brown base to reddish brown all over; receptacles tall, rounded on the ends; disk flowers reddish brown.
Fruit: Achenes produced by disk flowers only; flattened, enclosed in scales, about 1/16" tall; pappus scales tiny.
Bloom Period: Spring, summer fall.
Distribution: Cameron, Hidalgo, Willacy, and Starr counties.
Comments: The leaves are all more or less the same size. On the long headed coneflower, *R. peduncularis,* the leaves on the upper stems are noticeably smaller; also, the receptacle is usually taller.

Mexican Hat *Ratibida columnifera*

Long Headed Coneflower
Family: Asteraceae (Compositae)
Scientific Name: *Ratibida peduncularis*
Habit: Herbs up to 30" tall.
Leaves: Mostly crowded at the base, but alternate on the stems; blades usually once or twice lobed, up to 5½" long.
Flowers: Heads solitary, up to 2" broad; ray flowers variable, yellow to yellow with a reddish brown base to reddish brown all over; receptacles tall, rounded on the ends; disk flowers reddish brown.
Fruit: Achenes produced by disk flowers only; flattened, enclosed in scales, about 1/16" tall; pappus scales tiny.
Bloom Period: Spring, summer, fall.
Distribution: Cameron and Willacy counties.
Comments: The stem leaves are noticeably smaller

Long Headed Coneflower *Ratibida peduncularis*

than the lower leaves; this distinguishes this species from Mexican hat, *R. columnifera,* which also usually has a shorter receptacle.

Camphor Daisy

Family: Asteraceae (Compositae)
Scientific Name: *Rayjacksonia phyllocephala* [*Machaeranthera phyllocephala*]
Habit: Annuals, sometimes sticky, with a camphor odor when bruised.
Leaves: Opposite; blades mostly oblanceolate, up to 2" long, the margins sharp-toothed or lobed.
Flowers: Heads solitary, up to 1 5/8" broad; ray flowers yellow; disk flowers yellow, separated by scales.
Fruit: Achenes cylindrical, about 1/16" tall; pappus bristles 1/4" or longer.
Bloom Period: Spring, summer, fall.
Distribution: Cameron, Hidalgo, and Willacy counties.
Comments: This species is very abundant at or near the coast. It often grows with camphor weed, *Heterotheca subaxillaris*. Both have a camphor smell when the leaves are crushed. Camphor daisy has sharp-toothed leaves, while camphor weed leaves may have small teeth or none.

Camphor Daisy *Rayjacksonia phyllocephala*

Brown-Eyed Susan

Family: Asteraceae (Compositae)
Scientific Name: *Rudbeckia hirta*
Habit: Herbs up to 24" tall, all parts with long hairs; stems rough.
Leaves: Mostly alternate; blades ovate, up to 5" long, extending onto the petioles.
Flowers: Heads solitary, up to 2" broad; ray flowers infertile, yellow with reddish brown spots; disk flowers reddish brown, on conical receptacles.
Fruit: Achenes about 1/16" tall, enclosed in scales; pappus absent or of tiny teeth.
Bloom Period: Spring, summer.
Distribution: Hidalgo and Willacy counties.
Comments: This species prefers sandy soils. The genus is named for Olof Rudbeck, founder of Uppsala Botanic Garden in Sweden.

Brown-Eyed Susan *Rudbeckia hirta*

Yellow Sanvitalia

Family: Asteraceae (Compositae)
Scientific Name: *Sanvitalia ocymoides*
Habit: Annual herbs up to 8" tall, stems usually falling over and creeping along the ground.
Leaves: Opposite; blades ovate, rough with scales, up to 1 1/8" long.

Yellow Sanvitalia *Sanvitalia ocymoides*

Asteraceae

Flowers: Heads up to ½" broad; ray flowers pale yellow; disk flowers brown.
Fruit: Achenes about ³⁄₁₆" tall; pappus of ray flowers of 3 awns; pappus of disk flowers usually absent.
Bloom Period: Summer, fall.
Distribution: Cameron, Hidalgo, and Starr counties.
Comments: This species makes an attractive potted plant; seeds (achenes) are sometimes offered in seed catalogs. The flower heads contain some very sharp scales and can be painful to bare fingers.

Border Bonebract
Family: Asteraceae (Compositae)
Scientific Name: *Sclerocarpus uniserialis* var. *austrotexanus*
Habit: Annuals with weak and sprawling stems up to 39" long.
Leaves: Lower ones opposite, upper ones alternate; blades ovate, up to 3 ⅛" long, the margins usually toothed.
Flowers: Heads up to 1 ⅝" broad; ray flowers and disk flowers orange-yellow.
Fruit: Achenes about ³⁄₁₆" tall, each enclosed in a scale; pappus absent.
Bloom Period: Spring, summer, fall.
Distribution: Hidalgo, Willacy, and Starr counties.
Comments: This is an attractive butterfly nectar plant, preferring sandy soils. It resembles a large *Wedelia*. Noticeable characteristics of this species are the loose phyllaries and the relatively small number of ray flowers.

Border Bonebract *Sclerocarpus uniserialis* var. *austrotexanus*

Texas Groundsel
Family: Asteraceae (Compositae)
Scientific Name: *Senecio ampullaceus*
Habit: Annuals up to 32" tall.
Leaves: Alternate, hairy, the hairs sometimes falling away; lower blades obovate, up to 6" long, the margins often toothed; upper leaves smaller, clasping the stem.
Flowers: Heads up to 1 ⅛" broad; ray flowers and disk flowers yellow.
Fruit: Achenes about ⅛" tall; pappus bristles up to ¼" long.
Bloom Period: Spring.
Distribution: Cameron, Hidalgo, Willacy, and Starr counties.

Texas Groundsel *Senecio ampullaceus*

Comments: This species is distinguished from our other senecios by its broad, unlobed leaves. It was considered endemic to Texas until recently, when two populations were found in Oklahoma. The plants are toxic to cattle. The genus name is derived from the Latin word *senex*, meaning "old man," referring to the noticeable white pappus.

Broom Groundsel

Family: Asteraceae (Compositae)
Scientific Name: *Senecio riddellii* [*S. spartioides*]
Habit: Somewhat woody perennials up to 40" or taller.
Leaves: Alternate; blades somewhat succulent; up to 2 ¾" long, greatly dissected into linear lobes.
Flowers: Heads up to 1 ⅛" broad; ray flowers and disk flowers yellow.
Fruit: Achenes ⅛" or taller; pappus bristles about ¼" long.
Bloom Period: Spring, fall.
Distribution: Cameron, Willacy, and Starr counties.
Comments: This species is easily distinguished from our other Senecios by its highly dissected leaves. It grows mostly in sandy areas at or near the coast. The plants are beautiful and are worthy of cultivation. However, they are toxic to livestock. The genus name is derived from the Latin word *senex,* meaning "old man," referring to the noticeable white pappus.

Broom Groundsel *Senecio riddellii*

Bush Sunflower

Family: Asteraceae (Compositae)
Scientific Name: *Simsia calva*
Habit: Perennials up to 30" tall from a succulent root.
Leaves: Opposite, rough-hairy to the touch; blades mostly triangular, up to 1 ¾" long.
Flowers: Heads up to 2" broad; ray flowers and disk flowers orange-yellow.
Fruit: Achenes flattened, about 3/16" tall, with a small "wing" around the edge; pappus absent.
Bloom Period: Spring, summer, fall.
Distribution: Cameron, Hidalgo, Willacy, and Starr counties.
Comments: This plant makes an attractive perennial in the garden. It blooms almost continuously, as long as it receives water. It is a host plant to the Bordered Patch butterfly and is also a good butterfly nectar plant.

Bush Sunflower *Simsia calva*

Asteraceae

Tall Goldenrod

Family: Asteraceae (Compositae)
Scientific Name: *Solidago canadensis* [*S. altissima* var. *pluricephala, S. altissima* var. *altissima*]
Habit: Perennials up to 6' or taller, covered with tiny hairs.
Leaves: Alternate; blades linear or lanceolate, up to 2 ¾" long.
Flowers: Heads about 3/16" broad, crowded on one side of the inflorescence, the branches curling down; ray flowers yellow, small, not noticeable; disk flowers yellow.
Fruit: Achenes very tiny; pappus bristles about ⅛" long.
Bloom Period: Fall.
Distribution: Cameron County.
Comments: The curling tips of the one-sided inflorescence are characteristic of this genus. The short hairs on this species distinguish *S. canadensis* from seaside goldenrod, *S. sempervirens,* which is hairless.

Tall Goldenrod *Solidago canadensis*

Tall Goldenrod *S. canadensis* inflorescence

Seaside Goldenrod

Family: Asteraceae (Compositae)
Scientific Name: *Solidago sempervirens*
Habit: Perennials up to 6' or taller.
Leaves: Alternate; blades elliptic or grasslike, up to 7" long.
Flowers: Heads about 3/16" broad, crowded on one side of the inflorescence, the branches curling down; ray flowers yellow, small, not noticeable; disk flowers yellow.
Fruit: Achenes very tiny; pappus bristles about 3/16" long.
Bloom Period: Summer, fall.
Distribution: Cameron and Willacy counties.
Comments: This plant is typically found growing in coastal sands. It would be an attractive addition to a butterfly garden. The curling tips of the one-sided inflorescence are characteristic of this genus. The absence of hairs distinguishes this species from tall goldenrod, *S. canadensis. Ambrosia* (ragweed) and its relative, *Iva,* often share the same habitat with goldenrod. Their flowers are inconspicuous, but they produce much wind-blown pollen, which is allergenic. Probably they have given the goldenrods the bad reputation of causing hay fever. In fact, goldenrod pollen is too heavy to be spread effectively by wind.

Seaside Goldenrod *Solidago sempervirens*

Spiny Sow Thistle

Family: Asteraceae (Compositae)
Scientific Name: *Sonchus asper*
Habit: Annuals with hollow stems and milky sap, up to 40" or taller.
Leaves: Alternate; petioles with a pair of rounded basal "ears"; blades 6" or longer, the margins lobed and sharp-toothed.
Flowers: Heads up to 1" broad; flowers yellow, resembling ray flowers but with 5 small lobes.
Fruit: Achenes about 1/8" tall; pappus bristles about 1/4" tall.
Bloom Period: Spring.
Distribution: Cameron and Hidalgo counties.
Comments: The leaves of this species are pricklier than those of common sow thistle, *S. oleraceus*. Also, the paired "ears" at the base of the petiole are rounded rather than pointed. Both species are introduced from the Old World. Goldfinches eat the seeds.

Spiny Sow Thistle *Sonchus asper*

Common Sow Thistle

Family: Asteraceae (Compositae)
Scientific Name: *Sonchus oleraceus*
Habit: Annuals with hollow stems and milky sap, up to 40" or taller.
Leaves: Alternate; petioles with a pair of pointed basal "ears"; blades up to 6" long, the margins lobed and sharp-toothed.
Flowers: Heads up to 1" broad; flowers yellow, all resembling ray flowers, but with 5 small lobes.
Fruit: Achenes about 1/8" tall; pappus bristles about 1/4" long.
Bloom Period: Mostly spring and summer, less frequently in fall and winter.
Distribution: Cameron, Hidalgo, and Starr counties.
Comments: This plant is native to Eurasia. Goldfinches eat the seeds. The basal "ears" on the petioles are pointed. The similar spiny sow thistle, *S. asper*, has petiolar basal ears that are more or less rounded.

Common Sow Thistle *Sonchus oleraceus*

Asteraceae

Wireweed

Family: Asteraceae (Compositae)
Scientific Name: *Symphyotrichum subulatum* [*S. divaricatum, Aster subulatus*]
Habit: Annuals up to 40" or taller.
Leaves: Alternate; blades linear or tapering to a point, up to 6" long.
Flowers: Heads up to ¾" broad; ray flowers white or bluish; disk flowers yellow.
Fruit: Achenes up to ½" long; pappus bristles ⅛" or longer.
Bloom Period: Summer, fall.
Distribution: Cameron, Hidalgo, Willacy, and Starr counties.
Comments: This plant is very tenacious. Even if cut back severely, it will usually bloom in its season. Plants only a few inches tall have been observed in bloom.

Wireweed *Symphyotrichum subulatum*

Blue Boneset, Spring Mistflower

Family: Asteraceae (Compositae)
Scientific Name: *Tamaulipa azurea* [*Eupatorium azureum*]
Habit: Shrubs with sprawling stems up to 10' long.
Leaves: Opposite; blades triangular or ovate, up to 2 ⅜" long.
Flowers: Heads ⅜" tall; ray flowers absent; disk flowers bluish, rarely white.
Fruit: Achenes about 1/16" tall; pappus bristles about ⅛" long.
Bloom Period: Spring, summer, fall. The heaviest bloom is in spring.
Distribution: Cameron, Hidalgo, and Willacy counties.
Comments: This plant grows in the LRGV and in Mexico. It does not occur elsewhere in Texas. It is shade tolerant and is an important butterfly nectar plant. Species of *Chromolaena, Conoclinium, Fleischmannia,* and *Tamaulipa* were until fairly recently included in the single genus *Eupatorium*.

Blue Boneset, Spring Mistflower *Tamaulipa azurea*

Blue Boneset, Spring Mistflower *T. azurea* white-flowered form

Common Dandelion

Family: Asteraceae (Compositae)
Scientific Name: *Taraxacum officinale*
Habit: Annuals or perennials with a large taproot.
Leaves: All crowded basally; blades deeply lobed, up to 6" long.
Flowers: Heads about 1" or more broad, borne singly on stems about 8" tall; flowers all strap shaped, 5-lobed, yellow; disk flowers none.
Fruit: Achenes ribbed and beaked; pappus bristles very prominent, whitish.
Bloom Period: Spring.
Distribution: Hidalgo County.
Comments: This species is rare in southernmost Texas. The only population known is in McAllen. It was found in 1984 by the junior author and is still thriving. In earlier days, all parts of the plant were used as food. The flower heads are still used by some to make wine.

Common Dandelion *Taraxacum officinale*

Common Dandelion *T. officinale* fruiting head

Square Bud Daisy

Family: Asteraceae (Compositae)
Scientific Name: *Tetragonotheca repanda*
Habit: Perennials up to 16" tall.
Leaves: Opposite; blades ovate or triangular, up to 4 3/8" long.
Flowers: Heads solitary, up to 2 5/8" broad; ray flowers and disk flowers yellow.
Fruit: Achenes 4-sided, about 3/16" tall; pappus of tiny scales.
Bloom Period: Spring, summer, fall.
Distribution: Hidalgo, Willacy, and Starr counties.
Comments: This species is found growing in sandy soils. It is endemic to Texas. The genus name comes from the Latin *tetragonus*, meaning "with 4 angles." The square-shaped involucre of 4 large phyllaries gives the common name and a quick way to recognize this species.

Square Bud Daisy *Tetragonotheca repanda*

Square Bud Daisy *T. repanda* flowering head showing the 4 green phyllaries

Asteraceae

Texas Nerve Ray

Family: Asteraceae (Compositae)
Scientific Name: *Tetragonotheca texana*
Habit: Perennials; stems branching a few times, somewhat hairy.
Leaves: Opposite; blades often lobed, ovate to lanceolate in outline, up to 4" long, usually less.
Flowers: Heads fairly numerous, about 1 ½" broad; ray flowers yellow, about ½" long; disk flowers yellow.
Fruit: Achenes 4-sided, about ³⁄₁₆" tall; pappus absent or of very tiny scales.
Bloom Period: Spring and summer.
Distribution: Hidalgo and Starr counties.
Comments: This species, like square bud daisy, *T. repanda,* also has an involucre of 4 phyllaries, giving the buds a square shape. The leaves (at least some of them) are lobed, and the inflorescences are branching. The leaves of *T. repanda* are simple, and the inflorescences are not branching.

Texas Nerve Ray *Tetragonotheca texana*

Slender Leaf Fournerve

Family: Asteraceae (Compositae)
Scientific Name: *Tetraneuris linearifolia* [*Hymenoxys linearifolia*]
Habit: Annuals that arise from a short-lived basal rosette of leaves; stems erect or falling over.
Leaves: Hairy and sometimes glandular. Basal leaves spatulate or oblanceolate, entire or lobed; stem leaves mostly linear.
Flowers: Heads solitary or in groups; ray flowers yellow, about ⅝" long; disk flowers yellow.
Fruit: Achenes vaguely 4-angled and hairy, up to ⅛" tall; pappus of 5–8 scales.
Bloom Period: Spring, summer.
Distribution: Hidalgo and Starr counties.
Comments: The genus name is derived from two Greek words for "four" and "nerve," referring to the four nerves on the ray flowers. This species is quite widespread in Texas.

Slender Leaf Fournerve *Tetraneuris linearifolia*

Ambiguous Green Thread

Family: Asteraceae (Compositae)
Scientific Name: *Thelesperma ambiguum*
 [*Thelesperma megapotamicum* var. *ambiguum*]
Habit: Perennials up to 20" tall.
Leaves: Opposite; blades up to 6" long, dissected one, two, or three times, or sometimes simple.
Flowers: Heads up to 2" broad; ray flowers yellow but sometimes absent; disk flowers reddish brown.
Fruit: Achenes about ¼" tall; pappus of 2 small awns.
Bloom Period: Summer, fall.
Distribution: Hidalgo and Starr counties.
Comments: This species can be somewhat deceptive. Some populations either have no ray flowers or these fall early, presenting only the reddish brown disks; others are beautifully decorated with yellow rays. The pincushion daisy, *Gaillardia suavis,* also sometimes drops the ray flowers, but its inner phyllaries (involucral bracts) are separate, whereas *Thelesperma ambiguum* has inner phyllaries that are at least partially grown together.

Ambiguous Green Thread *Thelesperma ambiguum*

Nueces Green Thread

Family: Asteraceae (Compositae)
Scientific Name: *Thelesperma nuecense*
Habit: Annuals up to 3' or taller.
Leaves: Opposite; blades up to 5 ⅛" long, dissected, the segments threadlike.
Flowers: Heads up to 2 ⅛" broad; ray flowers gold with reddish brown basal spot, or sometimes almost all reddish brown; disk flowers purplish brown.
Fruit: Achenes about ¼" long; pappus of 2 tiny awns.
Bloom Period: Spring.
Distribution: Hidalgo, Willacy, and Starr counties.
Comments: This species prefers sandy soils. It is endemic to Texas. In our area, this species tends to resemble greatly *T. filifolium,* which occurs near the LRGV. The basal spot on the disk flowers can be so small that it may be overlooked. *T. filifolium* may have a basal streaked area but never a definite spot.

Nueces Green Thread *Thelesperma nuecense*

Asteraceae

Five Needle Dogweed

Family: Asteraceae (Compositae)
Scientific Name: *Thymophylla pentachaeta* [*Dyssodia pentachaeta*]
Habit: Small aromatic perennial herbs, grayish to green.
Leaves: Usually opposite, crowded, up to ⅝" long; blades lobed, the lobes linear or threadlike.
Flowers: Heads up to ⅝" broad; ray flowers yellow; disk flowers yellow. Upper parts of involucres dotted with glands.
Fruit: Achenes cylindrical, ⅛" or less tall; pappus of scales ⅛" or less tall.
Bloom Period: Spring, summer, fall.
Distribution: Cameron, Hidalgo, Willacy, and Starr counties.
Comments: Four varieties are recognized. The variety *pentachaeta* is found in Texas and northern Mexico. It is similar to tiny tim, *T. tenuiloba,* which has alternate leaves.

Five Needle Dogweed *Thymophylla pentachaeta*

Tiny Tim, Dogweed

Family: Asteraceae (Compositae)
Scientific Name: *Thymophylla tenuiloba* [*Dyssodia tenuiloba*]
Habit: Annual herbs, sometimes living several years; plants rounded, up to 8" tall.
Leaves: Alternate, crowded, up to ⅝" long, divided into linear segments.
Flowers: Heads about ⅝" broad; ray flowers yellow; disk flowers yellow. Upper parts of involucres dotted with glands.
Fruit: Achenes cylindrical, about ⅛" tall; pappus of fringed scales about the same length as the achene.
Bloom Period: Spring and fall.
Distribution: Cameron, Hidalgo, Willacy, and Starr counties.
Comments: Tiny Tim is an attractive native ornamental, endemic to Texas and northern Mexico. The leaves are aromatic when crushed. This species is very similar to five needle dogweed, *T. pentachaeta,* which has opposite leaves.

Tiny Tim, Dogweed *Thymophylla tenuiloba*

Ashy Dogweed

Family: Asteraceae (Compositae)
Scientific Name: *Thymophylla tephroleuca* [*Dyssodia tephroleuca*]
Habit: Gray hairy perennials up to 12" tall, the upper stems white-woolly.
Leaves: Alternate, usually linear, up to ⅝" long.
Flowers: Heads up to 1" broad; ray flowers yellow; disk flowers yellow. Upper parts of involucres dotted with glands.
Fruit: Achenes cylindrical, ⅛" or longer; pappus of 10 or more scales with points at the ends.
Bloom Period: Winter, spring, summer.
Distribution: Starr County, very rare.
Comments: This species is on the federal and state endangered lists. It is endemic to Texas. The only specimen from Starr County was collected by E. Clover in 1932, and the site has not been relocated. The species was rediscovered in Webb and Zapata counties in the 1960s. The gray color of the plants distinguishes ashy dogweed from the other *Thymophylla* species.

Ashy Dogweed *Thymophylla tephroleuca*

Wright's Bugheal

Family: Asteraceae (Compositae)
Scientific Name: *Trichocoronis wrightii*
Habit: Annuals with weak stems, mostly creeping along the ground.
Leaves: Mostly opposite; blades ovate or oblong, about 1" long, the margins toothed.
Flowers: Heads about ¼" tall; ray flowers absent; disk flowers white or bluish.
Fruit: Achenes ¹⁄₁₆" or taller; pappus absent or of tiny scales.
Bloom Period: Spring.
Distribution: Cameron, Hidalgo, and Willacy counties.
Comments: This small plant is rarely seen here but usually grows in moist places. It was named in honor of Charles Wright, one of the best known botanists in the United States in the 1800s.

Wright's Bugheal *Trichocoronis wrightii*

Asteraceae

Tridax, Coat Buttons

Family: Asteraceae (Compositae)
Scientific Name: *Tridax procumbens*
Habit: Hairy herbs with prostrate stems.
Leaves: Opposite; blades hairy, broadly lanceolate, about 2" long, the margins toothed.
Flowers: Heads ¾" broad; ray flowers few, white; disk flowers yellow.
Fruit: Achenes about 1/16" tall; pappus bristles ¼" long.
Bloom Period: Summer, fall.
Distribution: Cameron and Hidalgo counties.
Comments: This species was recently reported from Texas. It grows in disturbed areas, spreading quickly along roadsides and railroad tracks. Its range is continuous from Texas southward into Mexico. It is listed as a noxious weed in Florida and Hawaii.

Tridax, Coat Buttons *Tridax procumbens*

Trixis

Family: Asteraceae (Compositae)
Scientific Name: *Trixis inula* [*T. radialis*]
Habit: Shrubs up to 6' or taller, usually sprawling.
Leaves: Alternate; blades elliptic, up to 3 ½" long
Flowers: Heads up to ¾" tall; flowers yellow, all 2-lipped. There are no ray or disk flowers.
Fruit: Achenes about ¼" tall; pappus bristles about ⅜" long.
Bloom Period: All seasons.
Distribution: Cameron, Hidalgo, Willacy, and Starr counties.
Comments: *Trixis* grows in brush lands and is an attractive ornamental. This species is easily recognized by its yellow 2-lipped flowers. Western populations have narrower leaves than the eastern ones. In our area, only four other species of Compositae share the 2-lipped characteristic; all have flowers colored other than yellow.

Trixis *Trixis inula* narrow-leaf form

Trixis *T. inula* two-lipped flower

Saladillo

Family: Asteraceae (Compositae)
Scientific Name: *Varilla texana*
Habit: Somewhat woody perennials producing mounds about 12" tall.
Leaves: Alternate, dark green, succulent, rod-shaped, up to 1 ⅛" long.
Flowers: Heads about ⅜" tall and ⅝" broad; ray flowers absent; disk flowers yellow.
Fruit: Achenes ¹⁄₁₆" or slightly taller; pappus absent.
Bloom Period: Spring, summer, fall.
Distribution: Hidalgo and Starr counties.
Comments: The dark green mounds and yellow heads lacking ray flowers are important characteristics of this species. It grows in saline, gypseous soils. The Spanish common name, which translates as "salted," refers to the plant's saline habitat.

Saladillo *Varilla texana*

Cow Pen Daisy

Family: Asteraceae (Compositae)
Scientific Name: *Verbesina encelioides*
Habit: Annuals up to 36" tall.
Leaves: Opposite; blades triangular, up to 4 ⅜" long.
Flowers: Heads about 2" broad; ray flowers yellow; disk flowers yellow.
Fruit: Achenes often winged, up to ¼" tall; pappus of tiny awns
Bloom Period: Spring, fall.
Distribution: Cameron, Hidalgo, Willacy, and Starr counties.
Comments: This is an attractive, showy plant that would do well in the garden. It usually grows in disturbed areas and edges of cultivated fields. It is a host plant for the Bordered Patch butterfly.

Cow Pen Daisy *Verbesina encelioides*

Frostweed

Family: Asteraceae (Compositae)
Scientific Name: *Verbesina microptera*
Habit: Perennials with woody bases; stems winged, up to 6' or taller.
Leaves: Alternate; petioles clasping the stems; blades ovate or triangular, up to 5 ⅛" long, the margins toothed.
Flowers: Heads up to ⅝" broad; ray flowers and disk flowers white.

Frostweed *Verbesina microptera*. Note the winged stems on the plants in the foreground

Asteraceae

Fruit: Achenes about ³⁄₁₆" tall; pappus of tiny awns.
Bloom Period: Summer, fall.
Distribution: Cameron, Hidalgo, Willacy, and Starr counties.
Comments: A hard frost causes a white foamy substance to emerge from the stems; hence the common name. This an excellent butterfly nectar plant.

Frostweed *V. microptera* flower head

Skeleton Leaf Golden Eye
Family: Asteraceae (Compositae)
Scientific Name: *Viguiera stenoloba*
Habit: Shrubs up to 4' or taller.
Leaves: Alternate; blades up to 2" long, green above and whitish on lower surfaces, linear or divided into narrow linear segments.
Flowers: Heads about 1" broad; ray flowers yellow, disk flowers yellow.
Fruit: Achenes about ⅛" tall; pappus absent.
Bloom Period: Spring, summer, fall.
Distribution: Cameron, Hidalgo, and Starr counties.
Comments: This species makes an ideal garden ornamental, almost constantly in bloom until winter. It responds to vigorous pruning.

Skeleton Leaf Golden Eye *Viguiera stenoloba*

Hairy Wedelia
Family: Asteraceae (Compositae)
Scientific Name: *Wedelia acapulcensis* var. *hispida* [*W. texana, W. hispida, Zexmenia hispida*]
Habit: Weak-stemmed perennials, woody, rough-hairy, up to 3' or taller.
Leaves: Opposite; blades lanceolate, up to 2 ⅝" long, the margins usually toothed.
Flowers: Heads about 1 ⅛" broad; ray flowers yellow to orange; disk flowers yellow.
Fruit: Achenes about ¼" tall; pappus of a few awns up to ⅛" long.
Bloom Period: Spring, summer, fall.
Distribution: Cameron, Hidalgo, and Starr counties.
Comments: This species makes a reliable garden ornamental. It is a good butterfly nectar plant and is a host plant to the Bordered Patch butterfly.

Hairy Wedelia *Wedelia acapulcensis* var. *hispida*

Spiny Goldenweed

Family: Asteraceae (Compositae)

Scientific Name: *Xanthisma spinulosum* var. *spinulosum* [*Machaeranthera pinnatifida*, *Haplopappus spinulosus*]

Habit: Perennials with woody bases, up to 20" or taller.

Leaves: Alternate, up to 1 ¼" long, pointing upward, pinnately lobed with long spinelike extensions on the lobes.

Flowers: Heads about 1" broad; ray flowers and disk flowers yellow.

Fruit: Achenes about ⅛" tall, with pappus of bristles.

Bloom Period: Summer, fall.

Distribution: Starr County.

Comments: This plant looks like a miniature camphor daisy, *Rayjacksonia phyllocephala*. It lacks the camphor scent of that species. Both species were formerly included in the genus *Machaeranthera*. There have been many earlier name changes. Several varieties have been described.

Spiny Goldenweed *Xanthisma spinulosum* var. *spinulosum*

Eastern Sleepy Daisy

Family: Asteraceae (Compositae)

Scientific Name: *Xanthisma texanum* var. *orientale*

Habit: Erect annual herbs up to 32" tall.

Leaves: Alternate; basal leaves deeply lobed, soon dying; those on upper stems lanceolate to oblanceolate, up to 1" long, gradually becoming smaller; margins usually toothed.

Flowers: Heads up to 1 ¼" broad; ray flowers and disk flowers yellow to orange.

Fruit: Achenes flattened, about 1/16" tall; pappus of scales up to ¼" long.

Bloom Period: Spring, summer, fall.

Distribution: Hidalgo, Willacy, and Starr counties.

Comments: This is a plant of sandy soils. It is endemic to Texas. The common name comes from the late opening of the ray flowers. The genus name comes from the Greek word *xanthos*, meaning "yellow."

Eastern Sleepy Daisy *Xanthisma texanum* var. *orientale*

Asteraceae

Cocklebur

Family: Asteraceae (Compositae)
Scientific Name: *Xanthium strumarium*
Habit: Annuals up to 6" or taller.
Leaves: Mostly alternate; blades somewhat 3-lobed, up to 4 ¾" long, the bases heart shaped.
Flowers: Heads axillary, unisexual; ray flowers absent; disk flowers very tiny.
Fruit: Fruiting heads very spiny, cylindrical.
Bloom Period: Summer, fall.
Distribution: Cameron, Hidalgo, and Starr counties.
Comments: This species usually grows in low areas. It is poisonous to livestock. The seeds were eaten by the now-extinct Carolina Parakeet. The plant is easily recognized by the prominent prickly fruiting heads. The fruits (cockleburs) are dispersed by sticking onto animals' fur. The barbed ends of the bur spikes were the inspiration for Velcro. The leaves have been used to produce a yellow dye.

Cocklebur *Xanthium strumarium*

Asiatic Hawkweed

Family: Asteraceae (Compositae)
Scientific Name: *Youngia japonica*
Habit: Herbs with milky sap; stems 12" or taller, arising from a basal rosette.
Leaves: Mostly crowded at the base; blades lobed, variable, up to 6" or longer.
Flowers: Heads about ¼" or taller.
Fruit: Achenes about ⅛" long; pappus bristles ⅛" or longer.
Bloom Period: Fall, winter.
Distribution: Cameron and Hidalgo counties; probably others.
Comments: A native of Asia, this species was probably introduced with bedding plants for the garden. It is generally present only in cultivated areas.

Asiatic Hawkweed *Youngia japonica*

Desert Zinnia

Family: Asteraceae (Compositae)
Scientific Name: *Zinnia acerosa*
Habit: Small woody plants up to 7" tall.
Leaves: Opposite, needle shaped or linear, up to ¾" long.
Flowers: Heads about 1" or broader; ray flowers white; disk flowers yellow.
Fruit: Achenes ⅛" or taller; pappus of 2 or 3 awns.
Bloom Period: Summer, fall.
Distribution: Starr County.
Comments: Desert zinnia grows on gravelly or caliche hills, occasionally limestone sands. The freshly opened flowers are a yellowish color, turning white at maturity.

Desert Zinnia *Zinnia acerosa*

Black Mangrove

Family: Avicenniaceae
Scientific Name: *Avicennia germinans*
Habit: Shrubs up to 6 ½' or taller.
Leaves: Opposite; blades shiny, mostly elliptic, up to 3 ⅛" long.
Flowers: Inflorescences axillary or terminal; corollas white, 4-lobed.
Fruit: Capsule one-seeded, flattened, ovoid, up to ¾" long, capable of floating long distances.
Bloom Period: Summer.
Distribution: Cameron and Willacy counties.
Comments: The shiny leaves call attention to this plant. It grows near the water in coastal areas. The branches provide habitat for birds, and the pneumatophores (root branches that stick up from the mud, like straws) create a nursery area for a number of marine animals (see photo p. ix). Mangroves are more common in salty areas where there is less competition, but they can grow in fresh water. Our three other mangrove species are in different families.

Black Mangrove *Avicenniaceae germinans* flowers

Black Mangrove *A. germinans* fruit

Bataceae

Madeira Vine, Sacasile

Family: Basellaceae
Scientific Name: *Anredera leptostachys* [*Boussingaultia leptostachya*]
Habit: Perennial twining vines with red stems emerging from tuberous roots.
Leaves: Alternate, somewhat succulent; blades ovate to elliptic, up to 2 ⅜" long.
Flowers: Inflorescences branching, with many fragrant flowers; sepals greenish, less than ⅛" long; petals absent.
Fruit: Enclosed in the enlarged calyx (like a tiny Chinese lantern).
Bloom Period: Fall.
Distribution: Cameron, Hidalgo, and Starr counties; also Latin America.
Comments: This would be an attractive vine in the garden. The leaves and tubers have been used as food. Madeira vine is not known elsewhere in Texas.

Madeira Vine, Sacasile *Anredera leptostachys*

Vidrillos

Family: Bataceae
Scientific Name: *Batis maritima*
Habit: Woody, creeping plants, forming large mats.
Leaves: Opposite, yellowish, cylindrical, succulent, up to 1" long.
Flowers: Tiny, crowded in cylindrical spikes about ⅜" long.
Fruit: More or less ovoid, up to ¾" long.
Bloom Period: Mostly summer.
Distribution: Cameron and Willacy counties.
Comments: This is a coastal species, occurring in salt flats and salt marshes. The large yellowish mats are noticeable from quite a distance. The leaves have been used in salads. They vary in their salty taste.

Vidrillos *Batis maritima*

Vidrillos *B. maritima* young fruit

Catclaw Vine

Family: Bignoniaceae

Scientific Name: *Macfadyena unguis-cati* [*Doxantha unguis-cati*]

Habit: Vines with catclaw tendrils, growing from tuberous roots.

Leaves: Opposite, compound with 3 leaflets. Leaflets are more or less ovate, about 2" long or sometimes longer. The margins are entire, occasionally toothed or lobed. Often one of the side leaflets is replaced by a tendril that, instead of twining, ends in 3 small claws.

Flowers: Bilateral, yellow, up to 3" long.

Fruit: Long and slim, up to 20" long, containing seeds with papery wings.

Bloom Period: Spring.

Distribution: Cameron, Hidalgo, and Starr counties.

Comments: Native to tropical America, this species is spreading throughout the southeastern United States. The plants in bloom are beautiful but produce a smothering blanket of foliage. This species was seen by the junior author in 1984 near Santa Maria in Cameron County. William MacWhorter reports seeing the vines growing around Llano Grande (present site of the World Birding Center in Weslaco) in the 1930s.

Catclaw Vine *Macfadyena unguis-cati*

Catclaw Vine *M. unguis-cati* leaves with claws

Boraginaceae 141

Yellow Show
Family: Bixaceae (Cochleospermaceae)
Scientific Name: *Amoreuxia wrightii*
Habit: Herbs, sometimes woody basally, up to 20" tall, arising from a carrot-shaped tuber.
Leaves: Alternate, 5-lobed, up to 3 1/8" broad, the lobes toothed.
Flowers: Slightly bilateral or regular; petals 5, up to 1 3/8" long, orange to pinkish, with purple basal marks.
Fruit: An ovoid capsule, up to 2" long, inflated, resembling a Chinese lantern, the seeds visible inside.
Bloom Period: Spring, summer, fall.
Distribution: Cameron, Hidalgo, and Starr counties.
Comments: This plant is more common in Starr County than in the other two. It would be an excellent addition to a xeric garden.
A. wrightii was named in honor of Charles Wright, one of the best known botanists in the United States in the 1800s.

Yellow Show *Amoreuxia wrightii*

Yellow Show *A. wrightii* dried fruit

Wild Olive, Anacahuite, Anacahuita
Family: Boraginaceae
Scientific Name: *Cordia boissieri*
Habit: Small trees up to 26' tall
Leaves: Alternate; blades hairy, ovate, up to 6 3/4" long.
Flowers: Radial; corollas funnel shaped, white with yellow centers, up to 1 3/4" long.
Fruit: An ovoid, yellowish drupe up to 1 1/8" long.
Bloom Period: All seasons.
Distribution: Cameron, Hidalgo, and Starr counties.
Comments: This small tree is a popular native ornamental. Appreciation of its value is increasing, as evidenced by its increased use along highways and other public areas. It is an important nectar plant for hummingbirds and butterflies.

Wild Olive, Anacahuite, Anacahuita *Cordia boissieri*

Mexican Cat's Eye

Family: Boraginaceae
Scientific Name: *Cryptantha mexicana*
Habit: Usually low-growing, hairy annuals.
Leaves: Opposite at the base but alternate above; blades mostly oblong, up to 1 ¾" long, much smaller up the stem.
Flowers: Corollas white, more or less funnel shaped, about ⅛" long, in coils with linear bracts.
Fruit: Usually 4 nutlets (mericarps).
Bloom Period: Spring, summer.
Distribution: Cameron and Starr counties.
Comments: This species can be distinguished from Texas cat's eye, *C. texana*, by the linear bracts on the inflorescence and by the fruit usually consisting of 4 nutlets.

Mexican Cat's Eye *Cryptantha mexicana*

Texas Cat's Eye

Family: Boraginaceae
Scientific Name: *Cryptantha texana*
Habit: Leaning or erect hairy annuals.
Leaves: Opposite at the base but alternate above; blades spatulate, up to 2 ⅜" long.
Flowers: Corollas white, funnel shaped, about ⅛" long, in coils.
Fruit: Usually one nutlet (mericarp), occasionally more.
Bloom Period: Spring.
Distribution: Willacy and Starr counties.
Comments: Characteristics commonly seen in this family are a fruit of 4 nutlets, the inflorescence in one-sided coils, and the petals partially grown together. This species varies from the norm in having only one nutlet, the others aborting.

Texas Cat's Eye *Cryptantha texana*

Boraginaceae

Anacua

Family: Boraginaceae
Scientific Name: *Ehretia anacua*
Habit: Trees up to 49' tall.
Leaves: Alternate; blades ovate, up to 3 ⅛" long, with stiff, slanted, scratchy hairs.
Flowers: Inflorescences in tight one-sided coils; corollas white, funnel shaped, about ¼" long.
Fruit: Drupes yellow to orange, 4-seeded.
Bloom Period: Spring, summer, fall.
Distribution: Cameron, Hidalgo, Willacy, and Starr counties.
Comments: The fruit of this species is a 4-seeded drupe rather than the more common 4 nutlets. The fleshy fruit are an important food source for birds. The leaves are scratchy when rubbed backwards and have been used as sandpaper. When rubbed against the skin, the leaves produce a painful red mark, a popular playground activity. This native tree is commonly cultivated. Small trees are produced from the roots, and the seeds germinate and grow easily.

Anacua *Ehretia anacua*

Anacua *E. anacua* flowering branches

Anacua *E. anacua* fruiting branches

Scorpion's Tail

Family: Boraginaceae
Scientific Name: *Heliotropium angiospermum*
Habit: Erect herbs.
Leaves: Alternate, sometimes almost appearing to be opposite; blades ovate, up to 4 ⅜" long.
Flowers: Inflorescences in 1-sided coils; corollas white with yellow centers, about ⅛" broad.
Fruit: Usually 4 nutlets (mericarps)
Bloom Period: All seasons.
Distribution: Cameron, Hidalgo, Willacy, and Starr counties.
Comments: The genus name comes from two Greek words meaning "turning to the sun." It has been shown that plants in this genus do not actually follow the sun, although some other plants do. This is an excellent butterfly nectar plant.

Scorpion's Tail *Heliotropium angiospermum*

Crowded Heliotrope

Family: Boraginaceae
Scientific Name: *Heliotropium confertifolium*
Habit: Perennials; hairy, silver colored, prostrate or forming low mounds 4" tall or less.
Leaves: Mostly alternate; blades narrow, up to 5/16" long.
Flowers: Inflorescences in one-sided coils; corollas white with yellow centers, about 5/16" long.
Fruit: Up to 4 nutlets (mericarps)
Bloom Period: Spring, summer, fall.
Distribution: Cameron, Hidalgo, and Starr counties.
Comments: Noticeable characteristics of this species are its silvery color and its tendency to be prostrate. It grows in gravelly or gypseous soils, often with desert zinnia, *Zinnia acerosa*, and oreja de perro, *Tiquilia canescens*.

Crowded Heliotrope *Heliotropium confertifolium*

Boraginaceae

Seaside Heliotrope

Family: Boraginaceae
Scientific Name: *Heliotropium curassavicum*
Habit: Succulent herbs, prostrate except at the stem tips.
Leaves: Alternate; blades oblanceolate, up to 2 ⅛" long.
Flowers: Inflorescences in one-sided coils; corollas white with yellow centers, about ⅛" broad.
Fruit: Usually 4 nutlets (mericarps).
Bloom Period: All seasons.
Distribution: Cameron, Hidalgo, Willacy, and Starr counties.
Comments: This widespread species is common at the coast but also occurs inland on saline soils. It is often seen growing from cracks in sidewalks.

Seaside Heliotrope *Heliotropium curassavicum*

Turnsole, Indian Heliotrope

Family: Boraginaceae
Scientific Name: *Heliotropium indicum*
Habit: Erect hairy herbs up to 40" tall.
Leaves: Alternate; blades ovate to heart shaped, up to 4 ¾" long.
Flowers: Inflorescences coiled, the flowers on one side. Corollas blue or violet, funnel shaped, about ³⁄₁₆" broad.
Fruit: Usually 4 nutlets (mericarps).
Bloom Period: Summer, fall, winter.
Distribution: Cameron, Hidalgo, and Willacy counties.
Comments: *H. indicum* has been reported to be native to Asia. However, recent studies indicate that it is native to the American tropics (Diggs, Lipscomb, & O'Kennon, 1999), and possibly introduced to the Old World Tropics. We have sen it growing naturally, far from human influence. This is our only *Heliotropium* with blue flowers. It would make an attractive garden plant. It is typically found growing in low places but is not commonly seen in the LRGV. This is an excellent butterfly nectar plant.

Turnsole, Indian Heliotrope *Heliotropium indicum*

Trailing Heliotrope

Family: Boraginaceae
Scientific Name: *Heliotropium procumbens*
Habit: Erect or leaning herbs.
Leaves: Alternate, obovate or elliptic, rounded at the tips, up to 1 ⅛" long.
Flowers: Inflorescences in one-sided coils; corollas white with yellow centers, about ⅛" broad.
Fruit: Usually 4 nutlets (mericarps).
Bloom Period: All seasons.
Distribution: Cameron, Hidalgo, Willacy, and Starr counties.
Comments: The erect growth and rounded leaf tips help distinguish this plant from the other heliotropes.

Trailing Heliotrope *Heliotropium procumbens*

Coastal Plain Heliotrope

Family: Boraginaceae
Scientific Name: *Heliotropium racemosum*
Habit: Sprawling, hairy herbs.
Leaves: Alternate, mostly lanceolate, up to 1 ⅝" long.
Flowers: Corollas white, about ⅝" broad; coils not always noticeable.
Fruit: Usually 2 nutlets.
Bloom Period: Spring, summer, fall.
Distribution: Cameron and Hidalgo counties; probably Starr County.
Comments: Endemic to Texas, this species prefers sandy soils. The flowers are much larger than most of our other heliotropes. The Texas heliotrope, *H. texanum,* also has large flowers, but its fruit is 4-lobed whereas *H. racemosum* fruit is 2-lobed.

Coastal Plain Heliotrope *Heliotropium racemosum*

Boraginaceae 147

Texas Heliotrope

Family: Boraginaceae
Scientific Name: *Heliotropium texanum*
Habit: Annuals up to 18" tall.
Leaves: Alternate, hairy, oblanceolate to lanceolate, up to 1 ⅛" long.
Flowers: Corollas white, about ½" broad, in one-sided coils.
Fruit: Usually 4 nutlets.
Bloom Period: Summer, fall.
Distribution: Hidalgo and Willacy counties.
Comments: This species is usually found growing in sandy soils. The fruit contains 4 nutlets, and the corollas are slightly smaller than those of the coastal plain heliotrope, *H. racemosum*, which has a fruit of 2 nutlets.

Texas Heliotrope *Heliotropium texanum*

Texas Heliotrope *H. texanum* fruit of 4 nutlets

Narrow Leaf Heliotrope

Family: Boraginaceae
Scientific Name: *Heliotropium torreyi*
Habit: Small branching shrubs up to 20" tall
Leaves: Alternate, linear, up to 1 ⅛" long.
Flowers: Inflorescences at the branch tips; corollas yellow, up to 3/16" long.
Fruit: Usually 4 nutlets (mericarps) about ⅛" broad, slightly less in height.
Bloom Period: Spring, summer, fall.
Distribution: Hidalgo County.
Comments: Growing on limestone or caliche slopes, this species is rarely encountered in our area. It is distinguished from the other heliotropes by its yellow flowers.

Narrow Leaf Heliotrope *Heliotropium torreyi*

Narrow Leaf Puccoon

Family: Boraginaceae
Scientific Name: *Lithospermum incisum*
Habit: Erect perennials up to 12" tall.
Leaves: First leaves up to 4 ¾" long and ⅜" broad, clustered at the base; stem leaves alternate, linear, up to 2 ⅜" long, becoming smaller toward the top of the stem.
Flowers: Spring flowers sterile, yellow, trumpet-shaped, about 1" long and up to ¾" broad, the five lobes with crinkled margins. Late spring and summer flowers fertile, tiny, remaining closed, self-fertilizing.
Fruit: Usually 4 nutlets (mericarps) about ⅛" tall.
Bloom Period: Spring, summer.
Distribution: Hidalgo County.
Comments: This species grows on sandy soils. It is easy to recognize, with its linear leaves and yellow trumpet-shaped flowers with crinkled margins. The fertile flowers are self-fertilized without opening (cleistogamous). Because of their minute size, they are usually overlooked. The roots were used to make a red dye by Native Americans and early settlers (Diggs et al 1999). The common name is derived from the Powhatan word, *poughkone*.

Narrow Leaf Puccoon *Lithospermum incisum* sterile flowers

Rough Gromwell

Family: Boraginaceae
Scientific Name: *Lithospermum matamorense*
Habit: Hairy, prostrate annuals, sometimes erect.
Leaves: Alternate, hairy; blades mostly elliptic, up to 3 ⅛" long. The upper leaves are smaller.
Flowers: Solitary from the leaf axils; corollas white with yellow centers, up to ¼" broad.
Fruit: Usually 4 nutlets (mericarps)
Bloom Period: Spring, summer.
Distribution: Cameron, Hidalgo, and Willacy counties.
Comments: This species resembles *Heliotropium*. A magnifying lens is necessary to distinguish the two genera. *Heliotropium* has an ovary that is lobed only slightly, if at all; *Lithospermum* has a deeply lobed ovary.

Rough Gromwell *Lithospermum matamorense*

Boraginaceae

Oreja De Perro

Family: Boraginaceae
Scientific Name: *Tiquilia canescens* [*Coldenia canescens*]
Habit: Prostrate, hairy perennials.
Leaves: Alternate; blades ovate, up to 3/8" long.
Flowers: Solitary or few; corollas pinkish, about 3/16" broad.
Fruit: 4 nutlets that fit together to form a sphere.
Bloom Period: All seasons.
Distribution: Cameron, Hidalgo, and Starr counties.
Comments: The first collections of this genus were made in South America by the Spanish explorers Ruiz and Pavon. The genus name comes from the common name they recorded as "tiquil-tiquil." This was probably a misunderstanding of the Quechua word for flower, *t'ika* (phonetic spelling). In the Quechua language, repetition of a noun at least sometimes indicates the plural, suggesting that the name given to the explorers was simply "flowers." The Spanish common name translates as "dog's ear," a reference to the leaf shape and color.

Oreja De Perro *Tiquilia canescens*

Chiggery Grapes

Family: Boraginaceae
Scientific Name: *Tournefortia hirsutissima*
Habit: Sprawling and climbing woody plants.
Leaves: Alternate; blades elliptic to ovate, up to 6" long.
Flowers: Inflorescence fragrant, reminiscent of orange blossoms, in branching one-sided coils; corollas white, radial, about 3/8" broad.
Fruit: A white drupe up to 5/16" broad with one small black dot, containing 2–4 seeds.
Bloom Period: Spring.
Distribution: Cameron County, only one locality known, in San Benito.
Comments: Thanks to Mary Ann and David Sato for calling this plant to our attention. It was not planted on their property but suddenly appeared and grew without any care into a huge liana occupying a space about 20' by 20' and climbing 15' high into trees. Seeds could have been dropped by birds. The species is native to various parts of Mexico (including Llera and El Encino in Tamaulipas), Central and South America, and the West Indies and also grows in Florida. The common name comes from the practice in the West Indies of rubbing the leaves on the skin to remove chiggers (parasitic mites).

Chiggery Grapes *Tournefortia hirsutissima*

Chiggery Grapes *T. hirsutissima* fruit

Mexican Tournefortia, Googly-Eyed Vine

Family: Boraginaceae
Scientific Name: *Tournefortia volubilis*
Habit: Twining, somewhat woody vines.
Leaves: Alternate; blades mostly ovate, up to 3 ⅛" long.
Flowers: Inflorescences in one-sided coils; corollas white to greenish, about 3/16" broad.
Fruit: A white drupe with 1–3 black dots, containing 2–3 seeds. The black dots are reminiscent of crossed eyes.
Bloom Period: All seasons.
Distribution: Cameron and Hidalgo counties.
Comments: In Texas this species has not been reported outside our area. It also occurs in Florida, Mexico, Caribbean islands, and South America. It is a host plant for the Saucy Beauty moth. A pathogen has infected *Tournefortia* in the LRGV recently, causing the fruits to enlarge and become covered with a fuzzy skin.

Mexican Tournefortia, Googly-Eyed Vine *Tournefortia volubilis*

Mexican Tournefortia, Googly-Eyed Vine *T. volubilis* fruiting branch

Leaf Mustard

Family: Brassicaceae (Cruciferae)
Scientific Name: *Brassica juncea*
Habit: Annuals up to 4' tall, usually less.
Leaves: Alternate; lower blades lobed, up to 6" long; upper leaves smaller.
Flowers: Petals 4, yellow, about ⅜" long.
Fruit: Siliques (capsules divided into 2 parts) erect, up to 2 ⅜" long.
Bloom Period: Spring, summer.
Distribution: Cameron County.
Comments: This plant was abundant when rains came at the right time in 2003. It was not a noticeable part of the landscape before that time, nor has it been since. It is a native of Eastern Europe and Western Asia. The familiar vegetables cabbage, cauliflower, broccoli, and turnips are all in the genus *Brassica*.

Leaf Mustard *Brassica juncea*

Brassicaceae

Gulf Sea Rocket
Family: Brassicaceae (Cruciferae)
Scientific Name: *Cakile geniculata*
Habit: Fleshy herbs often forming large mounds.
Leaves: Alternate; blades mostly oblanceolate, up to 2 ¾" long.
Flowers: Petals 4, white to lavender, up to ¼" long.
Fruit: Siliques almost rod shaped, slightly larger in the middle, up to 1 ⅛" long, the partition transverse rather than longitudinal.
Bloom Period: Spring, summer, and fall.
Distribution: Cameron and Willacy counties, in sandy soils near the coast.
Comments: This species is easily distinguished from other members of the family by its siliques with transverse partitions. It is a host plant for the Great Southern White butterfly. The young leaves are used in salads and sandwiches.

Gulf Sea Rocket *Cakile geniculata*

Shepherd's Purse
Family: Brassicaceae (Cruciferae)
Scientific Name: *Capsella bursa-pastoris*
Habit: Herbs up to 20" tall.
Leaves: Mostly basal, lobed; stem leaves alternate, small.
Flowers: Petals 4, white, about 1/16" long.
Fruit: Siliques flattened, heart shaped on the tips, up to 5/16" long.
Bloom Period: Spring.
Distribution: Cameron and Hidalgo counties.
Comments: This species is a native of Europe but is now widespread. This is the only known plant that produces a carnivorous seed. The seed produces a chemical that attracts tiny organisms, which it digests. The fruits, with their heart-shaped tips, make this species easy to recognize. The common name and specific epithet come from the similarity of the fruit to a traditional shepherd's purse made from the scrotum of a sheep.

Shepherd's Purse *Capsella bursa-pastoris*

Western Tansy Mustard

Family: Brassicaceae (Cruciferae)
Scientific Name: *Descurainia pinnata*
Habit: Annuals up to 32" tall, with glandular hairs; also branched hairs.
Leaves: Alternate, compound with segments less than ⅛" broad.
Flowers: Radial; petals 4, yellow or rarely whitish, about 1/16" or slightly longer.
Fruit: Siliques slightly curved, about 3/16" long.
Bloom Period: Winter, spring.
Distribution: Cameron, Hidalgo, Willacy, and Starr counties.
Comments: The flowers are tiny but noticeable when present in large numbers. The plants are poisonous to livestock.

Western Tansy Mustard *Descurainia pinnata*

Wedge Leaf Draba, Whitlow Wort

Family: Brassicaceae (Cruciferae)
Scientific Name: *Draba cuneifolia*
Habit: Erect hairy annuals up to 12" tall.
Leaves: Basal leaves obovate, up to 2" long; stem leaves few.
Flowers: Petals 4, white, about 3/16" long or less, sometimes absent.
Fruit: Siliques flattened, more or less elliptic, up to ⅝" long.
Bloom Period: Spring.
Distribution: Cameron, Hidalgo, and Starr counties.
Comments: These small plants with tiny flowers are easy to overlook. They are more noticeable and easier to recognize when the siliques are aligned on the stems. "Whitlow" refers to a medical condition affecting fingernails. Before the advent of modern medicine, there was an old belief called the Doctrine of Signatures. It was believed that the Creator had placed healing plants on the earth and made some part of the plant to resemble the body part needing healing. Evidently some part of this plant, possibly the fruit, was thought to resemble fingernails. "Wort" is an old English word for "plant."

Wedge Leaf Draba, Whitlow Wort *Draba cuneifolia*

Brassicaceae

Southern Pepperweed

Family: Brassicaceae (Cruciferae)
Scientific Name: *Lepidium austrinum*
   [*L. lasiocarpum* var. *orbiculare*]
Habit: Erect, hairy herbs up to 24" tall.
Leaves: Alternate; basal leaves deeply lobed, up to 3 ½" long; upper ones smaller with margins toothed or entire.
Flowers: Tiny; petals 4, white.
Fruit: Siliques more or less ovoid, up to ³⁄₁₆" long, produced in a series as the flowers open and mature.
Bloom Period: Spring, summer.
Distribution: Cameron, Hidalgo, and Willacy counties.
Comments: The long string of siliques below the flowers is often mistaken for a row of leaves.

Southern Pepperweed *Lepidium austrinum*

Bicolored Greggia

Family: Brassicaceae (Cruciferae)
Scientific Name: *Nerisyrenia camporum*
   [*Greggia camporum*]
Habit: Perennials, white-hairy, usually woody basally;
Leaves: Alternate; blades oblanceolate, usually toothed, up to 2 ⅜" long.
Flowers: Petals 4, white to pale lavender, up to ½" long.
Fruit: Siliques white-hairy, linear or oblong, sometimes slightly curved, up to 1 ⅛" long; pedicels absent.
Bloom Period: Spring, summer.
Distribution: Starr County.
Comments: This species is very showy when in full bloom. It is easily confused with *Synthlipsis*. *Nerisyrenia* has siliques with length three or more times their width; *Synthlipsis* siliques are shorter.

Bicolored Greggia *Nerisyrenia camporum*

Big Flower Bladderpod

Family: Brassicaceae (Cruciferae)

Scientific Name: *Paysonia grandiflora* [*Lesquerella grandiflora*]

Habit: Annuals or biennials with star-shaped hairs; younger stems erect, the older ones sprawling.

Leaves: Alternate; petioles mostly absent; lower blades usually deeply cut, broad and lobed at the base; upper ones smaller.

Flowers: Petals 4, yellow, up to ⅜" long.

Fruit: Flowering pedicels curved or straight, only rarely S-shaped; fruit globose.

Bloom Period: Spring.

Distribution: Cameron, Hidalgo, Willacy, and Starr counties.

Comments: This species is endemic to Texas. The globose fruit and the broad leaf bases, combined with the lack of petioles, distinguish it from other *Paysonia* and *Physaria* species, which are quite similar.

Big Flower Bladderpod *Paysonia grandiflora*

Berlandier's Woolly Pod Bladderpod

Family: Brassicaceae (Cruciferae)

Scientific Name: *Paysonia lasiocarpa* subsp. *berlandieri* [*Lesquerella lasiocarpa* var. *berlandieri*]

Habit: Annuals, perennials, or biennials with star-shaped hairs; younger stems erect, the older ones sprawling.

Leaves: Alternate; petioles mostly absent; lower blades usually deeply cut; upper ones smaller.

Flowers: Petals 4, yellow, up to ¼" long.

Fruit: Fruiting pedicels curving downward; fruit flattened.

Bloom Period: Spring.

Distribution: Cameron, Hidalgo, and Willacy counties.

Comments: The flattened fruit distinguishes this species from our other species of *Paysonia* and *Physaria,* which are quite similar and are not all included here.

Berlandier's Woolly Pod Bladderpod *Paysonia lasiocarpa* subsp. *berlandieri*

Brassicaceae

Silver Bladderpod

Family: Brassicaceae (Cruciferae)
Scientific Name: *Physaria argyrea* [*Lesquerella argyraea*]
Habit: Perennials or biennials with star-shaped hairs; younger stems erect, the older ones sprawling.
Leaves: Alternate; petioles extremely short; lower blades usually deeply cut; upper ones smaller, toothed or entire.
Flowers: Petals 4, yellow, up to ¼" long.
Fruit: Flowering pedicels S-shaped; fruit globose or ellipsoid.
Bloom Period: Spring.
Distribution: Cameron, Hidalgo, Willacy, and Starr counties.
Comments: *Physaria* and *Paysonia* species are very similar. The S-shaped pedicels and the globose or ellipsoid fruit distinguish this species from the others.

Silver Bladderpod *Physaria argyraea*

Lindheimer's Bladderpod

Family: Brassicaceae (Cruciferae)
Scientific Name: *Physaria lindheimeri* [*Lesquerella lindheimeri*]
Habit: Annuals with star-shaped hairs; younger stems erect, the older ones sprawling.
Leaves: Alternate; petioles usually present; lower leaf blades up to 1 ⅝" long, tapering basally.
Flowers: Petals 4, yellow, about ¼" long; styles about ⅛" long.
Fruit: Globose, about 3/16" broad; pedicels pointing upward or S-shaped.
Bloom Period: Spring.
Distribution: Cameron, Hidalgo, and Willacy counties.
Comments: The similar silver bladderpod, *Physaria argyraea,* has styles usually greater than ⅛" long. *Paysonia grandiflora* also has S-shaped pedicels, but its leaves clasp the stems. Our *Physaria* and *Paysonia* species are very similar. *Physaria lindheimeri* was named in honor of Ferdinand Jacob Lindheimer, a German-born botanist prominent in Texas in the nineteenth century.

Lindheimer's Bladderpod *Physaria lindheimeri*

Zapata Bladderpod

Family: Brassicaceae (Cruciferae)
Scientific Name: *Physaria thamnophila*
  [*Lesquerella thamnophila*]
Habit: Silvery perennials covered with star-shaped hairs.
Leaves: Alternate; up to 4 ¾" long; margins entire or slightly toothed.
Flowers: Petals 4, yellow, about ¼" long.
Fruit: Fruiting pedicels curving downward; fruit more or less globose, up to ¼" broad.
Bloom Period: Spring, summer.
Distribution: Starr County.
Comments: This plant is rarely seen. It is endemic to Texas and is listed as rare and endangered. Our *Paysonia* and *Physaria* species are very similar.

Zapata Bladderpod *Physaria thamnophila*

Cylindrical Yellow Cress

Family: Brassicaceae (Cruciferae)
Scientific Name: *Rorippa teres*
Habit: Annuals or biennials somewhat hairy; stems erect or leaning, up to 12" long.
Leaves: Alternate; blades more or less oblong, deeply cut.
Flowers: Inflorescences densely flowered; petals 4, yellow, minute.
Fruit: Siliques cylindrical, sometimes slightly curved, up to ⅝" long.
Bloom Period: Mostly spring, some fall and winter.
Distribution: Cameron, Hidalgo, Willacy, and Starr counties.
Comments: The minute yellow flowers and the deeply cut leaves are major characteristics of this species.

Cylindrical Yellow Cress *Rorippa teres*

Brassicaceae

Rio Grande Valley Selenia

Family: Brassicaceae (Cruciferae)
Scientific Name: *Selenia grandis*
Habit: Annuals beginning with a basal rosette of leaves, then producing sprawling stems.
Leaves: Alternate, deeply cut, the lobes cut again; up to 8" long.
Flowers: Petals 4, yellow, up to ⅜" long, pleasantly aromatic.
Fruit: Oblong, up to ¾" long.
Bloom Period: Winter, spring.
Distribution: Cameron and Hidalgo counties.
Comments: This attractive plant is noticeable for its large flowers. It is endemic to Texas and is found around the edges of fields and in disturbed areas.

Rio Grande Valley Selenia *Selenia grandis*

Rio Grande Valley Selenia *S. grandis* fruit

Viereck's Winged Rockcress

Family: Brassicaceae (Cruciferae)
Scientific Name: *Sibara viereckii* [*S. runcinata*]
Habit: Annual herbs with slender stems up to 24" tall.
Leaves: Basal leaves deeply cut in rosettes, the upper ones alternate, smaller and usually entire.
Flowers: Petals 4, white to pale lavender, about ⁵⁄₁₆" long.
Fruit: Siliques linear, about ¾" long.
Bloom Period: Winter, spring.
Distribution: Cameron, Hidalgo, Willacy, and Starr counties.
Comments: The flowers are small and easily overlooked.

Viereck's Winged Rockcress *Sibara viereckii*

Rocket Mustard

Family: Brassicaceae (Cruciferae)
Scientific Name: *Sisymbrium irio*
Habit: Annual herbs up to 24" tall.
Leaves: Alternate, the lower ones deeply lobed, the upper ones smaller and nearly entire.
Flowers: Petals 4, yellow, about 3/16" long.
Fruit: Siliques cylindrical, up to 2" long, less than 1/16" broad.
Bloom Period: Mostly spring.
Distribution: Cameron, Hidalgo, and Starr counties.
Comments: This is one of the earliest of our spring wildflowers. It is native to Europe.

Rocket Mustard *Sisymbrium irio*

Gregg's Keelpod

Family: Brassicaceae (Cruciferae)
Scientific Name: *Synthlipsis greggii*
Habit: Annuals or biennials, very hairy and weak-stemmed, often sprawling.
Leaves: Alternate, deeply toothed, sometimes entire, up to 5½" long.
Flowers: Petals 4, white to lavender, up to ½" long.
Fruit: Siliques mostly ellipsoid, up to ½" or slightly longer.
Bloom Period: Winter, spring.
Distribution: Hidalgo and Starr counties.
Comments: This species grows in Texas from Hidalgo County to the Trans-Pecos; it also occurs in Mexico. It is easily confused with *Nerisyrenia,* which has siliques with length three or more times the width; *Synthlipsis* siliques are shorter.

Gregg's Keelpod *Synthlipsis greggii*

Buddlejaceae

Shinners' Rocket

Family: Brassicaceae (Cruciferae)
Scientific Name: *Thelypodiopsis shinnersii*
Habit: Hairless erect annuals.
Leaves: Alternate, the lower ones with winged petioles; blades obovate, up to 4" long.
Flowers: Radial; petals 4, white, ¼" long.
Fruit: Siliques rounded in cross section, up to 2 ⅜" long.
Bloom Period: Spring.
Distribution: Cameron County.
Comments: In Texas this rare species is known only from Cameron County. It also occurs in Mexico. A small population was found in Harlingen recently. To our knowledge, this is the first photograph of a living plant of *T. shinnersii*.

Shinners' Rocket *Thelypodiopsis shinnersii*

Rio Grande Butterfly Bush, Tepozán

Family: Buddlejaceae (Loganiaceae)
Scientific Name: *Buddleja sessiliflora*
Habit: Shrubs up to 6 ½' tall.
Leaves: Opposite, somewhat hairy, silvery gray; blades ovate or elliptic, up to 4" long.
Flowers: In dense clusters from the leaf axils; corollas greenish, up to 3⁄16" long.
Fruit: Capsules about 3⁄16" tall.
Bloom Period: Spring, summer.
Distribution: Cameron and Hidalgo counties.
Comments: In Texas this species occurs only in the Lower Rio Grande Valley, never occurring far from the banks of the Rio Grande. The flowers smell like the inside of a dog's ears.

Rio Grande Butterfly Bush, Tepozán *Buddleja sessiliflora*

Polly Prim

Family: Buddlejaceae (Loganiaceae)
Scientific Name: *Polypremum procumbens*
Habit: Erect, leaning, or prostrate annuals or perennials.
Leaves: Opposite; blades linear, up to ¾" long.
Flowers: Corollas white, 4-lobed, about ⅛" long.
Fruit: Capsules more or less globose, about ¹⁄₁₆" broad.
Bloom Period: Spring, summer.
Distribution: Cameron, Hidalgo, and Willacy counties.
Comments: Polly prim tends to grow in sandy soils. It has a great resemblance to prairie bluets, *Hedyotis nigricans,* which has an inferior ovary. Polly prim has a superior ovary.

Polly Prim *Polypremum procumbens*

Night-Blooming Cereus, Barbed-Wire Cactus

Family: Cactaceae
Scientific Name: *Acanthocereus tetragonus* [*A. pentagonus, Cereus pentagonus*]
Habit: Stems branching, 3-, 4-, or rarely 5- to 7-ribbed, growing 10' or longer, sometimes climbing in trees; spines whitish, up to 1 ½" long.
Leaves: None.
Flowers: Radial, about 8" broad, sepals and petals numerous, intergrading; petals white, fragrant, opening at night. They are pollinated by Hawk moths and bats.
Fruit: Fleshy, red when ripe, spiny, about 3 ⅛" long.
Bloom Period: Spring, summer.
Distribution: Cameron, Hidalgo, and Willacy counties.
Comments: Much of its wild brush habitat is now gone, but this species is often cultivated. It is neither invasive nor as difficult to handle as the prickly pear. The young stems and ripe fruit are edible, after spines are removed. The barbed, brittle glochids typical of prickly pear are not present in this species. It will freeze to the ground in some years, regrowing from roots and underground stems. The stems are eaten by rats and rabbits. Birds, tortoises, and coyotes eat the ripe fruit.

Night-Blooming Cereus, Barbed-Wire Cactus *Acanthocereus tetragonus*

Night-Blooming Cereus, Barbed-Wire Cactus
A. tetragonus fruit

Fishhook Cactus, Root Cactus

Family: Cactaceae
Scientific Name: *Ancistrocactus scheeri*
  [*Echinocactus scheeri, Sclerocactus scheeri*]
Habit: Stems up to 6" long, the spines ending in curved hooks. Stems grow from a long white fleshy tuberous root.
Leaves: None.
Flowers: Radial, greenish yellow, up to 1 ⅛" broad, sepals and petals intergrading.
Fruit: Green, about 1 ⅜" long, somewhat fleshy but drying at maturity.
Bloom Period: Winter, spring
Distribution: Hidalgo, Willacy, and Starr counties.
Comments: The spines can vary in color from light yellow to brown. They often prevent the flowers from opening fully. This is one of our earliest flowering cacti. The common name (Root Cactus) comes from the characteristic of producing a fleshy tuberous root.

Fishhook Cactus, Root Cactus *Ancistrocactus scheeri*

Sea Urchin Cactus, Sand Dollar Cactus, Star Cactus

Family: Cactaceae
Scientific Name: *Astrophytum asterias*
  [*Echinocactus asterias*]
Habit: Stems green, roundish and spineless, up to 6" broad, dotted with small clumps of whitish hairs.
Leaves: None.
Flowers: Radial, yellow with orange centers, up to 2" broad, sepals and petals intergrading.
Fruit: Fleshy, green or pinkish, up to ½" tall.
Bloom Period: Spring and summer, flowering sporadically after rains.
Distribution: Starr County.
Comments: This is federally listed as endangered. Starr County is the only currently recorded Texas locality for the star cactus. It is also known in Mexico but is not common there. The small dots of whitish hairs on the stems distinguish this species from peyote, *Lophophora williamsii*, which also lacks spines. Seed-grown plants are now available from nurseries. Plants are threatened by habitat destruction, collecting, and foraging by cottontail rabbits. Observations by the Nature Conservancy of Texas suggest that lack of other forage during drought forces cottontails to concentrate on sand dollar cactus, drastically reducing the already threatened populations.

Sea Urchin Cactus, Sand Dollar Cactus, Star Cactus
Astrophytum asterias

Runyon's Coryphantha

Family: Cactaceae

Scientific Name: *Coryphantha macromeris* var. *runyonii* [*C. runyonii, Mammillaria runyonii*]

Habit: Much branched from a fleshy root, forming mats or mounds up to 40" broad. Stem hemispheric, or older ones cylindrical, with large, conspicuous knobs (tubercles) with spines at the tips up to 2 1/8" long, usually shorter. Spine and tubercle length are variable.

Leaves: None.

Flowers: Apical, radial, about 2" broad, sepals and petals numerous, intergrading; petals rose colored, some with stripes.

Fruit: Green, ovoid, up to 1 1/8" tall.

Bloom Period: Spring, summer, fall, occasionally into winter.

Distribution: Cameron and Starr counties.

Comments: According to *Flora of North America*, vol. 4 (2003), this group of plants should not have varietal status. Since there is not full agreement on this, we are being conservative and retaining the variety *runyonii*. An extreme form of this species, with very short tubercles and spines, was collected in Starr County by Viola and Harry Sunderland. It was introduced into the nursery trade as *Coryphantha sunderlandii*.

Runyon's Coryphantha *Coryphantha macromeris* var. *runyonii*

Junior Tom Thumb Cactus, Runyon's Pincushion Cactus

Family: Cactaceae

Scientific Name: *Coryphantha pottsiana* [*C. robertii, Escobaria runyonii, Mammillaria robertii*]

Habit: Stems more or less cylindrical, up to 3 1/4" tall, forming clumps, with many conspicuous knobs (tubercles) about 5/16" long; spines almost hiding the stems. The spines are variable from white to dark brown.

Leaves: None.

Flowers: Radial, about 1" broad, sepals and petals numerous, intergrading; petals bronze to tan in color.

Fruit: Red when ripe, up to 3/8" long.

Bloom Period: Early spring.

Distribution: Hidalgo and Starr counties.

Comments: According to Powell and Weedin's *Cacti of the Trans-Pecos and Adjacent Areas* (2004), *Coryphantha robertii* is the same as

Junior Tom Thumb Cactus, Runyon's Pincushion Cactus *Coryphantha pottsiana*

C. pottsiana. Since *C. pottsiana* is the older name, *C. robertii* is reduced to a synonym of *C. pottsiana*. Runyon's pincushion cactus is sometimes confused with the hair covered cactus, *Mammillaria prolifera*, which has dense hairs among the spines, covering the stems. *Coryphantha pottsiana* does not.

Cactaceae

Horse Crippler, Devil's Head

Family: Cactaceae

Scientific Name: *Echinocactus texensis* [*Homalocephala texensis*]

Habit: Stems hemispheroidal and ribbed, up to 12" broad. Spines reddish, turning white with age, variable in length.

Leaves: None.

Flowers: Radial, up to 2 ¼" broad, lavender, pink or white with reddish centers; sepals and petals numerous, intergrading.

Fruit: Red, spiny, up to 1½" long.

Bloom Period: Spring.

Distribution: Cameron, Hidalgo, Willacy, and Starr counties.

Comments: Like many cacti, this one is well camouflaged and often overlooked. The spines are sturdy and can penetrate a shoe or a tire. Easily grown from seed, this cactus makes an attractive rock garden addition. At one time, this was one of the barrel cacti used to make Mexican cactus candy.

Horse Crippler, Devil's Head *Echinocactus texensis*

Berlandier's Alicoche, Blank's Alicoche

Family: Cactaceae

Scientific Name: *Echinocereus berlandieri* [*E. blankii*]

Habit: Young stems erect, then falling over, growing up to 14" long and 1 ⅝" broad, sprouting from underground rhizomes. Spines up to 1 ⅛" long, dark brown, rarely yellow.

Leaves: None.

Flowers: Radial, purple with darker centers, up to 5" broad; sepals and petals numerous, intergrading.

Fruit: Green, up to 1" long.

Bloom Period: Spring, summer.

Distribution: Cameron, Hidalgo, and Willacy counties.

Comments: The taxonomy of this species is confused. (1) The cactus we formerly called *E. berlandieri* has been combined with *E. pentalophus* as an extremely variable species, called *E. pentalophus*. (2) Then the name *E. berlandieri* was given to what had previously been known *E. blankii;* and then (3) *E. blankii* was eliminated from our flora. This is according to the *Flora of North America,* vol. 4 (2003). In the past, there have been statements that the flowers of *E. berlandieri*

Berlandier's Alicoche *Echinocereus berlandieri*

do not open fully. We have observed this on cloudy days, especially with cool temperatures. In general, however, the flowers open normally as in other cacti. The plants grow on well-drained clay soils. *E. berlandieri* was named in honor of Jean Louis Berlandier, who did extensive plant collecting in Mexico and Texas.

Strawberry Cactus, Strawberry Pitaya

Family: Cactaceae

Scientific Name: *Echinocereus enneacanthus*

Habit: Stems erect to sprawling, 3" to 16" or longer; ribs about 7–10.

Leaves: None.

Flowers: Radial, about 3" broad, sepals and petals numerous, intergrading; petals purple.

Fruit: Pale yellow or dull red, up to 1 1/8" tall.

Bloom Period: Spring.

Distribution: Cameron, Hidalgo, and Starr counties.

Comments: The fruits are edible, tasting like strawberries.

Strawberry Cactus, Strawberry Pitaya *Echinocereus enneacanthus*

Small Papillosus, Yellow Flowered Alicoche

Family: Cactaceae

Scientific Name: *Echinocereus papillosus* var. *angusticeps* [*E. angusticeps*]

Habit: Plants forming clumps. Stems spiny, erect or leaning, up to 8" long, up to 2" broad or more, with ribs divided into nodules (tubercles).

Leaves: None.

Flowers: Radial, up to 4" broad; sepals and petals numerous, intergrading; petals yellow, basally orange-red.

Fruit: Greenish, spiny.

Bloom Period: Spring.

Distribution: Cameron, Hidalgo, and Starr counties.

Comments: The Cameron County locality is doubtful. Based on reports of intermediate forms, some taxonomists do not recognize any varieties of *E. papillosus*. Others recognize var. *angusticeps* (small stem size and large number of stems) and var. *papillosus* (larger stem size and few stems per cluster). Var. *angusticeps* is endemic to Texas, limited to red sandy soils of northwestern Hidalgo County and northeastern Starr County. The var. *papillosus* is more widespread in eroded limestone and caliche.

Small Papillosus, Yellow Flowered Alicoche *Echinocereus papillosus* var. *angusticeps*

Cactaceae

Lady Finger Cactus

Family: Cactaceae
Scientific Name: *Echinocereus pentalophus*
Habit: Stems prostrate, branching, up to 6" long and 1 ⅛" broad, with 4–5 ribs.
Leaves: None.
Flowers: Radial, 4" or broader, sepals and petals numerous, intergrading; petals purple with light centers.
Fruit: Green, up to 1" tall.
Bloom Period: Spring.
Distribution: Cameron, Hidalgo, and Starr counties; probably also Willacy County.
Comments: Western populations have longer, thicker stems than more eastern populations. Some forms of this species have been mistakenly called *Echinocereus berlandieri* (see Berlandier's alicoche).

Lady Finger Cactus *Echinocereus pentalophus*

Pencil Cactus, Sacasil, Dahlia Cactus

Family: Cactaceae
Scientific Name: *Echinocereus poselgeri* [*Wilcoxia poselgeri*]
Habit: Stems arising from tuberous roots, up to ⅝" broad and 57" long, erect when young, later clambering in shrubs; spines about ¼" long, appressed against the stem.
Leaves: None.
Flowers: Radial, up to 2" broad; sepals and petals numerous, intergrading; petals deep pink.
Fruit: Green to brownish, with white-hairy spots around the spines, up to ¾" tall.
Bloom Period: Spring.
Distribution: Cameron, Hidalgo, Willacy, and Starr counties.
Comments: This cactus has for many years been known as *Wilcoxia*. Only recently has it been placed into the genus *Echinocereus*. It is sometimes called dahlia cactus because the tubers resemble those of dahlias. The plants are inconspicuous except when in bloom.

Pencil Cactus, Sacasil, Dahlia Cactus *Echinocereus poselgeri*

Fitch's Rainbow Cactus

Family: Cactaceae
Scientific Name: *Echinocereus reichenbachii* var. *fitchii* [*E. fitchii*]
Habit: Stems up to 6" tall, ribbed, with spiny nodules.
Leaves: None.
Flowers: Radial, up to 4" broad, sepals and petals numerous, intergrading; petals pink.
Fruit: Green, oval to roundish, up to 1" long.
Bloom Period: Spring.
Distribution: Starr County
Comments: The spine colors can vary from white to brown to pink.

Fitch's Rainbow Cactus *Echinocereus reichenbachii* var. *fitchii*

Fitch's Rainbow Cactus *Echinocereus reichenbachii* var. *fitchii* in habitat

Lower Rio Grande Valley Barrel Cactus

Family: Cactaceae
Scientific Name: *Ferocactus hamatacanthus* var. *sinuatus* [*Echinocactus sinuatus*]
Habit: Stems spheroidal when young, elongating with age, up to 24" or taller and 8" broad.
Leaves: None.
Flowers: Radial, about 3" broad, sepals and petals numerous, intergrading; petals yellow.
Fruit: Green, ovoid, about 1" tall.
Bloom Period: Summer.
Distribution: Cameron, Hidalgo, and Starr counties.
Comments: This is our largest barrel-type cactus, growing in saline and gypseous clay soils. Populations on the coastal lomas and along the Arroyo Colorado are disjunct from the western Hidalgo and Starr County populations. Exceptional individuals have been reported to be 60" tall. Many of the older plants were removed and sold for cultivation. This species is easily confused with twisted rib, *Hamatocactus bicolor,* which has small red fruit and a yellow flower with a red center.

Lower Rio Grande Valley Barrel Cactus *Ferocactus hamatacanthus* var. *sinuatus*

Cactaceae

Dog Cholla

Family: Cactaceae
Scientific Name: *Grusonia schottii* [*Opuntia schottii*]
Habit: Stems prostrate or leaning, up to 3" long. Spines 6–12 in a cluster, up to 2" long.
Leaves: Very small, cylindrical, curved, produced during new growth and falling quickly.
Flowers: Radial, up to 2 ¾" broad; sepals and petals numerous, intergrading; petals yellow.
Fruit: Yellow, up to 1 ⅝" tall.
Bloom Period: Late spring, early summer.
Distribution: Cameron, Hidalgo, and Starr counties.
Comments: The Cameron County populations have not been located recently. The stem joints easily detach and become attached to animals, providing a convenient means of dispersion. The spines can easily penetrate a tennis shoe.

Dog Cholla *Grusonia schottii*

Dog Cholla *G. schottii* fruit

Twisted Rib, Hedgehog Cactus

Family: Cactaceae
Scientific Name: *Hamatocactus bicolor* [*Hamatocactus setispinus, Thelocactus setispinus, Ferocactus setispinus*]
Habit: Ovoid when young, growing cylindroid with the ribs twisted.
Leaves: None.
Flowers: Radial, up to 2 ⅛" broad, sepals and petals numerous, intergrading; petals yellow, red basally.
Fruit: Red, spherical to ovoid, up to ¾" broad.
Bloom Period: Spring or early fall.
Distribution: Cameron, Hidalgo, and Starr counties.
Comments: The stem sizes vary from one population to another. At one time, two separate varieties based on size were recognized. The larger stem size is sometimes confused with Lower Rio Grande Valley barrel cactus, *Ferocactus hamatacanthus* var. *sinuatus*, which has all-yellow flowers. *Hamatocactus bicolor* flowers have red centers.

Twisted Rib, Hedgehog Cactus *Hamatocactus bicolor*

Peyote

Family: Cactaceae

Scientific Name: *Lophophora williamsii*

Habit: Stems almost globular, or hemispherical, up to 5" broad, usually less. Most of the stem is underground, especially during drought. Spines absent.

Leaves: None.

Flowers: Radial, up to 1" broad, sepals and petals numerous, intergrading; petals pink.

Fruit: Pale pink, up to ¾" tall.

Bloom Period: Spring.

Distribution: Hidalgo and Starr counties.

Comments: The fruit can take a year or more to mature. The plants have no spines, but tufts of woolly hairs emerge where spines would be. It is well known that peyote plants contain hallucinogenic substances. Historical records indicate locations in western Hidalgo County. Recent attempts to relocate the Hidalgo County populations have been unsuccessful. Populations are decreasing because of root plowing and collecting for its hallucinogenic properties. Multiheaded specimens may indicate past harvesting.

Peyote *Lophophora williamsii*

Little Chiles, Nipple Cactus, Pincushion Cactus

Family: Cactaceae

Scientific Name: *Mammillaria heyderi*

Habit: Stems hemispherical, green or blue green, up to 5" broad; tubercles in spiral rows. Stems usually single, sometimes in clusters.

Leaves: None.

Flowers: Radial, ⅝–1 ⅛" broad, sepals and petals numerous, intergrading; petals pinkish to almost white, with midstripes.

Fruit: Bright red, ⅜–1 ⅜" tall.

Bloom Period: Spring.

Distribution: Cameron, Hidalgo, Willacy, and Starr counties.

Comments: The flowers and fruit occur in rings near the top of the stem. The fruit are edible, described as "sweet and tart."

Little Chiles, Nipple Cactus, Pincushion Cactus *Mammillaria heyderi*

Little Chiles, Nipple Cactus, Pincushion Cactus *M. heyderi* plant in fruit

Cactaceae

Hair Covered Cactus

Family: Cactaceae

Scientific Name: *Mammillaria prolifera* var. *texana* [*M. multiceps*]

Habit: Stems in large clusters, globose or egg shaped, up to 3 ½" tall, with a whitish appearance caused by hairs and bristles. Tubercles present.

Leaves: None.

Flowers: Radial, about ¾" broad, sepals and petals numerous, intergrading; petals yellow to tan in color, with midstripes.

Fruit: Red, up to ¾" long.

Bloom Period: Spring.

Distribution: Cameron and Hidalgo counties.

Comments: The spine colors vary from yellow to dark brown. The photo shows the brown-spined variation (left) and the yellow-spined variation (right). This species is similar to junior Tom Thumb cactus, *Coryphantha pottsiana*, which lacks the hairs and bristles. The multiple stems develop very rapidly, hence the specific epithet.

Hair Covered Cactus *Mammillaria prolifera* var. *texana*

Yellow Flowered Pincushion Cactus

Family: Cactaceae

Scientific Name: *Mammillaria sphaerica* [*M. longimamma, Dolicothele sphaerica*]

Habit: Plants sometimes forming clusters, arising from fleshy roots, more or less globose, up to 3 ⅛" broad, with tubercles.

Leaves: None.

Flowers: Radial, about 2½" broad, sepals and petals numerous, intergrading; petals yellow.

Fruit: Maroon to almost white, up to 5/16" long.

Bloom Period: Summer.

Distribution: Cameron, Hidalgo, and Starr counties.

Comments: The plants are usually found growing in the partial shade of shrubs. The flowers are fragrant. They do not occur in a ring as in other mammillarias.

Yellow Flowered Pincushion Cactus *Mammillaria sphaerica*

Texas Prickly Pear

Family: Cactaceae

Scientific Name: *Opuntia engelmannii* var. *lindheimeri* [*O. lindheimeri* var. *lindheimeri*]

Habit: Stems flattened, jointed, growing 10' or taller. Spines yellow, in clusters, some sturdy and some (glochids) shorter, brittle, and barbed.

Leaves: Very small, cylindrical, curved, produced during new growth and falling quickly.

Flowers: Radial; up to 4" broad, sepals and petals numerous, intergrading; petals yellow in the upper Valley, but in the Brownsville area and in the Arroyo Colorado brush, near Harlingen, they occur in all shades from red to yellow.

Fruit: Purple, up to 2 ¾" tall.

Bloom Period: Spring.

Distribution: Cameron, Hidalgo, Willacy, and Starr counties.

Comments: This species was named in honor of Georg Engelmann and Ferdinand Jacob Lindheimer, both prominent Texas botanists. The blooming season is short, but these are among the most magnificent flowers with their brilliance and color. Spines discourage approaching too closely. The glochids are the most irritating type of spines, since they are barbed and brittle, breaking off and remaining in one's flesh. Flower color is highly variable in Cameron County, ranging from shades of red to orange, pink, or yellow. Elsewhere the flowers are mostly yellow unless transplanted from Cameron County. Texas prickly pear is an important food plant for wildlife. This and some other cactus species are sometimes infected with colonies of scale insects, which appear white and cottony on the outer part. The inner part contains the red cochineal dye that has been used for centuries to dye feathers, textiles, and other items.

Texas Prickly Pear *Opuntia engelmannii* var. *lindheimeri*, red-flowering form

Texas Prickly Pear *O. engelmannii* var. *lindheimeri*, yellow-flowering form

Cactaceae

Desert Christmas Cactus, Tasajillo

Family: Cactaceae

Scientific Name: *Opuntia leptocaulis* [*Cylindropuntia leptocaulis*]

Habit: Stems cylindrical, erect to leaning or falling over. Spine length is variable, some individuals with long spines, others nearly spineless.

Leaves: Very small, falling away early.

Flowers: Radial, up to ⅞" broad; sepals and petals numerous, intergrading; petals greenish to yellow or bronze.

Fruit: Red, rarely yellow, up to 1" long.

Bloom Period: Spring, some summer and fall.

Distribution: Cameron, Hidalgo, Willacy, and Starr counties.

Comments: This cactus often forms huge clumps. The stems can clamber up into nearby trees. Around Christmas time, the fruit become bright red, hence the common name. A large clump can be very attractive. This species is unusual in that the stems often branch from the fruit. A plant with yellow fruit was found in Bexar County.

Desert Christmas Cactus, Tasajillo *Opuntia leptocaulis* showing stems growing out of the fruit

Desert Christmas Cactus, Tasajillo *O. leptocaulis* flower

Low Growing Prickly Pear

Family: Cactaceae

Scientific Name: *Opuntia macrorhiza* [*O. leptocarpa*]

Habit: Stems flattened, jointed, erect at first but falling over with 3 or more joints. Spines yellow or tan, some sturdy and some (glochids) brittle and barbed.

Leaves: Very small, cylindrical, curved, produced during new growth and falling quickly.

Flowers: Radial, up to 3" broad, sepals and petals numerous, intergrading; petals yellow or orange, reddish basally.

Fruit: Red, about 3 ½" tall.

Bloom Period: Spring.

Distribution: Cameron, Hidalgo, Willacy, and Starr counties.

Comments: This species occurs in sandy soils near the coast and also inland. The reddish center of the flower distinguishes this species from the more common Texas prickly pear, *O. engelmannii* var. *lindheimeri*.

Low Growing Prickly Pear *Opuntia macrorhiza*

Lemon Vine

Family: Cactaceae

Scientific Name: *Pereskia aculeata*

Habit: Vinelike, clambering stems with two kinds of spines. Primary spines (the first formed) are 2 per areole, curved like a cat's claw, an aid to the clambering habit. Secondary spines are up to 25 per areole, straight, from 3/8" to 1 3/8" long.

Leaves: Alternate, thick and succulent, lanceolate, ovate, or oblong, up to 4 3/8" long.

Flowers: Fragrant, radial, sepals and petals intergrading, variable color from white to yellow or greenish or pinkish, 1 1/2–2" broad.

Fruit: Juicy, yellow to orange when ripe, roundish, about 1 5/8" tall.

Bloom Period: Summer, fall.

Distribution: Willacy County.

Comments: Native to the South American tropics, this species has been cultivated and has subsequently escaped. It is found growing naturally in Texas, Florida, Mexico, and Central America. It was first reported to be growing naturally in Willacy County by Joe Ideker in 1996. This is our only cactus with persisting leaves.

Lemon Vine *Pereskia aculeata*

Queen of the Night, Moonlight Cactus

Family: Cactaceae

Scientific Name: *Selenicereus spinulosus*

Habit: Stems ribbed, up to 3/4" broad, clambering, producing aerial roots; spines 1/16" long or less.

Leaves: None.

Flowers: Opening at night, radial, white or occasionally pinkish, about 3 1/2" broad or more.

Fruit: An elliptical or ovoid, yellow or red spiny berry up to 2" long. The spines brush away easily when the fruit ripens.

Bloom Period: Spring, summer.

Distribution: Cameron County.

Comments: This species has not been seen outside cultivation in recent years and has been omitted from some floras. William MacWhorter, in Weslaco, reports finding it in brush near the mouth of the Rio Grande and taking cuttings from that plant. The plant in the photograph was grown from one of those cuttings. All the plants we have observed in Texas have yellow fruit. We have observed some in Mexico with red fruit.

Queen of the Night, Moonlight Cactus *Selenicereus spinulosus*

Queen of the Night, Moonlight Cactus *S. spinulosus* fruit

Glory of Texas

Family: Cactaceae
Scientific Name: *Thelocactus bicolor*
Habit: Stems ovoid, becoming longer with age; tubercles present, at least in younger parts; spines white, brownish, or red.
Leaves: None.
Flowers: Radial, about 2 ⅜" broad, sepals and petals numerous, intergrading; petals rose colored, dark basally.
Fruit: Green.
Bloom Period: Spring, summer, fall.
Distribution: Starr County.
Comments: Unlike most cacti, this cactus blooms sporadically throughout the spring, summer, and fall.

Glory of Texas *Thelocactus bicolor*

Annual Waterwort

Family: Callitrichaceae
Scientific Name: *Callitriche terrestris*
Habit: Tiny creeping herbs.
Leaves: Opposite, more or less ovate, about ⅛" long.
Flowers: Axillary, highly reduced; male flowers of a single anther; female flowers of a single pistil.
Fruit: More or less heart shaped, less than 1/16" tall.
Bloom Period: Spring.
Distribution: Cameron and Willacy counties.
Comments: This tiny herb grows at the edges of resacas and other wet places.

Annual Waterwort *Callitriche terrestris*

Berlandier's Lobelia

Family: Campanulaceae
Scientific Name: *Lobelia berlandieri*
Habit: Annuals, erect or leaning, up to 20" tall.
Leaves: Alternate; blades more or less ovate, up to 1 ⅜" long, the margins entire or toothed.
Flowers: Bilateral, 2-lipped, blue with white centers, occasionally white, up to ½" long.
Fruit: Capsules up to ¼" long, splitting top to bottom.
Bloom Period: Spring.
Distribution: Cameron, Hidalgo, and Willacy counties.
Comments: This plant prefers sandy soils. The plants contain nicotine alkaloids and are toxic to livestock. Consumption can result in death. *Lobelia*, with its bilateral flowers, is easily distinguished from *Triodanis*, which has radial flowers.

Berlandier's Lobelia *Lobelia berlandieri*

Berlandier's Lobelia *L. berlandieri* white-flowered form

Small Venus' Looking Glass

Family: Campanulaceae
Scientific Name: *Triodanis perfoliata* var. *biflora* [*T. biflora*]
Habit: Annuals up to 24" tall.
Leaves: Alternate; blades ovate, about ¾" long
Flowers: Radial; corollas blue to purplish, up to ⅜" long and ¾" broad.
Fruit: Capsules up to ⅜" long, the seeds released through 2 or 3 pores.
Bloom Period: Spring.
Distribution: Cameron and Willacy counties.
Comments: Recently the two species *T. perfoliata* and *T. biflora* were merged into two varieties of *T. perfoliata*. *Triodanis,* with its radial flowers, is easily distinguished from *Lobelia,* which has bilateral flowers.

Small Venus' Looking Glass *Triodanis perfoliata* var. *biflora*

Vara Blanca, Hoary Caper

Family: Capparaceae (Capparidaceae)
Scientific Name: *Capparis incana*
Habit: Shrubs; stems and leaves covered with white and red star-shaped hairs.
Leaves: Alternate, lanceolate, up to 2" or longer.
Flowers: Radial; petals 4, white or yellowish.
Fruit: Capsules one-seeded. Seeds red.
Bloom Period: Spring, summer.
Distribution: Cameron and Hidalgo counties.
Comments: Our plants are a northern occurrence of a wide distribution in Tamaulipas, Mexico. In the area around San Fernando, Tamaulipas, *C. incana* is abundant in the brush. The single plant in Hidalgo County (found by James Everitt) is no longer living. Lindley and Fleet Lentz found the only known United States population, located in Brownsville. Several plants are now in cultivation. The capers sold as condiments come from the buds of *Capparis spinosa,* which grows around the Mediterranean. The Spanish common name translates as "white stem."

Vara Blanca, Hoary Caper *Capparis incana*

Vara Blanca, Hoary Caper *C. incana* fruit

Capparaceae

Spider Wisp

Family: Capparaceae (Capparidaceae)
Scientific Name: *Cleome gynandra*
Habit: Annuals, erect growing, with glandular hairs.
Leaves: Alternate, palmate with 3 leaflets.
Flowers: Somewhat bilateral; petals 4, white, up to ¾" long, the stamens exserted well beyond the corollas.
Fruit: Capsules up to 4" long.
Bloom Period: Summer, fall.
Distribution: Cameron and Willacy counties.
Comments: This species is introduced from Africa. It is found occasionally in Texas.

Spider Wisp *Cleome gynandra*

Narrow Leaf Rhombopod

Family: Capparaceae (Capparidaceae)
Scientific Name: *Cleomella angustifolia*
Habit: Hairless bushy annuals up 24" or much taller.
Leaves: Alternate, compound, 3-foliate, the leaflets mostly linear, 2" or longer.
Flowers: Somewhat bilateral; petals 4, yellow, ¼" long.
Fruit: Capsules triangular, ⅜" wide.
Bloom Period: Summer, fall.
Distribution: Brooks County
Comments: This species has not previously been reported from any area near the Valley. It is included because of its presence in the sandy soils of Brooks County, just north of Hidalgo County. It could easily be present in the sandy soils of Hidalgo County, only a short distance away. Because of the regular, 4-petaled flowers, this species could be mistaken for a member of the Brassicaceae family.

Narrow Leaf Rhombopod *Cleomella angustifolia*

Allthorn, Junco

Family: Capparaceae (Capparidaceae, Koeberliniaceae)
Scientific Name: *Koeberlinia spinosa*
Habit: Shrubs up to 5' tall; stems green, highly branched, the ends all sharp pointed.
Leaves: Tiny or absent.
Flowers: Radial; petals 4, greenish white, up to ¼" long.
Fruit: A black berry, more or less round, about ¼" broad.
Bloom Period: All seasons.
Distribution: Cameron, Hidalgo, Willacy, and Starr counties.
Comments: This shrub has an appropriate common name. It indeed seems to be "all thorns," with its apparently leafless, thorny green stems where photosynthesis occurs. The flowers are sometimes malodorous. The chocolate brown heartwood is prized for knife handles and gun stocks.

Allthorn, Junco *Koeberlinia spinosa* with fruit

Allthorn, Junco *K. spinosa* flowers

Clammy Weed

Family: Capparaceae (Capparidaceae)
Scientific Name: *Polanisia dodecandra* subsp. *riograndensis*
Habit: Annuals about 24" tall, glandular viscid, with an offensive odor.
Leaves: Alternate, palmate with 3 lobes, each lobe about 1 ½" long.
Flowers: Bilateral; petals 4, purple or pink, up to ⅝" long, the stamens usually exserted well beyond the corollas.
Fruit: Capsules about 1 ¼" or longer.
Bloom Period: Spring, summer, fall.
Distribution: Cameron, Hidalgo, Willacy, and Starr counties.
Comments: This species is attractive but not pleasant to touch. The glands release glue-like material when touched, and bruising the leaves releases an offensive odor. The flowers attract nectaring butterflies.

Clammy Weed *Polanisia dodecandra* subsp. *riograndensis*

Caryophyllaceae

Short Gland Clammy Weed
Family: Capparaceae (Capparidaceae)
Scientific Name: *Polanisia erosa* subsp. *breviglandulosa*
Habit: Annuals, glandular viscid, with an unpleasant odor, slender and branching, up to 24" tall.
Leaves: Alternate, palmate with 3 lobes, each lobe about 1 ⅝" long.
Flowers: Bilateral; petals usually yellow, rarely white or pinkish, about ⅜" long.
Fruit: Capsules up to 2 ⅜" long.
Bloom Period: Spring, summer, fall.
Distribution: Hidalgo, Willacy, and Starr counties.
Comments: This plant is endemic to Texas. It prefers loose sandy soil.

Short Gland Clammy Weed *Polanisia erosa* subsp.*breviglandulosa*

South Texas Nailwort
Family: Caryophyllaceae
Scientific Name: *Paronychia jonesii*
Habit: Annuals, mostly prostrate.
Leaves: Opposite; blades hairy, up to 1 ⅝" long.
Flowers: Radial; sepals 5, white, about ¹⁄₁₆" long; petals minute.
Fruit: Papery, with one seed.
Bloom Period: Spring, summer, fall.
Distribution: Hidalgo County.
Comments: This plant is endemic to Texas and is usually found growing in loose sand. It may be confused with *Chamaesyce* (Euphorbiaceae) but is easily distinguished by breaking a stem. *Chamaesyce* bleeds milky sap when damaged; *Paronychia* does not.

South Texas Nailwort *Paronychia jonesii*

Saltmarsh Sandspurry
Family: Caryophyllaceae
Scientific Name: *Spergularia salina* [*S. marina*]
Habit: Erect branching annuals with stems up to 14" long, with glandular hairs.
Leaves: Opposite, linear, up to 1 ⅝" long.
Flowers: Petals 5, but variable in number and size, pink to white, about ⅛" long.
Fruit: Capsules about ⅛" tall, splitting into 3 parts.
Bloom Period: Spring, summer.
Distribution: Hidalgo and Willacy counties.
Comments: This is an introduced species, a native of Europe. It prefers salty soils.

Saltmarsh Sandspurry *Spergularia salina*

Common Chickweed
Family: Caryophyllaceae
Scientific Name: *Stellaria media*
Habit: Somewhat hairy annuals, prostrate, forming mats.
Leaves: Opposite; ovate to elliptic, up to 1 ⅝" long.
Flowers: Radial; petals white, 5, but divided and seemingly 10, about ³⁄₁₆" long.
Fruit: Capsules up to ¼" tall.
Bloom Period: Spring.
Distribution: Hidalgo County.
Comments: This species is a Eurasian native, now widespread throughout Texas.

Common Chickweed *Stellaria media*

Prostrate Starwort
Family: Caryophyllaceae
Scientific Name: *Stellaria cuspidata* subsp. *prostrata*
Habit: Hairy annuals or sometimes perennials, prostrate or sprawling.
Leaves: Opposite; blades triangular or broadly ovate, about ¾" long.
Flowers: Radial; petals 5, white, about ³⁄₁₆" long.
Fruit: Capsules with few to many seeds.
Bloom Period: Winter, spring, summer.
Distribution: Cameron and Hidalgo counties.
Comments: This plant usually prefers some shade.

Prostrate Starwort *Stellaria cuspidata* subsp. *prostrata*

Leatherleaf, Guttapercha
Family: Celastraceae
Scientific Name: *Maytenus phyllanthoides* [*M. texana*]
Habit: Erect or prostrate shrubs; stems minutely hairy.
Leaves: Alternate; blades thick and leathery, obovate, up to 1 ¾" long
Flowers: Unisexual, male and female on different plants; petals 4, greenish, about ¹⁄₁₆" long.
Fruit: An ovoid capsule about ½" long; seeds red.
Bloom Period: Spring, summer, fall.
Distribution: Cameron, Hidalgo, and Willacy counties.
Comments: In the salty areas, the shrubs are mostly low or prostrate; in better soils, they can grow 2' or taller.

Leatherleaf, Guttapercha *Maytenus phyllanthoides*

Celastraceae

Afinador, Mortonia

Family: Celastraceae
Scientific Name: *Mortonia greggii*
Habit: Shrubs up to 6' or taller with pinkish stems.
Leaves: Alternate, bright green, oblanceolate, up to 1" long.
Flowers: Radial; petals 5, white, about 1/16" long
Fruit: Achenes about 3/16" long.
Bloom Period: Winter, spring.
Distribution: Hidalgo and Starr counties.
Comments: In Texas this species is reported only from Hidalgo and Starr counties. It is not commonly seen. Mortonia's preferred habitat is gravelly hills.

Afinador, Mortonia *Mortonia greggii*

Desert Yaupon

Family: Celastraceae
Scientific Name: *Schaefferia cuneifolia*
Habit: Shrubs reaching over 6' in height, usually shorter.
Leaves: Alternate or in clusters; blades spatulate, often notched at the tips, up to 5/8" long.
Flowers: Radial, unisexual, male and female flowers on different plants; petals 4, greenish white, about 1/8" long.
Fruit: Fleshy, up to 3/16" broad, orange or red when mature.
Bloom Period: Spring, summer, fall.
Distribution: Cameron, Hidalgo, Willacy, and Starr counties.
Comments: The specific epithet *cuneifolia,* Latin for "wedge-shaped leaves," emphasizes the leaf shape. The common name yaupon is used for other species in different families.

Desert Yaupon *Schaefferia cuneifolia*

Desert Yaupon *S. cuneifolia* plant in fruit

Submerged Hornwort

Family: Ceratophyllaceae
Scientific Name: *Ceratophyllum demersum*
Habit: Rootless herbs forming large masses submerged in water.
Leaves: Whorled and finely dissected, up to ⅝" long.
Flowers: Minute, not noticeable.
Fruit: Achenes up to ¼" long, with 2 basal spines.
Bloom Period: Summer.
Distribution: Cameron County.
Comments: The floating plants are often found in local resacas.

Submerged Hornwort *Ceratophyllum demersum*

Spiny Fruit Saltbush

Family: Chenopodiaceae
Scientific Name: *Atriplex acanthocarpa*
Habit: Shrubs forming large mounds up to 40" tall.
Leaves: Alternate; blades more or less arrowhead shaped, silvery in color, up to 2 ⅜" long.
Flowers: Very small, male and female on separate plants.
Fruit: Utricles (slightly inflated, one-seeded) enclosed in bracts that are spiny on the margins and sides.
Bloom Period: Spring, summer, fall.
Distribution: Cameron, Willacy, and Starr counties.
Comments: The specific epithet *acanthocarpa* is derived from two Greek words meaning "spiny" and "fruit." This species (at least the female plant) is easily recognized by its noticeably spiny fruit. It prefers alkaline soils.

Spiny Fruit Saltbush *Atriplex acanthocarpa*

Four Wing Saltbush

Family: Chenopodiaceae
Scientific Name: *Atriplex canescens*
Habit: Erect woody perennials up to 24" or much taller.
Leaves: Alternate; blades narrowly spatulate to narrowly oblong, gray, up to 2" long.
Flowers: Very small, male and female on separate plants.
Fruit: Utricles (slightly inflated, one-seeded) enclosed in bracts producing 4 wings.
Bloom Period: Spring, summer, fall.
Distribution: Starr County.

Four Wing Saltbush *Atriplex canescens*

Comments: Many species in this family are difficult to identify, requiring a microscope and the use of keys. The four wings on the fruit make this species (at least the female plant) easy to recognize.

Chenopodiaceae

Matamoros Saltbush

Family: Chenopodiaceae
Scientific Name: *Atriplex matamorensis*
Habit: Erect perennials, woody at the base.
Leaves: Mostly opposite, dense, the blades narrow, silvery colored, about ½" long.
Flowers: Very small, male and female flowers on separate plants.
Fruit: Utricles (slightly inflated, one-seeded) enclosed in bracts that are spiny along the margins.
Bloom Period: Spring, summer, fall.
Distribution: Cameron and Willacy counties.
Comments: The narrow leaves are a distinguishing characteristic of this species. It prefers alkaline soils.

Matamoros Saltbush *Atriplex matamorensis*

Sand Atriplex

Family: Chenopodiaceae
Scientific Name: *Atriplex pentandra* [*A. arenaria*]
Habit: Perennials, sometimes annuals, leaning or falling over, woody at the base.
Leaves: Alternate; blades various shapes, silver colored, up to 1 ¼" long.
Flowers: Very small, male and female flowers on the same plant.
Fruit: Utricles (slightly inflated, one-seeded) enclosed in bracts that are spiny on the margins and sides.
Bloom Period: Spring, summer, fall.
Distribution: Cameron, Hidalgo, and Willacy counties.
Comments: The leaning or falling over habit plus the fruit characteristics define this species. It prefers alkaline soils. The leaves of many species of *Atriplex* have a bad taste, especially if the plants were under water stress. Sand atriplex leaves are more tender and have been used in salads.

Sand Atriplex *Atriplex pentandra*

Pitseed Goosefoot, Stinkweed

Family: Chenopodiaceae
Scientific Name: *Chenopodium berlandieri*
Habit: Erect annuals, usually bad-smelling.
Leaves: Alternate; blades ovate to oblong, up to 1 ½" long, the younger leaves often with mealy surfaces.
Flowers: Tiny, mealy, in small clusters on the inflorescences, which bear small leaves.
Fruit: Utricles (slightly inflated, one-seeded).
Bloom Period: Spring, summer, fall.
Distribution: Cameron, Hidalgo, Willacy, and Starr counties.
Comments: This species was named for Jean Louis Berlandier, who made extensive plant collections in Mexico and Texas. The bad smell, like the air from a tire inner tube, often defines this species, although there are some who cannot smell it.

Pitseed Goosefoot, Stinkweed *Chenopodium berlandieri*

Pitseed Goosefoot, Stinkweed *C. berlandieri* inflorescence

Nettle Leaf Goosefoot

Family: Chenopodiaceae
Scientific Name: *Chenopodium murale*
Habit: Erect succulent annuals.
Leaves: Alternate; blades ovate, with coarse teeth, up to 3 ⅛" long.
Flowers: Tiny, mealy, in clusters on the inflorescences.
Fruit: Utricles (slightly inflated, one-seeded).
Bloom Period: Spring, summer, fall, winter.
Distribution: Cameron, Hidalgo, Willacy, and Starr counties.
Comments: The lack of a bad odor separates this species from pitseed goosefoot, *C. berlandieri*.

Nettle Leaf Goosefoot *Chenopodium murale*

Chenopodiaceae

Desert Goosefoot

Family: Chenopodiaceae
Scientific Name: *Chenopodium pratericola*
 [*C. desiccatum* var. *leptophylloides*]
Habit: Annual erect herbs up to 32" tall, often mealy looking.
Leaves: Alternate; blades lanceolate to elliptic, up to 2 3/8" long. The older leaf blades usually have a pair of basal lobes.
Flowers: Tiny, mealy, in clusters.
Fruit: Utricles (slightly inflated, one-seeded).
Bloom Period: Summer, early fall.
Distribution: Hidalgo and Starr counties.
Comments: The mealiness of the leaves and flowers combined with the basal lobes on the older leaf blades separate this from the other *Chenopodium* species.

Desert Goosefoot *Chenopodium pratericola*

Keeled Goosefoot

Family: Chenopodiaceae
Scientific Name: *Dysphania carinata*
 [*Chenopodium carinatum*]
Habit: Annual herbs; stems erect or falling over.
Leaves: Alternate, aromatic; blades ovate, up to 1 1/8" long, the margins with broad rounded teeth.
Flowers: Tiny, in clusters in the leaf axils.
Fruit: Achenes ovoid, roughened and keeled, less than 1/16" tall.
Bloom Period: Summer and fall.
Distribution: Hidalgo County.
Comments: This is a new record for the Lower Rio Grande Valley. In Texas it has been known only in far East Texas. Native to Australia, New Zealand, and New Caledonia, it has been introduced to other parts of the world with wool. After arrival in Texas, it could have been distributed with plant nursery stock or perhaps through some natural means. The crushed leaves have a citrus scent.

Keeled Goosefoot *Dysphania carinata*

Annual Saltwort

Family: Chenopodiaceae
Scientific Name: *Salicornia bigelovii*
Habit: Erect succulent annuals.
Leaves: Opposite, scalelike, joined basally.
Flowers: Tiny, slightly protruding from behind the scale leaves.
Fruit: Utricles hidden by the scale leaves.
Bloom Period: Spring, summer, fall, and winter.
Distribution: Cameron and Willacy counties.
Comments: This species occurs in salt flats and salt marshes. A similar species grows on the coast with *Salicornia bigelovii*. It was for many years misidentified as *Salicornia virginica,* but recent studies have shown that it is perennial saltwort, *Sarcocornia utahensis* (synonym: *Salicornia utahensis*).

Annual Saltwort *Salicornia bigelovii*

Annual Saltwort *S. bigelovii* inflorescence

Tumbleweed, Russian Thistle

Family: Chenopodiaceae
Scientific Name: *Salsola tragus* [*S. australis, S. kali*]
Habit: Annual herbs; stems green with purple stripes.
Leaves: Alternate, linear, up to 1 ⅛" long, with sharp tips.
Flowers: Axillary, with white calyxes up to ⅜" broad.
Fruit: Utricles (slightly inflated, one-seeded).
Bloom Period: Summer and fall.
Distribution: Cameron, Hidalgo, and Starr counties.
Comments: This species was introduced to the United States from Eurasia. It is especially prevalent in disturbed areas and agricultural fields. The tumbling movement of the wind-blown plants disperses the seeds over a broad area.

Tumbleweed, Russian Thistle *Salsola tragus*

Chenopodiaceae

Perennial Saltwort

Family: Chenopodiaceae
Scientific Name: *Sarcocornia utahensis*
Habit: Perennials, erect or falling over.
Leaves: Opposite, scalelike, joined basally.
Flowers: Tiny, protruding from behind the scale leaves.
Fruit: Utricles, hidden by the scale leaves.
Bloom Period: Spring, summer, and fall.
Distribution: Cameron and Willacy counties.
Comments: This species occurs in salt flats and salt marshes. It has for many years been called *Salicornia virginica*. Recent studies have shown that to be incorrect. The plants are ordinarily green but sometimes turn a reddish color. There is a great similarity to *Salicornia bigelovii*, another coastal plant, which is an annual.

Perennial Saltwort *Sarcocornia utahensis*

Tufted Sea Blite

Family: Chenopodiaceae
Scientific Name: *Suaeda conferta*
Habit: Woody plants forming dense mats, the stem ends sometimes bending upward.
Leaves: Alternate, succulent, crowded at the tips, linear to ellipsoid, up to ⅝" long.
Flowers: Radial, usually perfect, tiny, about ⅛" broad, forming clusters at the nodes; petals absent.
Fruit: Small utricles (slightly inflated, one-seeded).
Bloom Period: Spring, summer, and fall.
Distribution: Starr and Cameron counties.
Comments: This mat-forming plant vegetatively resembles *Lenophyllum* (*Sedum*) in the Crassulaceae family. We have observed it growing abundantly in Starr and Zapata counties in saline, gypseous soils. It is reported for the coastal areas of Cameron County, but we have not seen it there. It also occurs in Mexico and the West Indies.

Tufted Sea Blite *Suaeda conferta*

Tufted Sea Blite *Suaeda conferta* inflorescence

Annual Seepweed
Family: Chenopodiaceae
Scientific Name: *Suaeda linearis*
Habit: Hairless, succulent herbs with weak stems.
Leaves: Alternate; blades linear, succulent, up to 1 ½" long, green, seasonally turning purple.
Flowers: Tiny, difficult to examine.
Fruit: Utricles (slightly inflated, one-seeded).
Bloom Period: Spring, summer, fall.
Distribution: Cameron and Willacy counties.
Comments: We have several species of *Suaeda;* they are difficult to distinguish without the use of a microscope and keys. This species is found in alkaline soils.

Annual Seepweed *Suaeda linearis*

Tampico Seepweed
Family: Chenopodiaceae
Scientific Name: *Suaeda tampicensis*
Habit: Erect, green and woody perennials with hairy stems.
Leaves: Alternate; blades cylindrical, up to ⅝" long.
Flowers: Tiny, difficult to examine.
Fruit: Utricles (slightly inflated, one-seeded).
Bloom Period: Spring, summer, fall.
Distribution: Cameron and Starr counties.
Comments: The hairy stems distinguish this species from annual seepweed, *S. linearis.* We have several species of *Suaeda;* they are difficult to distinguish without the use of a microscope and keys. This is a plant of alkaline soils.

Tampico Seepweed *Suaeda tampicensis*

Georgia Rockrose
Family: Cistaceae
Scientific Name: *Helianthemum georgianum*
Habit: Perennials up to 16" tall.
Leaves: Alternate, hairy; blades narrowly elliptic to spatulate, green above and whitish below, up to 1 ⅜" long.
Flowers: Two types: cleistogamous (self-pollinating, never opening); and chasmogamous (opening normally), with 5 yellow petals up to ⅝" long.
Fruit: Capsules up to 2 ⅛" tall.
Bloom Period: Spring, summer.
Distribution: Cameron, Hidalgo, and Willacy counties.
Comments: Plants of this species prefer sandy soils. The flowers remain open for only a short time during the day.

Georgia Rockrose *Helianthemum georgianum*

Clusiaceae

San Saba Pinweed
Family: Cistaceae
Scientific Name: *Lechea san-sabeana*
Habit: Perennials up to 14" tall, with hairs appressed against the stem.
Leaves: Alternate and opposite; blades linear, up to ½" long.
Flowers: Radial, in arching inflorescences, all the flowers on one side; sepals 5, the 2 outer ones narrow, the 3 inner ones with a green keel in fruit; petals 3, about 1/16" long, red.
Fruit: Capsules 6-seeded, about ⅛" tall.
Bloom Period: Spring, summer.
Distribution: Hidalgo and Willacy counties.
Comments: This species is endemic to Texas. We have two species of *Lechea* in Texas. In *L. san-sabeana,* the flower pedicels bend to 90 degrees or downward; in *L. tenuifolia* (not in our area), the flower pedicels bend upward. Both occur in sandy soils.

San Saba Pinweed *Lechea san-sabeana*

Few Flowered St. John's Wort
Family: Clusiaceae (Hypericaceae)
Scientific Name: *Hypericum pauciflorum*
Habit: Somewhat succulent herbs up to 28" tall, with woody bases.
Leaves: Opposite or whorled; blades narrow or linear, up to 1 ⅛" long.
Flowers: Radial, solitary; petals 5, orange-yellow, up to ⅜" long.
Fruit: Capsules cylindrical, up to ¼" tall.
Bloom Period: Spring, summer, fall.
Distribution: Cameron, Hidalgo, and Willacy counties.
Comments: Plants of this species prefer sandy soils. The flowers are not long lasting. Two different forms were seen in the same population, one with larger leaves and larger flowers, the other with smaller leaves and smaller flowers.

Few Flowered St. John's Wort *Hypericum pauciflorum*

Button Mangrove

Family: Combretaceae
Scientific Name: *Conocarpus erectus*
Habit: Small trees or shrubs up to 27' tall.
Leaves: Alternate, elliptic to oblanceolate, more or less leathery, up to 3½" long.
Flowers: In globose heads; corollas absent; calyxes greenish, about 1/16" long.
Fruit: Small, about 1/16" tall.
Bloom Period: Intermittently throughout the year.
Distribution: Cameron and Willacy counties.
Comments: This is a tropical mangrove, found from Mexico to Brazil, Africa, and the Galapagos Islands. In the United States it is found in Florida and rarely on the southern Texas coast. We also have the unrelated black mangrove, *Avicennia germinans,* and red mangrove, *Rhizophora mangle.* Button mangrove was discovered on Padre Island by Tom Patterson. *Conocarpus* is often placed into the Myrtaceae family.

Button Mangrove *Conocarpus erectus*

Field Bindweed

Family: Convolvulaceae
Scientific Name: *Convolvulus arvensis*
Habit: Stems trailing or twining, 40" or longer.
Leaves: Alternate; blades up to 2 ¾" long, variable in shape, commonly with a pair of lobes at the base.
Flowers: Radial, funnel shaped, white to pink, about ¾" long. The petals are fused together.
Fruit: Capsules with 2–4 seeds.
Bloom Period: Spring, summer, fall.
Distribution: Hidalgo and Starr counties.
Comments: This is an introduced species, native to Eurasia. The genus name comes from the Latin *convolvulo,* meaning "to twine around."

Field Bindweed *Convolvulus arvensis*

Carr's Bindweed

Family: Convolvulaceae
Scientific Name: *Convolvulus carrii*
Habit: Hair perennials, stems twining, prostrate or climbing onto shrubs.
Leaves: Alternate; blades up to 2 ¾" long and 1 ⅛" broad, with basal lobes rounder or sometimes narrowing and turned outward; margins irregularly toothed.
Flowers: Arising, usually singly, from the leaf axils; corollas radial, funnel shaped (the petals grown together), white with purple centers, up to 1 ⅝" broad.
Fruit: Capsules with 2–4 seeds; seed coats smooth or nearly so.
Bloom Period: Spring, summer.
Distribution: Hidalgo County, also in Brooks County very near Starr County.
Comments: *Convolvulus carrii* grows in deep sandy soils. It has been collected by several different taxonomists but misidentified as *C. equitans,* which has similar coloration but much smaller flowers. Only recently was *C. carri* recognized as an undescribed species by Billie L. Turner (Turner, 2009) at the University of Texas at Austin. Previously, only two species of *Convolvulus* were known to be native to North America. All three species are native to the Lower Rio Grande Valley. The flowers of this species are very like those of *Merremia dissecta,* which has distinctly lobed leaves. The leaf margins of *C. carrii* are only toothed. The third *Convolvulus* species, *C. arvensis,* is distinguished from *C. carrii* by its smaller corollas, which lack the purple center.

Carr's Bindweed *Convolvulus carrii*

Texas Bindweed

Family: Convolvulaceae
Scientific Name: *Convolvulus equitans*
Habit: Stems trailing or twining, up to 6' or longer.
Leaves: Alternate; blades hairy, arrowhead shaped, with toothed or lobed margins, up to 2 ⅛" long.
Flowers: Radial, funnel shaped, white to pinkish, often with red centers, up to 1 ⅜" long. The petals are fused together.
Fruit: Capsules about 5/16" long.
Bloom Period: Spring, summer.
Distribution: Cameron, Hidalgo, and Starr counties.
Comments: Texas bindweed is usually found growing in sandy soils.

Texas Bindweed *Convolvulus equitans*

Leafless Cressa

Family: Convolvulaceae
Scientific Name: *Cressa nudicaulis*
Habit: Gray, hairy perennials up to 8" tall.
Leaves: Alternate, scalelike, about ⅛" long.
Flowers: Radial, funnel shaped, white, about ⅜" long. The petals are fused together.
Fruit: Capsules with 1–4 seeds.
Bloom Period: All seasons.
Distribution: Cameron and Willacy counties.
Comments: These are plants of the coastal sands and clay hills. The plants look like gray, narrow sticks. Leaves are not noticeable.

Leafless Cressa *Cressa nudicaulis*

Silky Alkaliweed

Family: Convolvulaceae
Scientific Name: *Cressa truxillensis* [*C. depressa*]
Habit: Low, branching, hairy perennials up to 10" tall.
Leaves: Alternate; blades gray-hairy, elliptic to lanceolate, up to ½" long.
Flowers: Radial, funnel shaped, white, about ⅜" long.
Fruit: Capsules with 1–4 seeds.
Bloom Period: Spring, summer.
Distribution: Hidalgo County.
Comments: This species has been reported from West Texas and along the Coastal Bend. It has not been previously reported for this area. It grows in alkaline, saline soils. We have seen it growing in abundance at La Sal del Rey.

Silky Alkaliweed *Cressa truxillensis*

Convolvulaceae

Carolina Ponyfoot

Family: Convolvulaceae
Scientific Name: *Dichondra carolinensis*
Habit: Perennials growing prostrate, rooting at the nodes.
Leaves: Alternate; blades kidney shaped or roundish, up to ¾" long and broad.
Flowers: Radial, funnel shaped, greenish, about 1/16" broad. The pedicels are straight.
Fruit: Capsules about 1/16" broad.
Bloom Period: Spring.
Distribution: Cameron County.
Comments: These plants make an attractive ground cover in a slightly damp area. Because the pedicels are straight, the flowers are not hidden, as in Asian ponyfoot, *D. micrantha*.

Carolina Ponyfoot *Dichondra carolinensis*

Asian Ponyfoot

Family: Convolvulaceae
Scientific Name: *Dichondra micrantha*
Habit: Perennials growing prostrate, rooting at the nodes.
Leaves: Alternate; blades kidney shaped or roundish, about ¾" long and broad.
Flowers: Radial, funnel shaped, greenish, up to ⅛" long. Because the pedicels are curved, the flowers are usually hidden on the lower surface of the plants.
Fruit: Capsules about 1/16" broad.
Bloom Period: Spring.
Distribution: Cameron, Hidalgo, and Starr counties.
Comments: These small plants make an attractive ground cover. The flowers are hidden below the leaves, unlike those of Carolina ponyfoot, *D. carolinensis,* which are seen above the leaves.

Asian Ponyfoot *Dichondra micrantha*

Slender Evolvulus, Ojo De Víbora

Family: Convolvulaceae
Scientific Name: *Evolvulus alsinoides* var. *angustifolius*
Habit: Small, hairy perennials, usually prostrate.
Leaves: Alternate; blades elliptic, up to ¾" long.
Flowers: Radial, funnel shaped, blue, sometimes white or whitish, up to ⅜" broad. The petals are grown together.
Fruit: Capsules with 1–4 seeds.
Bloom Period: All seasons.
Distribution: Cameron, Hidalgo, Willacy, and Starr counties.
Comments: This plant would make a nice hanging basket. The genus name comes from a Latin word meaning "untwist," referring to the nontwining habit of this plant (in contrast to many members of this family). The Spanish common name translates as "adder's eye." As is true with many common names, the connection is obscure.

Slender Evolvulus, Ojo De Víbora *Evolvulus alsinoides* var. *angustifolius*

Hairy Evolvulus

Family: Convolvulaceae
Scientific Name: *Evolvulus nuttallianus*
Habit: Hairy perennials with stems erect or falling over.
Leaves: Alternate; blades lanceolate to elliptic, up to ¾" long.
Flowers: Radial, funnel shaped, lavender to whitish, up to ½" broad. The petals are grown together.
Fruit: Capsules with 1–4 seeds.
Bloom Period: Spring, summer.
Distribution: Hidalgo County.
Comments: This species has not previously been reported this far south in Texas. It is easily distinguished from our other species of *Evolvulus* by its silvery appearance, with dense hairs on both surfaces of the leaves. Silky evolvulus, *E. sericeus,* is also hairy, but it has white flowers.

Hairy Evolvulus *Evolvulus nuttallianus*

Convolvulaceae

Silky Evolvulus

Family: Convolvulaceae
Scientific Name: *Evolvulus sericeus*
Habit: Small, hairy perennials, usually prostrate.
Leaves: Alternate; blades narrowly elliptic to linear, up to ¾" long.
Flowers: Radial, funnel shaped, white, about ⅜" broad, sometimes more. The petals are grown together.
Fruit: Capsules with 1–4 seeds.
Bloom Period: Spring, summer, fall.
Distribution: Cameron and Hidalgo counties.
Comments: This species usually grows in sandy soils. The leaves have a silky appearance with their dense, appressed hairs. The specific epithet, derived from a Latin word for "silky," comes from this characteristic. The seeds and leaves are eaten by wildlife.

Silky Evolvulus *Evolvulus sericeus*

Moonflower

Family: Convolvulaceae
Scientific Name: *Ipomoea alba*
Habit: Twining vines, with stems that are sometimes prickly.
Leaves: Alternate; blades heart shaped, up to 6 ¾" long.
Flowers: Radial, funnel shaped, white, about 3 ⅛" broad with a long tube about 4" long. The petals are grown together.
Fruit: A brown, globose capsule subtended by the dried sepals.
Bloom Period: Summer, fall.
Distribution: Cameron and Hidalgo counties.
Comments: This species is native to the tropics, a cultivar that has become naturalized, especially in moist places.

Moonflower *Ipomoea alba*

Red Center Morning Glory

Family: Convolvulaceae
Scientific Name: *Ipomoea amnicola*
Habit: Twining vines.
Leaves: Alternate; blades heart shaped, sometimes 3-lobed, up to 3 ½" long.
Flowers: Radial, funnel shaped, white to purplish with a purple center, up to 1 ½" long. The petals are grown together.
Fruit: A brown, globose capsule subtended by the dried sepals.
Bloom Period: All seasons.
Distribution: Cameron, Hidalgo, Willacy, and Starr counties.
Comments: This is an introduced plant, a native of Paraguay. This species has calyx lobes that are blunt or rounded, not tapering to a point.

Red Center Morning Glory *Ipomoea amnicola*

Bush Morning Glory

Family: Convolvulaceae
Scientific Name: *Ipomoea carnea* subsp. *fistulosa* [*I. fistulosa*]
Habit: Shrubs up to 13' or taller, usually not more than 6' tall.
Leaves: Alternate; blades heart shaped, up to 8" long.
Flowers: Radial, funnel shaped, pinkish lavender with dark centers, sometimes white, up to 3 ½" long. The petals are grown together.
Fruit: A brown, globose capsule subtended by the dried sepals.
Bloom Period: All seasons.
Distribution: Cameron, Hidalgo, and Willacy counties.
Comments: This is an introduced plant, a native of the tropics. It has been widely cultivated and has subsequently become naturalized, preferring moist localities.

Bush Morning Glory *Ipomoea carnea* subsp. *fistulosa*

Convolvulaceae

Tie Vine

Family: Convolvulaceae
Scientific Name: *Ipomoea cordatotriloba* var. *torreyana* [*I. trichocarpa* var. *torreyana*]
Habit: Twining vines.
Leaves: Alternate; blades heart shaped, often 3-to 5-lobed, up to 3 ½" long.
Flowers: Radial, funnel shaped, pinkish to reddish purple with dark centers, about 1 ⅛" long.
Fruit: A brown, globose capsule subtended by the dried sepals.
Bloom Period: Summer, fall.
Distribution: Cameron, Hidalgo, and Willacy counties.
Comments: There are several *Ipomoea* species that are vines with purplish flowers. This species has calyx lobes that are more or less equal and taper to a point.

Tie Vine *Ipomoea cordatotriloba* var. *torreyana*

Crest-Rib Morning Glory

Family: Convolvulaceae
Scientific name: *Ipomoea costellata* var. *edwardsensis*
Habit: Annuals growing erect when young; older stems with some twining.
Leaves: Alternate; petioles ¼"-1" long; blades divided into narrow segments up to 1" long.
Flowers: Peduncles rising from the leaf axils with 1 or 2 flowers; corollas white, about ½" long, radial, the lower part tubular, the upper part 5-lobed, spreading nearly flat.
Fruit: Capsules about ¼" broad, with 4 seeds.
Bloom Period: Summer, fall.
Distribution: Starr County.
Comments: This is the first report of *I. costellata* for the Lower Rio Grande Valley. Populations occur in West Texas (var. *costellata*) and again in Central Texas (var. *edwardsensis*). One collection is reported from Webb County (var. *costellata*). *I. costellata* var. *costellata* has flowers with 10 lobes, and the colors can range from white to yellow, blue, or pale lavender. Crest-Rib Morning Glory is easily recognized by its smaller white flowers and its highly dissected leaves.

Crest-Rib Morning Glory *Ipomoea costellata* var. *edwardsensis*

Blue Morning Glory

Family: Convolvulaceae

Scientific Name: *Ipomoea hederacea*

Habit: Twining vines.

Leaves: Alternate; blades more or less ovate, usually 3-lobed, up to 6" long, usually less.

Flowers: Radial, funnel shaped, blue with white centers, 1 ¼–2" long.

Fruit: A brown, globose capsule subtended by the dried sepals.

Bloom Period: Summer, fall.

Distribution: Cameron and Hidalgo counties.

Comments: It is easy to recognize this species because it is our only morning glory with blue flowers. There are some cultivars with larger flowers.

Blue Morning Glory *Ipomoea hederacea*

Beach Morning Glory

Family: Convolvulaceae

Scientific Name: *Ipomoea imperati* [*I. stolonifera*]

Habit: Prostrate vinelike perennials; stems rooting at the nodes.

Leaves: Alternate, leathery or succulent; blades notched apically or palmately 3–5 lobed, up to 1 ⅝" long.

Flowers: Radial, funnel shaped, white with yellow centers, up to 2 ¾" long. The petals are grown together.

Fruit: A brown, globose capsule subtended by the dried sepals.

Bloom Period: Spring, summer, fall.

Distribution: Cameron and Willacy counties.

Comments: This is a plant of the coastal sands. It is important in stabilizing the dunes.

Beach Morning Glory *Ipomoea imperati*

Convolvulaceae

Railroad Vine

Family: Convolvulaceae
Scientific Name: *Ipomoea pes-caprae* subsp. *brasiliensis*
Habit: Prostrate vinelike perennials; stems rooting at the nodes.
Leaves: Alternate, thick and leathery; blades notched apically, almost circular in outline, up to 3 ½" long and 4 ¼" broad.
Flowers: Radial, funnel shaped, purple, up to 2 ¾" long. The petals are grown together.
Fruit: A brown, globose capsule subtended by the dried sepals.
Bloom Period: Summer, fall.
Distribution: Cameron and Willacy counties.
Comments: This species can produce a tremendous spurt of growth in ideal conditions. The growth of one stem was measured and was found to average almost 10" a day. In peak times, plants of this species put on quite a show at the beaches. It is common in coastal sands.

Railroad Vine *Ipomoea pes-caprae* subsp. *brasiliensis*

Cliff Morning Glory

Family: Convolvulaceae
Scientific Name: *Ipomoea rupicola*
Habit: Trailing or twining vines.
Leaves: Alternate; blades heart shaped to arrowhead shaped, up to 3 ⅛" long; the margins are sometimes toothed.
Flowers: Radial, funnel shaped, reddish to pinkish purple with dark centers, up to 3 ⅛" long. The petals are grown together.
Fruit: A brown, globose capsule subtended by the dried sepals.
Bloom Period: Summer, fall.
Distribution: Cameron, Hidalgo, Willacy, and Starr counties.
Comments: This species has calyx lobes that are unequal and taper to a point.

Cliff Morning Glory *Ipomoea rupicola*

Saltmarsh Morning Glory, Arrow Leaf Morning Glory

Family: Convolvulaceae
Scientific Name: *Ipomoea sagittata*
Habit: Perennial vines with twining stems.
Leaves: Alternate; blades arrowhead shaped, up to 4" long.
Flowers: Radial, petals grown together; corollas funnel shaped, reddish purple, up to 4" broad.
Fruit: A brown, globose capsule subtended by the dried sepals.
Bloom Period: Spring, summer, fall.
Distribution: Cameron County.
Comments: This species is mainly found growing on the beaches and dunes along the Gulf Coast. In Texas it has previously been reported to grow only as far south as Refugio County. The Cameron County locality, South Padre Island, was reported to us by Christina Mild.

Saltmarsh Morning Glory, Arrow Leaf Morning Glory *Ipomoea sagittata*

Alamo Vine, Woodrose

Family: Convolvulaceae
Scientific Name: *Merremia dissecta* [*Ipomoea sinuata*]
Habit: Twining vines.
Leaves: Alternate; 5- to 7-lobed, more or less ovate in outline, up to 4" or longer, the margins toothed.
Flowers: Radial, funnel shaped, white with purple centers, up to 2" long. The petals are grown together.
Fruit: A brown, globose capsule subtended by the dried sepals.
Bloom Period: Spring, summer, fall.
Distribution: Cameron, Hidalgo, Willacy, and Starr counties.
Comments: Alamo vine is closely related to the *Ipomoea* vines and is sometimes hard to distinguish from one of them; the white flowers with purple centers combined with the lobed leaves are important (though not unique) characteristics. The dry fruit resembles a flower and is commonly called "woodrose."

Alamo Vine, Woodrose *Merremia dissecta*

Crassulaceae

Narrow Leaf Kalanchoe

Family: Crassulaceae
Scientific Name: *Kalanchoe delagoensis*
 [*K. tubiflora, Bryophyllum tubiflorum*]
Habit: Erect succulent perennials up to 40" tall.
Leaves: Opposite, or in threes, pencil shaped, mottled, up to 3" long, forming plantlets at the tips.
Flowers: In clusters held high above the plant; corollas tubular, about 1 ¼" long, reddish pink, hanging downward.
Fruit: Dry, many-seeded, opening along one side.
Bloom Period: Fall, winter.
Distribution: Cameron County.
Comments: This species is native to Africa and Madagascar. It was introduced because of its attractive flowers and ease of culture. It can survive and reproduce in the natural setting, outcompeting native species. When the older stems fall over, the leaves form plantlets that can start a colony a few feet from the parent plants. New plants also come up from the base of the parent plant. On occasion, plants have been thrown into the brush to die but instead continued to thrive. Large colonies thrive in coastal clay soils. The plants are poisonous if consumed.

Narrow Leaf Kalanchoe *Kalanchoe delagoensis*

Devil's Backbone

Family: Crassulaceae
Scientific Name: *Kalanchoe xhoughtonii*
Habit: Succulent perennials up to 3' tall.
Leaves: Opposite; blades lanceolate, up to 10" long, the margins toothed and producing plantlets.
Flowers: In branched inflorescences; corollas tubular, about 1" long, purplish, hanging downward.
Fruit: Dry, many-seeded, opening along one side.
Bloom Period: Fall, winter.
Distribution: Cameron County.
Comments: *K. xhoughtonii* is considered to be a hybrid between *K. daigremontiana* (*Bryophyllum daigremontianum*) and *K. delagoensis,* both of which are native to Madagascar. The "x" preceding the specific epithet (*houghtonii*) indicates hybrid origin. It is not known where and when the hybridization took place, but the hybrid is just as invasive as the parents. Seeds are only occasionally produced, but the plants reproduce profusely from the leaves. A single leaf can produce 20 or more plantlets. Among different plants, there is considerable variation in leaf size. The plants are often simply referred to by their genus name, *Kalanchoe.* The parent species were introduced through the gardening trade. They are invasive, forming hundreds of small plants by vegetative propagation. Although the flowers are beautiful, the plants can become pests and are difficult to eliminate because of their profuse vegetative reproduction. The plants are poisonous if consumed.

Devil's Backbone *Kalanchoe xhoughtonii*

Devil's Backbone *K. xhoughtonii* leaves

Texas Stonecrop

Family: Crassulaceae
Scientific Name: *Lenophyllum texanum* [*Sedum texanum*]
Habit: Mostly creeping, succulent perennials.
Leaves: Opposite; blades obovate, about 1" long.
Flowers: Radial, inflorescences at the stem ends; petals yellow with purple, about 3/16" long.
Fruit: Separating into several erect, many-seeded follicles.
Bloom Period: Winter.
Distribution: Cameron and Starr counties.
Comments: This plant is attractive in a succulent garden or grown as a potted plant. The leaves fall easily and will root, forming new plants. Texas stonecrop is a host plant for the Xami Hairstreak butterfly. In Cameron County this species grows on the clay lomas; in Starr County it grows in cracks in limestone at the tops of mesas. Rabbits and tortoises consume the plant.

Texas Stonecrop *Lenophyllum texanum*

Texas Stonecrop *L. texanum* inflorescence

Globe Berry

Family: Cucurbitaceae
Scientific Name: *Ibervillea lindheimeri*
Habit: Perennial vines with tendrils, growing from a swollen root.
Leaves: Alternate; blades variable, mostly 3-lobed, up to 3 1/8" long.
Flowers: Unisexual, male and female flowers growing on separate plants; male flowers in clusters, the female flowers single. Corollas radial, yellowish, about 1/2" broad.
Fruit: Spherical, up to 1 3/8" broad, red when ripe.
Bloom Period: Spring, summer.
Distribution: Cameron, Hidalgo, Willacy, and Starr counties.
Comments: Ibervilleas are often grown in containers with the swollen roots exposed. These vines have tendrils. Some other vines have twining stems instead. Other vinelike plants have neither tendrils nor twining stems; they simply sprawl on the ground or manage to climb onto various supports. Injured swollen roots release the odor of horseradish.

Globe Berry *Ibervillea lindheimeri*

I. lindheimeri was named in honor of Ferdinand Jacob Lindheimer, a prominent botanist in the nineteenth century.

Meloncito

Family: Cucurbitaceae
Scientific Name: *Melothria pendula*
Habit: Perennial vines with tendrils; stems slightly hairy.
Leaves: Alternate; blades heart shaped in outline; margins entire to 5-lobed, up to 2 ¾" long.
Flowers: Unisexual; male and female flowers on the same plant; male flowers in clusters; female flowers single or sometimes in clusters. Corollas yellow, about ⅛" broad.
Fruit: Cylindrical, up to ¾" long; green with white stripes, turning black when mature. The green stage resembles miniature watermelons.
Bloom Period: Spring, summer, fall.
Distribution: Cameron, Hidalgo, and Starr counties.
Comments: One way of distinguishing vines is their means of climbing. These vines have tendrils; some other vines have twining stems instead. The fruit are said to be poisonous. The Spanish common name translates as "little melon."

Meloncito *Melothria pendula*

Cusp Dodder

Family: Cuscutaceae
Scientific Name: *Cuscuta cuspidata*
Habit: Parasitic, with orange color, twining stems.
Leaves: None.
Flowers: Radial, white, about ⅛" broad.
Fruit: Capsules roundish, about ⅛" or less broad.
Bloom Period: Spring, fall.
Distribution: Cameron and Starr counties.
Comments: This species is usually seen parasitizing members of the Asteraceae (Compositae). As in slender dodder, *C. leptantha,* the flowers have pedicels, but they are much shorter than the pedicels of that species. This character, and the deeply divided calyx, are identifying markers of *C. cuspidata*.

Cusp Dodder *Cuscuta cuspidata*

Cuscutaceae

Slender Dodder

Family: Cuscutaceae
Scientific Name: *Cuscuta leptantha*
Habit: Parasitic, twining vines; stems orange in color.
Leaves: None.
Flowers: Radial, white, 1/16" or slightly longer.
Fruit: Capsules globose, with 2 to 4 seeds, separating from the base.
Bloom Period: Spring, summer.
Distribution: Cameron and Hidalgo counties.
Comments: The genus name comes from an Arabic word meaning "a tangled tuft of hair." The long flower pedicels of this species distinguish it from cusp dodder, *C. cuspidata*. The plant in the photograph is parasitizing a *Chamaesyce* sp.

Slender Dodder *Cuscuta leptantha*

Smooth Five Angled Dodder

Family: Cuscutaceae
Scientific Name: *Cuscuta pentagona* var. *glabrior* [*C. glabrior*]
Habit: Parasitic rootless twining vines; stems orange in color
Leaves: None.
Flowers: Radial, white, about 1/8" long. The fast-growing fruit makes the flower seem to be larger in diameter.
Fruit: Capsules, more or less globose, 1/8" or larger.
Bloom Period: Spring, summer.
Distribution: Cameron, Hidalgo, and Starr counties.
Comments: We have six species of *Cuscuta* in our area: *C. cuspidata*, *C. indecora* (not shown), *C. leptantha*, *C. pentagona*, *C. runyonii*, and *C. umbellata* (not shown). They are difficult to distinguish without a key and microscope or hand lens. This species was observed parasitizing pink smartweed, *Persicaria pensylvanica*.

Smooth Five Angled Dodder *Cuscuta pentagona* var. *glabrior*

Runyon's Dodder

Family: Cuscutaceae
Scientific Name: *Cuscuta runyonii*
Habit: Parasitic rootless twining vines, stems orange in color.
Leaves: None.
Flowers: Radial, whitish, about 1/8" long; sepals appear to be spiny.
Fruit: Capsules, more or less globose, about 1/8" broad, enclosed in the old flower.
Bloom Period: Spring, summer, fall.
Distribution: Hidalgo and Starr counties.
Comments: *Cuscuta runyonii* was named in honor or Robert Runyon, a self-taught botanist who was active in the Lower Rio Grande Valley. This species is notable for its "spiny" calyx. It was seen parasitizing *Thelesperma, Chamaesyce,* and *Dalea* species.

Runyon's Dodder *Cuscuta runyonii*

Dwarf Sundew

Family: Droseraceae
Scientific Name: *Drosera brevifolia* [*D. annua, D. leucantha*]
Habit: Tiny annual insectivorous herbs.
Leaves: Red colored, with glandular hairs; blades roundish, up to 5/16" broad.
Flowers: Radial, petals pinkish, up to 3/8" long.
Fruit: Capsules egg shaped, up to 1/8" broad.
Bloom Period: Spring, summer.
Distribution: Jim Hogg County.
Comments: The dimensions given are the maximum sizes. The plants we observed were 1/4" to 3/8" in total breadth. The LRGV has only three carnivorous species: The dwarf sundew; the aquatic bladderwort, *Utricularia gibba;* and seeds of shepherd's purse, *Capsella bursa-pastoris.* Sundews capture tiny insects with the sticky material on the ends of the glandular hairs of the leaves. When the insect is captured, the hairs slowly bend toward the center of the leaf blade, where the insect is digested. Although the plants were observed in Jim Hogg County, the locality was only two to three miles north of the Starr County border, with similar habitats ranging farther south. It is likely that they also grow in Starr County and possibly nearby Hidalgo County and have been overlooked because of their small size.

Dwarf Sundew *Drosera brevifolia*

Euphorbiaceae

Chapote, Texas Persimmon, Mexican Persimmon

Family: Ebenaceae
Scientific Name: *Diospyros texana*
Habit: Shrubs or small trees up to 19' tall with peeling bark.
Leaves: Alternate; blades obovate, up to 2 ⅜" long; underside hairy; margins usually curled back.
Flowers: Radial; male and female flowers on separate plants; petals 5, white, up to ½" long, partially united basally.
Fruit: A globose berry about ¾" broad, black when ripe.
Bloom Period: Spring.
Distribution: Cameron, Hidalgo, Willacy, and Starr counties.
Comments: The fruit is edible and is used to make jams and pastries. It must be fully ripe; otherwise it has an offensive, astringent taste. The fruit is also used to make a dye, said to produce colors from brown to nearly black. The tree is attractive for the landscape and an important wildlife food source.

Chapote, Texas Persimmon, Mexican Persimmon *Diospyros texana*

Round Copperleaf

Family: Euphorbiaceae
Scientific Name: *Acalypha monostachya*
Habit: Sprawling, hairy perennials.
Leaves: Alternate, simple; blades with straight hairs, more or less round, about ⅝" long; margins shallowly lobed.
Flowers: Male and female inflorescences usually on separate plants or occasionally on the same plant. Male spikes longer, red; female flowers with prominent red styles.
Fruit: A 3-seeded capsule.
Bloom Period: Spring, summer, fall.
Distribution: Cameron, Hidalgo, and Starr counties.
Comments: The rounded leaves with shallowly cut margins separate this from the other *Acalypha* species in our area. The genus name comes from an ancient Greek name for a nettle. Linnaeus gave the plant this name because he thought the plant resembled a nettle.

Round Copperleaf *Acalypha monostachya*

Poiret's Copperleaf
Family: Euphorbiaceae
Scientific Name: *Acalypha poirettii*
Habit: Erect hairy annuals up to 14" tall.
Leaves: Alternate, simple; blades ovate, about 2" long; margins with rounded teeth.
Flowers: Male and female flowers on the same spike, very small.
Fruit: A 3-seeded capsule.
Bloom Period: Summer, fall.
Distribution: Cameron, Hidalgo, and Starr counties
Comments: The erect habit and the leaf blades with coarse, often rounded teeth separate this from other *Acalypha* species in our area.

Poiret's Copperleaf *Acalypha poirettii*

Cardinal Feather
Family: Euphorbiaceae
Scientific Name: *Acalypha radians*
Habit: Sprawling perennials.
Leaves: Alternate, simple; blades roundish, up to ⅜" long, the margins deeply lobed.
Flowers: Male and female flowers on separate plants; male spikes reddish; female spikes with prominent red styles.
Fruit: A 3-seeded capsule.
Bloom Period: Spring, fall.
Distribution: Cameron, Hidalgo, Willacy, and Starr counties.
Comments: The middle photo illustrates the female inflorescences, from which the common name is derived. The deeply lobed, roundish leaves identify this species of *Acalypha*.

Cardinal Feather *Acalypha radians* with female flowers

Cardinal Feather *A. radians* male inflorescences

Euphorbiaceae

Vasey's Adelia

Family: Euphorbiaceae
Scientific Name: *Adelia vaseyi*
Habit: Erect shrubs up to 10' tall, with peeling bark. Stem nodes bumpy, enlarged.
Leaves: Alternate or in clusters; blades mostly spatulate, up to 1 5/8" long, but usually smaller; margins entire.
Flowers: Male and female flowers small, on separate plants; petals absent.
Fruit: A 3-seeded capsule about 5/16" broad.
Bloom Period: Spring, summer.
Distribution: Cameron, Hidalgo, Willacy, and Starr counties.
Comments: The erect branches often have a wandlike appearance, giving the shrub a distinctive look. It makes an excellent landscape plant and is a host plant for the Mexican Bluewing butterfly. Vasey's adelia is especially abundant in the Arroyo Colorado brush. The fruit pops open when ripe, expelling seeds, which are eaten by doves. Parrots eat the green fruit.

Vasey's Adelia *Adelia vaseyi*

Vasey's Adelia *A. vaseyi* 3-lobed fruit

Oreja De Ratón

Family: Euphorbiaceae
Scientific Name: *Bernardia myricifolia*
Habit: Shrubs up to 8' tall, with star-shaped hairs.
Leaves: Alternate, simple; blades ovate, up to 2" long; margins crinkled and toothed.
Flowers: Male and female flowers on separate plants, very small; petals absent.
Fruit: A 3-seeded capsule about ⅜" broad.
Bloom Period: Spring, summer, fall.
Distribution: Cameron, Hidalgo, Willacy, and Starr counties.
Comments: The leaves are considered good browse for deer and livestock. The shrubs make good landscape plants, especially in the xeriscape. *Bernardia* is a host plant for Lacy's Scrub-hairstreak butterfly. The common name translates as "rat's ear."

Oreja De Ratón *Bernardia myricifolia*

White Margined Sand Mat

Family: Euphorbiaceae
Scientific Name: *Chamaesyce albomarginata* [*Euphorbia albomarginata*]
Habit: Prostrate, hairless plants, often rooting at the nodes.
Leaves: Opposite; blades elliptic to rounded, up to ⅜" long; margins entire.
Flowers: Tiny, reduced; 4 white glandular appendages and a circular "crown" are the only noticeable floral parts.
Fruit: A 3-lobed, 3-seeded capsule.
Bloom Period: Spring, summer, fall.
Distribution: Cameron, Hidalgo, and Starr counties.
Comments: The apparent flower is actually a container for several highly reduced flowers. The white margin mentioned in the common name evidently refers to the white "crown" surrounding the tiny flowers; the leaves are not white-margined. Identifying characteristics: The plants are prostrate and hairless, with stipules united on both sides of stem and male flowers numbering less than 12 per cluster.

White Margined Sand Mat *Chamaesyce albomarginata*

Euphorbiaceae

Ashy Sand Mat
Family: Euphorbiaceae
Scientific Name: *Chamaesyce cinerascens*
   [*Euphorbia cinerascens*]
Habit: Prostrate perennials with woody taproots, the stems sometimes arching.
Leaves: Opposite; blades ovate to oblong, about ⅜" long; margins entire.
Flowers: Tiny, reduced; glandular appendages 4 or absent, red or white, attached to red glands, are the only noticeable floral parts.
Fruit: A 3-lobed, 3-seeded capsule.
Bloom Period: Spring, summer, fall.
Distribution: Starr County.
Comments: The tiny, bright red floral glands are its most noticeable characteristic. The genus name comes from an ancient Greek name for a prostrate or creeping plant. Identifying characteristics: plants are prostrate and green or greenish, with short stem hairs and hairy capsules.

Ashy Sand Mat *Chamaesyce cinerascens*

Heart Leaf Sand Mat
Family: Euphorbiaceae
Scientific Name: *Chamaesyce cordifolia*
   [*Euphorbia cordifolia*]
Habit: Plants small, prostrate, hairless, bleeding milky sap when damaged.
Leaves: Opposite; blades elliptic, up to ⅜" long; margins entire.
Flowers: Tiny, reduced; 4 white glandular appendages are the only noticeable floral parts.
Fruit: A 3-lobed, 3-seeded capsule.
Bloom Period: Spring, summer.
Distribution: Cameron and Willacy counties; probably also Starr County.
Comments: This species is commonly found in loose sands, at the coast and also inland. When stressed, the leaves turn a reddish color. Identifying characteristics: stems are prostrate and hairless, stipules are not united on both sides of stem, and leaf margins are finely toothed.

Heart Leaf Sand Mat *Chamaesyce cordifolia*

Ridge Seed Sand Mat

Family: Euphorbiaceae

Scientific Name: *Chamaesyce glyptosperma* [*Euphorbia glyptosperma*]

Habit: Plants small, prostrate, hairless, bleeding milky sap when damaged.

Leaves: Opposite; blades oblong to elliptic, up to ¼" long; margins minutely toothed apically.

Flowers: Tiny, reduced; 4 white glandular appendages are the only noticeable floral part.

Fruit: A 3-lobed, 3-seeded capsule.

Bloom Period: Spring, summer, fall.

Distribution: Cameron, Hidalgo, Willacy, and Starr counties.

Comments: What seems to be a flower is actually a container for several highly reduced flowers. Identifying characteristics: plants are green or greenish, sometimes turning red, prostrate, and hairy or hairless; capsules are hairless, stipules are not united on both sides of stem, and leaf margins are smooth.

Ridge Seed Sand Mat *Chamaesyce glyptosperma*

Graceful Sand Mat

Family: Euphorbiaceae

Scientific Name: *Chamaesyce hypericifolia* [*Euphorbia hypericifolia*]

Habit: Erect, hairless plants up to 39" tall, usually shorter, bleeding milky sap when damaged.

Leaves: Opposite; blades elliptical to lanceolate, up to 1 ⅛" long; margins finely toothed.

Flowers: Tiny, reduced; 4 white glandular appendages are the only noticeable floral part.

Fruit: A 3-lobed, 3-seeded capsule.

Bloom Period: Spring, summer, fall.

Distribution: Cameron, Hidalgo, and Starr counties.

Comments: This species is widespread in North America, frequently found in Texas. Identifying characteristics: plants are erect and hairless, with flower cups in tight clusters.

Graceful Sand Mat *Chamaesyce hypericifolia*

Euphorbiaceae

Laredo Sand Mat
Family: Euphorbiaceae
Scientific Name: *Chamaesyce laredana* [*Euphorbia laredana*]
Habit: Prostrate, gray-hairy annuals.
Leaves: Opposite; blades hairy, oblong to ovate, up to ⅜" or longer.
Flowers: Tiny, reduced; 4 pinkish white glandular appendages are the only noticeable floral part.
Fruit: A 3-lobed, 3-seeded capsule.
Bloom Period: Spring, summer, fall.
Distribution: Hidalgo and Starr counties.
Comments: The gray color of the plants and the hairiness of this species distinguish it from most of our other *Chamaesyce* species.

Laredo Sand Mat *Chamaesyce laredana*

Eyebane
Family: Euphorbiaceae
Scientific Name: *Chamaesyce nutans* [*Euphorbia nutans*]
Habit: Plants up to 19", sometimes taller, often minutely hairy at the stem tips, bleeding milky sap when damaged.
Leaves: Opposite; blades oblanceolate to oblong, sometimes slightly curved, up to 1 3/16" long; margins toothed.
Flowers: Tiny, reduced; 4 white glandular appendages are the only noticeable floral part.
Fruit: A 3-lobed, 3-seeded capsule.
Bloom Period: Spring, summer, fall.
Distribution: Cameron and Hidalgo counties.
Comments: This species commonly turns up uninvited in flower beds. Identifying characteristics: plants are erect, hairy or hairless, with flower cups not in tight clusters.

Eyebane *Chamaesyce nutans*

Prostrate Sand Mat

Family: Euphorbiaceae
Scientific Name: *Chamaesyce prostrata* [*Euphorbia prostrata*]
Habit: Prostrate, moderately hairy plants, bleeding milky sap when damaged.
Leaves: Opposite; blades ovate to elliptical, up to 5/16" long; margins often finely toothed.
Flowers: Tiny, reduced; 4 white glandular appendages are the only noticeable floral part.
Fruit: A 3-lobed, 3-seeded capsule.
Bloom Period: All seasons.
Distribution: Cameron, Hidalgo, Starr counties.
Comments: This species grows commonly in disturbed areas. It is fairly common in the state except for West Texas. Identifying characteristics: plants are prostrate, green or greenish, with long shaggy hairs and hairy capsules.

Prostrate Sand Mat *Chamaesyce prostrata*

Hierba De La Golondrina

Family: Euphorbiaceae
Scientific Name: *Chamaesyce serpens*
Habit: Prostrate hairless annuals, often rooting at the nodes, bleeding milky sap when damaged.
Leaves: Opposite; blades ovate to elliptic, up to 3/16" long.
Flowers: Tiny, reduced; 4 white glandular appendages are the only noticeable floral part.
Fruit: A 3-lobed, 3-seeded capsule.
Bloom Period: All seasons.
Distribution: Cameron, Hidalgo, Willacy, and Starr counties.
Comments: This species is widespread in North America and the tropics. In Texas it is commonly found in calcareous soil. Identifying characteristics: plants are prostrate, rooting along the stems, and hairless, with stipules united on both sides of stem; male flowers number less than 12 per cluster. The common name translates as "the swallow's plant."

Hierba De La Golondrina *Chamaesyce serpens*

Euphorbiaceae 213

Bull Nettle

Family: Euphorbiaceae

Scientific Name: *Cnidoscolus texanus*

Habit: Perennials up to 20" or taller, with stinging hairs, bleeding milky sap when damaged. The plants grow from a thick fleshy root, which can grow up to 8" broad and several feet in length.

Leaves: Alternate, palmately 3- to 5-lobed, about 6" broad.

Flowers: Male and female flowers in clusters together; tepals grown together basally, usually 5-lobed, white; lobes about ½" or longer.

Fruit: A 3-lobed, 3-seeded capsule about ¾" tall.

Bloom Period: Spring, summer, fall.

Distribution: Hidalgo, Willacy, and Starr counties.

Comments: This is a plant of sandy soils. The seeds are edible but are well defended by the stinging hairs. Bull nettle resembles young plants of the tropical species *Cnidoscolus urens* (called *mala mujer* in Mexico), which grows into a tree.

Bull Nettle *Cnidoscolus texanus*

Silver July Croton

Family: Euphorbiaceae

Scientific Name: *Croton argenteus* [*Julocroton argenteus*]

Habit: Erect annuals up to 47" tall, with star-shaped hairs.

Leaves: Alternate, simple; blades ovate, sometimes triangular, up to 4 ⅜" long; margins usually toothed.

Flowers: Male and female flowers tiny, in clusters together; calyx lobes of male flowers equal; those of female flowers with 3 large and 2 small lobes.

Fruit: A 3-seeded capsule about 3/16" tall.

Bloom Period: Summer, fall, winter.

Distribution: Cameron and Hidalgo counties.

Comments: The unequal female calyx lobes with fringed margins are useful characteristics for identifying this species.

Silver July Croton *Croton argenteus*

Silver Croton

Family: Euphorbiaceae

Scientific Name: *Croton argyranthemus*

Habit: Perennial herbs from creeping underground stems, sometimes woody, up to 12" tall.

Leaves: Alternate; blades obovate to elliptic, up to 2" long; both surfaces scaly, the lower surfaces silver in color.

Flowers: Male and female flowers on the same spike; calyxes 5-lobed, the lower portions grown together; petals 5, small.

Fruit: Capsules 3-seeded, about ¼" long.

Bloom Period: All seasons.

Distribution: Hidalgo, Willacy, and Starr Counties.

Comments: This is a plant of sandy soils. The silver color of the lower leaf surfaces and the stems make this *Croton* easy to identify. Occasionally this plant will emit a skunklike odor.

Silver Croton *Croton argyranthemus*

Woolly Croton

Family: Euphorbiaceae

Scientific Name: *Croton capitatus* var. *lindheimeri*

Habit: Annuals up to 39" tall with star-shaped hairs.

Leaves: Alternate, simple; blades lanceolate or ovate, up to 3" long, with abundant star-shaped hairs on both sides.

Flowers: Male and female flowers tiny, in clusters together; calyxes of female flowers 6-lobed, sometimes more.

Fruit: A 3-seeded capsule about ⅜" tall.

Bloom Period: Spring, summer, fall.

Distribution: Cameron, Hidalgo, Willacy, and Starr counties.

Comments: This species is typically seen growing in sandy soils at the coast and also inland. The variety was named in honor of Ferdinand Jacob Lindheimer, a prominent Texas botanist.

Woolly Croton *Croton capitatus* var. *lindheimeri*

Euphorbiaceae

Mexican Croton

Family: Euphorbiaceae
Scientific Name: *Croton ciliatoglandulifer*
Habit: Shrubs up to 39" tall with star-shaped hairs.
Leaves: Alternate, simple; blades ovate, up to 3 1/8" long, with star-shaped hairs on both sides; margins with noticeable stalked glands 1/16" or longer.
Flowers: Male and female flowers tiny, in clusters together, sometimes on separate plants.
Fruit: A 3-seeded capsule about 1/4" tall.
Bloom Period: All seasons.
Distribution: Willacy and Starr counties.
Comments: This plant is uncommon in our area but more abundant in Mexico. The glandular leaf edges make it easy to recognize. The glandular secretions can be irritating to the eyes or skin.

Mexican Croton *Croton ciliatoglandulifer*

Cortés Croton, Palillo

Family: Euphorbiaceae
Scientific Name: *Croton cortesianus*
Habit: Shrubs up to 6' tall with star-shaped hairs.
Leaves: Alternate, simple; blades elliptic, up to 4" long, with star-shaped hairs on the lower surfaces; margins usually entire.
Flowers: Male and female flowers tiny, on separate plants.
Fruit: A 3-seeded capsule about 1/4" long.
Bloom Period: All seasons.
Distribution: Cameron, Hidalgo, and Starr counties.
Comments: Cortés croton grows at the edges of brush lands. It is often planted as an ornamental. Fourteen croton species have been reported for our area. This species can usually be recognized by its larger green leaves and its preference for shaded localities. The Spanish common name translates as "toothpick" or "matchstick." As in many common names, the meaning is obscure.

Cortés Croton, Palillo *Croton cortesianus* male inflorescence

Cortés Croton, Palillo *C. cortesianus* 3-lobed fruit

Cory's Croton

Family: Euphorbiaceae
Scientific Name: *Croton coryi*
Habit: White-hairy annuals up to 40" tall.
Leaves: Alternate; blades ovate, up to 2 ¾" long, covered with white, star-shaped hairs.
Flowers: Male and female flowers tiny, on the same inflorescence; female calyxes with 6 or 7 lobes; male petals with silvery scales.
Fruit: A 3-seeded capsule about 3/16" tall.
Bloom Period: Summer, fall.
Distribution: Southern Kenedy, Brooks, and Jim Hogg counties, in sandy soil.
Comments: Since Cory's croton occurs in sandy soils of the three adjacent counties to the north of the Lower Rio Grande Valley, chances are high that it also occurs in similar habitat in the northern parts of Willacy, Hidalgo, and Starr counties. It is endemic to Texas.

Cory's Croton *Croton coryi*

Bristly Tropic Croton

Family: Euphorbiaceae
Scientific Name: *Croton glandulosus* var. *pubentissimus*
Habit: Annuals up to 22" tall with star-shaped hairs.
Leaves: Alternate, simple; blades ovate or elliptic, up to 1" long, with star-shaped hairs on both surfaces; margins toothed.
Flowers: Male and female flowers tiny, in clusters together.
Fruit: A 3-seeded capsule about 3/16" tall.
Bloom Period: All seasons.
Distribution: Cameron, Hidalgo, and Willacy counties.
Comments: This is a plant of sandy soils. This and most other crotons are host plants for Leaf Wing butterflies. Identification of this species requires the use of keys and a good magnifying lens.

Bristly Tropic Croton *Croton glandulosus* var. *pubentissimus*

Euphorbiaceae

Low Croton, Berlandier Croton

Family: Euphorbiaceae
Scientific Name: *Croton humilis*
Habit: Shrubs up to 6' tall, usually shorter, with star-shaped hairs.
Leaves: Alternate, simple; blades ovate, up to 3 ⅛" long, with star-shaped hairs on both surfaces.
Flowers: Male and female flowers tiny, usually in the same cluster, sometimes on separate plants.
Fruit: A 3-seeded capsule ¼" tall.
Bloom Period: All seasons.
Distribution: Cameron, Hidalgo, Willacy, and Starr counties.
Comments: Crotons get their name from a Greek word meaning "tick." This is derived from the appearance of the seeds, although some imagination is needed to see the similarity. The leaves turn orange in color during stress. Crotons are planted as ornamentals and to attract butterflies.

Low Croton, Berlandier Croton *Croton humilis*

Torrey's Croton

Family: Euphorbiaceae
Scientific Name: *Croton incanus* [*C. torreyanus*]
Habit: Shrubs up to 6' tall, usually shorter, with star-shaped hairs.
Leaves: Alternate, simple; blades oblong, sometimes ovate or narrowly elliptic, up to 2 ⅜" long, with star-shaped hairs on both surfaces.
Flowers: Male and female flowers tiny, in clusters together on the same inflorescence.
Fruit: A 3-seeded capsule.
Bloom Period: Spring, summer, fall.
Distribution: Cameron, Hidalgo, and Starr counties.
Comments: This plant is excellent for landscaping. It is our most common woody croton. During stress, the leaves have an attractive orange color. Identification of the crotons often requires use of keys and a good magnifying lens. The specific epithet is a Latin word meaning "gray."

Torrey's Croton *Croton incanus*

White Leaf Croton

Family: Euphorbiaceae
Scientific Name: *Croton leucophyllus*
Habit: Annuals up to 20" tall with star-shaped hairs.
Leaves: Alternate, simple; blades ovate, up to 3" long, with star-shaped hairs on both surfaces.
Flowers: Male and female flowers tiny, in the same cluster. Calyx lobes on the female flowers are unequal, with 3 large and 2 small.
Fruit: A 3-seeded capsule about 3/16" tall.
Bloom Period: Spring, summer, fall.
Distribution: Cameron, Hidalgo, and Starr counties.
Comments: Identification of the crotons is often difficult, but the irregular calyx lobes of the female flowers are helpful in identifying this species. The specific epithet comes from Latin words for "white leaf." Many crotons have whitish leaves.

White Leaf Croton *Croton leucophyllus*

Three Seed Croton

Family: Euphorbiaceae
Scientific Name: *Croton lindheimerianus*
Habit: Annuals up to 20" tall.
Leaves: Alternate; blades lanceolate to elliptic, up to 3 1/8" long, covered with star-shaped hairs, especially on the upper surfaces.
Flowers: Male and female flowers tiny, on the same inflorescence.
Fruit: A 3-seeded capsule about 3/16" tall.
Bloom Period: Spring, fall.
Distribution: Cameron, Hidalgo, and Starr counties.
Comments: The common name of this species is not very useful: most crotons have three seeds. Identification of the crotons is often difficult. The presence of male and female flowers on the same inflorescence is helpful, although some other crotons share this characteristic. *C. lindheimerianus* was named in honor of Ferdinand Jacob Lindheimer, a prominent Texas botanist in the nineteenth century.

Three Seed Croton *Croton lindheimerianus*

Euphorbiaceae

Park's Croton

Family: Euphorbiaceae
Scientific Name: *Croton parksii*
Habit: Annuals up to 32" tall.
Leaves: Alternate; blades lanceolate, up to 3 ⅛" long, with star-shaped hairs on both surfaces.
Flowers: Male and female flowers tiny, on separate plants.
Fruit: A 3-seeded capsule about 3/16" tall.
Bloom Period: Spring, summer, fall.
Distribution: Hidalgo and Willacy counties.
Comments: This species is endemic to Texas, growing in sandy soils. The main identifying characteristics of the species are the star-shaped hairs on both leaf surfaces, and the leaf blade length about 4 times the width. The leaves are fragrant when bruised.

Park's Croton *Croton parksii*

Beach Croton

Family: Euphorbiaceae
Scientific Name: *Croton punctatus*
Habit: Perennials, sometimes woody, up to 28" tall with star-shaped hairs.
Leaves: Alternate, simple; blades ovate, up to 2 ⅜" long, with star-shaped hairs on both surfaces; lower surfaces with tiny spots and usually brown veins.
Flowers: Male and female flowers tiny, usually on separate plants.
Fruit: A 3-seeded capsule about ⅜" tall.
Bloom Period: Spring, summer, fall.
Distribution: Cameron, Hidalgo, and Willacy counties.
Comments: This coastal dune species is easy to identify, with the dots on the lower leaf surfaces. The lower surfaces also usually have a brownish appearance because of the brown veins. The Hidalgo County population was found growing on sandy banks of the Rio Grande.

Beach Croton *Croton punctatus*

Beach Croton *C. punctatus* lower leaf surface

Texas Croton

Family: Euphorbiaceae

Scientific Name: *Croton texensis*

Habit: Annuals up to 32" tall.

Leaves: Alternate, simple; blades narrowly ovate to narrowly lanceolate, up to about 1 ⅜" long, with dense star-shaped hairs, more so on the lower surfaces.

Flowers: Male and female flowers tiny, on separate plants.

Fruit: A 3-seeded, bumpy capsule about ¼" tall.

Bloom Period: Summer, fall.

Distribution: Cameron and Hidalgo counties.

Comments: This *Croton* is recognizable by the warty capsules, and the male and female flowers on separate plants. It is fairly widespread in the state but has not previously been reported specifically for our area.

Texas Croton *Croton texensis*

Low Wild Mercury

Family: Euphorbiaceae

Scientific Name: *Ditaxis humilis* [*Argythamnia humilis*]

Habit: Leaning or trailing perennials with T-shaped hairs as seen with a magnifying lens.

Leaves: Alternate, simple; blades elliptical or spatulate, up to 2" long; margins entire or toothed.

Flowers: Male and female flowers small, in the same cluster; petals 5, up to ⅛" long; glands on the female flowers linear, less than 1/16" long.

Fruit: A 3-seeded capsule.

Bloom Period: Spring, summer.

Distribution: Cameron Hidalgo, and Starr counties.

Comments: In this species of *Ditaxis,* the inflorescences are shorter than the leaves, and the glands on the female flowers are linear. The specific epithet comes from a Latin word meaning "low growing."

Low Wild Mercury *Ditaxis humilis*

Euphorbiaceae

Hairy Silverbush

Family: Euphorbiaceae
Scientific Name: *Ditaxis pilosissima* [*Argythamnia mercurialina* var. *pilosissima*]
Habit: Perennial hairy herbs up to 28" tall.
Leaves: Alternate, usually without petioles; blades mostly elliptic, up to 1 ¾" long, the margins entire.
Flowers: Male and female flowers very small, in the same cluster; petals about ⅛" long but usually absent on the male flowers.
Fruit: A 3-seeded capsule.
Bloom Period: Spring, summer.
Distribution: Willacy, Starr, and Southern Brooks counties; probably also Hidalgo County.
Comments: In this species of *Ditaxis,* the inflorescences are longer than the leaves. This is a plant of sandy soils.

Hairy Silverbush *Ditaxis pilosissima*

Hairy Silverbush *D. pilosissima* male inflorescences

Common Silverbush

Family: Euphorbiaceae
Scientific Name: *Ditaxis neomexicana* [*Argythamnia neomexicana*]
Habit: Perennials with T-shaped hairs; stems trailing or leaning.
Leaves: Alternate, simple; blades elliptical or spatulate, up to 2" long.
Flowers: Male and female flowers on the same inflorescence; glands on female flowers tongue shaped.
Fruit: A 3-seeded capsule.
Bloom Period: Spring, summer.
Distribution: Hidalgo and Starr counties.
Comments: The inflorescences are shorter than the leaves, and the glands on the female flowers are tongue shaped.

Common Silverbush *Ditaxis neomexicana*

Ditaxis

Family: Euphorbiaceae
Scientific Name: *Ditaxis* sp.
Habit: Prostrate perennials with T-shaped hairs.
Leaves: Alternate, simple; blades ovate to obovate, 1 ¾" or longer.
Flowers: Tiny, male and female flowers on separate plants; glands on female flowers tongue shaped.
Fruit: A 3-seeded capsule
Distribution: Hidalgo and Starr counties.
Comments: This is an undescribed species pointed out by Bill Carr, who noted that, unlike in our other prostrate-growing *Ditaxis* species, male and female flowers are on separate plants. It grows on sandy soils.

Ditaxis *Ditaxis* sp.

Candelilla, Wax Plant

Family: Euphorbiaceae
Scientific Name: *Euphorbia antisiphyllitica*
Habit: Perennials forming clumps; stems flexible, wax-covered, up to 20" or taller.
Leaves: Alternate, about ⅛" long, falling early.
Flowers: Tiny, male and female flowers in cuplike structures located in the upper parts of the stems.
Fruit: A 3-compartment, 2-seeded capsule.
Bloom Period: After rains.
Distribution: Starr County.
Comments: In the early 1900s, populations of this species in the Trans-Pecos and in Mexico were greatly diminished by harvesting and extracting the wax from the stems. The high quality wax had many uses, including making of candles; hence the common name, which translates as "small candle." The Starr County candelillas are a darker green than those of Trans-Pecos Texas.

Candelilla, Wax Plant *Euphorbia antisiphyllitica*

Euphorbiaceae

Wild Poinsettia

Family: Euphorbiaceae
Scientific Name: *Euphorbia cyathophora*
  [*E. heterophylla* var. *cyathophora*]
Habit: Erect herbs up to 24" tall, bleeding milky sap when damaged.
Leaves: Mostly alternate; blades highly variable, linear to ovate, up to 4 ⅜" long; margins entire to toothed or lobed. Blades nearest the flowers colored reddish at the base.
Flowers: Tiny, male and female flowers in cuplike structures located in the upper parts of the plant.
Fruit: A 3-lobed capsule about ³⁄₁₆" tall.
Bloom Period: All seasons.
Distribution: Cameron and Hidalgo counties.
Comments: This species is sometimes cultivated for its miniature poinsettia-like appearance. In some individuals, the leaves remain green.

Wild Poinsettia *Euphorbia cyathophora*

Grassleaf Spurge

Family: Euphorbiaceae
Scientific Name: *Euphorbia graminea*
Habit: Erect annuals bleeding milky sap when damaged. Stems more or less square.
Leaves: Blades and petioles minutely hairy; lower blades alternate, broadly ovate, margins smooth or occasionally toothed; upper blades opposite, narrowly ovate, the margins smooth.
Flowers: Tiny, male and female flowers together in cuplike structures in the upper parts of the plant, the cup bordered by two white petal-like glandular appendages. Occasionally the cups contain only male flowers.
Fruit: A 3-seeded, 3-lobed capsule; seeds mottled.
Bloom Period: Spring, fall.
Distribution: Hidalgo County.
Comments: Grassleaf spurge is native to Mexico, Central America, and northern South America and known to have been introduced in Hawaii, California, and Florida. It was reported to be present in the LRGV by Everitt, Lonard and Little in *Weeds in South Texas and Northern Mexico* (2007). Found mainly in nursery areas or gardens, it has likely been spread through the nursery trade.

Grassleaf Spurge *Euphorbia graminea*

Heller's Spurge

Family: Euphorbiaceae
Scientific Name: *Euphorbia helleri*
 [*Poinsettia helleri*]
Habit: Hairless erect annuals up to 14" tall, bleeding milky sap when damaged.
Leaves: Alternate; blades spatulate up to 1 ¼" long.
Flowers: Tiny, male and female flowers in cuplike structures in the upper parts of the plant.
Fruit: A 3-seeded capsule; seeds white and smooth.
Bloom Period: Spring, summer.
Distribution: Cameron and Hidalgo counties.
Comments: Some of the euphorbias are very similar. Because of the small size of the flowers, optical aid and keys are necessary to distinguish the species.

Heller's Spurge *Euphorbia helleri*

Velvet Spurge

Family: Euphorbiaceae
Scientific Name: *Euphorbia innocua*
Habit: Perennials with prostrate or leaning, densely hairy stems.
Leaves: Alternate on lower stems, opposite on the upper ones; blades more or less ovate, up to 1 ¼" long, the margins entire.
Flowers: Tiny, male and female flowers in cuplike structures in the upper parts of the plant.
Fruit: A 3-seeded capsule.
Bloom Period: Late summer, fall, early winter.
Distribution: Willacy County
Comments: Velvet spurge is endemic to sandy soils of the Texas coast. It grows from a fleshy tuberous root. The genus *Euphorbia* is very large. The greatly reduced flowers make it a difficult family to classify, resulting in differences of opinion. For example, the genus *Chamaesyce* was considered a valid, separate genus, then it was merged with *Euphorbia*. Subsequently it was again separated, and we again have two genera rather than one.

Velvet Spurge *Euphorbia innocua*

Euphorbiaceae

Jicamilla

Family: Euphorbiaceae
Scientific Name: *Jatropha cathartica* [*J. berlandieri*]
Habit: Perennials forming a large tuber at the base. Stems up to 12" long.
Leaves: Usually alternate; blades up to 4" broad, deeply lobed palmately, with 5–7 lobes; the lobes cut into segments.
Flowers: Male and female flowers on the same plant; corollas red, up to ⅜" broad.
Fruit: A 3-seeded capsule about ½" tall.
Bloom Period: Spring, summer, fall.
Distribution: Cameron and Starr counties.
Comments: This species is often grown in a pot with the tuber exposed; however, when growing naturally, the tuber is underground. The plants are toxic to livestock and can cause death. However, they are unpalatable and are eaten only if the animal is starving. The Spanish common name probably refers to the similarity of the basal tuber to the edible root *jícama*.

Jicamilla *Jatropha cathartica*

Jicamilla *J. cathartica* flowers

Leather Stem, Sangre De Drago

Family: Euphorbiaceae
Scientific Name: *Jatropha dioica*
Habit: Thick-rooted perennials with rubbery stems up to 24" tall, bleeding red juice when damaged. The leaves fall during dry periods.
Leaves: Crowded in groups along the stem; blades simple, mostly spatulate, up to 2 3/8" long. Occasionally the blades are palmately lobed.
Flowers: Male and female flowers on separate plants; corollas whitish, bell shaped, about 5/16" long.
Fruit: A 1- or 2-seeded capsule up to 5/8" tall.
Bloom Period: Spring, summer, fall.
Distribution: Cameron, Hidalgo, Willacy, and Starr counties.
Comments: The plants are toxic to livestock but not fatal. A single plant can produce a colony from its roots. Fruit are seldom seen. The Spanish common name, which translates as "dragon's blood," refers to some similarity of this plant to a tree of the upper Amazon called *sangre de drago*. It is also in the Euphorbiaceae, and its sap is believed to have various healing properties. This is in contrast to the many examples of poisonous products of members of the Euphorbiaceae.

Leather Stem, Sangre De Drago *Jatropha dioica*

Walker's Manihot, Walker's Manioc

Family: Euphorbiaceae
Scientific Name: *Manihot walkerae*
Habit: Perennials with carrot-shaped roots; stems erect or prostrate, dying back seasonally. All parts smell of cyanide, especially if bruised.
Leaves: Alternate; blades palmately 5-lobed, rarely 3- or 7-lobed, the middle lobe up to 2" long.
Flowers: Male and female flowers on the same plant; corollas white, up to 1/2" long.
Fruit: A 3-seeded capsule.
Bloom Period: Spring, summer, and early fall.
Distribution: Hidalgo and Starr counties.
Comments: This species is on the state and federal lists of endangered plants. Initially only one plant was known to be growing wild in Texas; recently, however, several small populations have been found. They grow on caliche hills. The tubers are dug up and eaten by feral hogs.

Walker's Manihot, Walker's Manioc *Manihot walkerae*

Euphorbiaceae

Knotweed Leaf-Flower

Family: Euphorbiaceae
Scientific Name: *Phyllanthus polygonoides*
Habit: Perennial, nearly prostrate growing herbs.
Leaves: Alternate, simple; blades elliptic or spatulate, up to ½" long.
Flowers: Male and female flowers usually on the same plant; greenish and inconspicuous, usually hanging downward.
Fruit: A 3-compartmented capsule about ⅛" broad, with up to 2 seeds per compartment.
Bloom Period: All seasons.
Distribution: Cameron, Hidalgo, Willacy, and Starr counties.
Comments: Often the flowers are not seen unless the stems are lifted and the undersurface of the plant is examined.

Knotweed Leaf-Flower *Phyllanthus polygonoides*

Tender Leaf-Flower

Family: Euphorbiaceae
Scientific Name: *Phyllanthus tenellus*
Habit: Annuals up to 20" tall.
Leaves: Alternate, more or less elliptical, up to 1" long.
Flowers: Male and female flowers on the same plant; petals absent; sepals 5, small.
Fruit: Capsules roundish, less than ⅛" broad.
Bloom Period: All seasons.
Distribution: Hidalgo County.
Comments: This is a plant introduced from Asia. It has become a yard weed, possibly becoming dispersed through bedding plants from nurseries.

Tender Leaf-Flower *Phyllanthus tenellus*

Castor Bean

Family: Euphorbiaceae

Scientific Name: *Ricinus communis*

Habit: Large annual herbs up to 16' tall, usually less.

Leaves: Alternate; blades palmately lobed, roundish in outline, 8–24" broad.

Flowers: Inflorescences terminal, the yellowish male flowers above the reddish female flowers.

Fruit: Capsules 3-seeded, about ⅝" broad; seeds mottled, about ⅜" long.

Bloom Period: All seasons.

Distribution: Cameron and Hidalgo counties.

Comments: This is an introduced species that spreads easily along canals and ditches. It was formerly grown as an agricultural crop. Oil from the seeds was used in industry and also for medicinal purposes, such as castor oil. The seeds contain the poison ricin. The genus name comes from a Latin word for "tick," a reference to the ticklike appearance of the seeds.

Castor Bean *Ricinus communis*

Queen's Delight

Family: Euphorbiaceae

Scientific Name: *Stillingia sylvatica*

Habit: Slightly branched perennial herbs up to 28" tall, with milky sap.

Leaves: Alternate; blades narrowly ovate, up to 2 ¾" long, the margins toothed (sometimes minutely so).

Flowers: Inflorescences terminal, with tiny, greenish male and female flowers together; male calyxes 2-lobed, female calyxes 3-lobed; petals absent.

Fruit: Capsules 3-seeded, about ¼" tall.

Bloom Period: Spring.

Distribution: Hidalgo and Willacy counties.

Comments: This is a plant of sandy soils. The seeds are eaten by various birds.

Queen's Delight *Stillingia sylvatica*

Queen's Delight *S. sylvatica* fruit

Euphorbiaceae

Trecul's Queen's Delight

Family: Euphorbiaceae
Scientific Name: *Stillingia treculiana*
Habit: Perennials with milky sap, up to 16" tall.
Leaves: Alternate, simple; blades ovate, up to 1" long; margins coarsely toothed.
Flowers: Male and female flowers on the same plant, about ⅛" broad.
Fruit: A 3-lobed, 3-seeded capsule, 3/16–½" tall.
Bloom Period: Spring, summer, winter.
Distribution: Cameron, Hidalgo, and Starr counties.
Comments: The very short styles (about 1/16") distinguish this species from the other *Stillingia* species, which have styles about 3/16" long. The stems die back annually, reemerging from perennial rootstock.

Trecul's Queen's Delight *Stillingia treculiana*

Brush Noseburn

Family: Euphorbiaceae
Scientific Name: *Tragia glanduligera*
Habit: Perennials often growing on shrubs, with trailing or often twining stems up to 3' long, armed with stinging hairs.
Leaves: Alternate, simple; blades mostly ovate, up to 2" long; margins toothed.
Flowers: Male and female flowers on the same plant, about ⅛" broad.
Fruit: A 3-seeded capsule, about ⅛" tall.
Bloom Period: Spring, summer, fall.
Distribution: Cameron, Hidalgo, Willacy, and Starr counties.
Comments: If this plant is touched, the stinging hairs produce quite a bite. This species, which has glandular hairs on its inflorescences, is very similar to catnip noseburn, *T. ramosa*, which does not have glandular hairs on its inflorescences. Brush noseburn is a host plant for Blue-eyed Sailor, Common Mestra, Gray Cracker, and Red Rim butterflies.

Brush Noseburn *Tragia glanduligera*

Catnip Noseburn

Family: Euphorbiaceae
Scientific Name: *Tragia ramosa*
Habit: Perennials with trailing, rarely twining stems, armed with stinging hairs.
Leaves: Alternate, simple; blades ovate to linear, up to 2" long; margins toothed.
Flowers: Male and female flowers on the same plant, about ⅛" broad.
Fruit: A 3-seeded capsule about ⅛" tall.
Bloom Period: Spring, summer, fall.
Distribution: Hidalgo and Starr counties.
Comments: This is a new record for the LRGV. This species does not have glandular hairs on its inflorescences. It is otherwise very similar to brush noseburn, *T. glanduligera,* which does have glandular hairs on its inflorescences.

Fabaceae (Leguminosae)

Catnip Noseburn *Tragia ramosa*

The arrangement of this family (commonly called the legume family) requires some explanation. In the past, it has been segregated into three separate families. The same groups have also been called subfamilies of the Fabaceae. Regardless of the classification, these groups are easily distinguished, with their own floral characteristics. They are separated here, not to promote a taxonomic scheme but to group legumes with similar flowers for convenience in identification. The Caesalpinoideae have larger, somewhat bilateral flowers. The Mimosoideae have tiny radial flowers in globe- or rod-shaped clusters. The Papilionoideae have strongly bilateral, pea-shaped flowers.

   Members of this family have alternate leaves with stipules and the fruit is a legume.

Gray Nicker, Prickly Caesalpinia

Family: Fabaceae (Leguminosae: Caesalpinoideae)
Scientific Name: *Caesalpinia bonduc*
Habit: Sprawling large shrubs up to 6' tall, covered with curved prickles on leaves, stems, and fruit.

Gray Nicker, Prickly Caesalpinia *Caesalpinia bonduc*

Fabaceae

Leaves: Alternate, with curved prickles, twice compound; pinnae up to 7 pairs; leaflets up to 8 pairs per pinna, about 2" long. The paired, disk-shaped stipules are conspicuous.
Flowers: Inflorescences from the leaf axils; corollas slightly bilateral, about ¾" broad, petals 5, yellow, one petal with a reddish blotch.
Fruit: Legumes up to 2 ¾" long
Bloom Period: Fall, winter.
Distribution: Cameron and Willacy counties.
Comments: This primarily tropical seashore species was first brought to our attention by Tom Patterson, who discovered this plant growing on Padre Island. Subsequently, Ken and Selena King found a small population growing at Laguna Atascosa National Wildlife Refuge on the Laguna Madre. The seeds (called seabeans, along with those of several other species) are spread by ocean currents.

Tailed Rushpea

Family: Fabaceae (Leguminosae: Caesalpinoideae)
Scientific Name: *Caesalpinia caudata*
Habit: Low shrubs up to 20" tall, spreading from rhizomes, forming clumps.
Leaves: Alternate, twice compound; pinnae 3–11, the terminal one the largest, up to 4" long, with up to 11 pairs of leaflets; leaflets ⅛" to ½" long.
Flowers: from the leaf axils; corollas slightly bilateral, about ¾" broad but variable; petals light yellow, one of them darker; yellow glands on the outer surfaces.
Fruit: Legumes up to 1 ⅝" long, with reddish glands.
Bloom Period: Spring to early summer.
Distribution: Starr County.
Comments: This species has a somewhat limited distribution in Texas, in sandy soils from Starr County to Webb County.

Tailed Rushpea *Caesalpinia caudata*

Mexican Caesalpinia

Family: Fabaceae (Leguminosae: Caesalpinoideae)

Scientific Name: *Caesalpinia mexicana*

Habit: Trees with brittle branches, up to 19' or sometimes taller, without spines.

Leaves: Alternate, twice compound; pinnae up to 9; leaflets on each pinna up to 5 pairs, each up to 1" long.

Flowers: Slightly bilateral, bright yellow, sometimes with red; petals up to ½" long, one of them slightly different.

Fruit: Legumes brown, up to 2 ⅜" long.

Bloom Period: Spring, summer.

Distribution: Cameron, Hidalgo, and Starr counties.

Comments: This species is a host plant for the Curve-Wing Metalmark butterfly. The tree is attractive and is often cultivated. When the legumes ripen, in the heat of the day they suddenly pop open, throwing the seeds some distance. The seeds are thrown with such force that when they hit a building, it sounds as if someone were throwing gravel.

Mexican Caesalpinia *Caesalpinia mexicana*

Woodland Sensitive Pea

Family: Fabaceae (Leguminosae: Caesalpinoideae)

Scientific Name: *Chamaecrista calycioides* [*Cassia aristellata*]

Habit: Erect and leaning perennials; stems with short hairs.

Leaves: Alternate, once compound; leaflets about ⅝" long, up to 8 pairs, occasionally more.

Flowers: Solitary, from the leaf axils; corollas slightly bilateral; petals pale yellow, about ⅜" long, wilting by noon.

Fruit: Legumes linear, slightly curved, up to 1 ⅝" long.

Bloom Period: Summer, fall.

Distribution: Hidalgo, Willacy, and Starr counties.

Comments: This plant is separated from the other herbaceous *Chamaecrista* species by its flowers borne singly from the leaf axils and its deep swollen roots. It grows on sandy soils.

Woodland Sensitive Pea *Chamaecrista calycioides*

Fabaceae

Partridge Pea

Family: Fabaceae (Leguminosae: Caesalpinoideae)
Scientific Name: *Chamaecrista fasciculata* [*Cassia fasciculata* var. *ferrisiae*]
Habit: Annuals up to 50" tall with reddish stems, usually much shorter and spread out.
Leaves: Alternate, once compound; leaflets up to 15 pairs, about ⅛" broad.
Flowers: Inflorescences with up to 7 flowers; corollas slightly bilateral; petals yellow, up to ⅝" long, 4 of them with a dark basal spot.
Fruit: Legumes brown or reddish, slightly curved, up to 2 ¾" long.
Bloom Period: Spring, summer, fall.
Distribution: Cameron, Hidalgo, and Willacy counties.
Comments: This is a species of sandy soils, especially near the coast. It is a host plant for the Cloudless Sulphur and Little Yellow butterflies.

Partridge Pea *Chamaecrista fasciculata*

Texas Senna

Family: Fabaceae (Leguminosae: Caesalpinoideae)
Scientific Name: *Chamaecrista flexuosa* var. *texana* [*Cassia texana*]
Habit: Prostrate perennials, sometimes with curved hairs.
Leaves: Alternate, once compound; leaflets up to 16 pairs, about 1/16" broad.
Flowers: Mostly solitary, bilateral; petals yellow, up to ½" long.
Fruit: Legumes brown, up to 1 ⅝" long, with slanted divisions between the seeds.
Bloom Period: Spring, summer.
Distribution: Cameron, Hidalgo, Willacy, and Starr counties.
Comments: This species is usually found in sandy soils.

Texas Senna *Chamaecrista flexuosa* var. *texana*

Indian Rush Pea

Family: Fabaceae (Leguminosae: Caesalpinoideae)

Scientific Name: *Hoffmanseggia glauca*

Habit: Perennial herbs up to 12" tall, with swollen roots.

Leaves: Alternate, twice compound; pinnae up to 11; leaflets up to 11 pairs, about 3/16" long.

Flowers: Inflorescences with up to 15 flowers, with stalked glands; petals yellow, one with red marks, up to 1/2" long.

Fruit: Legumes brown, curved, up to 1 5/8" long.

Bloom Period: Spring, summer.

Distribution: Cameron, Hidalgo, Willacy, and Starr counties.

Comments: This species is especially found in saline areas. It produces new plants from long underground runners.

Indian Rush Pea *Hoffmanseggia glauca*

Retama

Family: Fabaceae (Leguminosae: Caesalpinoideae)

Scientific Name: *Parkinsonia aculeata*

Habit: Trees up to 32' tall, with green branches, with curved spines.

Leaves: Alternate, twice compound; pinnae usually 1 pair up to 12" long, usually less, the rachises flattened; leaflets widely spaced, up to 3/16" long, often falling away in dry times.

Flowers: Slightly bilateral; petals yellow, one of them with reddish marks, about 1/2" long.

Fruit: Legumes brown, up to 4" long, constricted between the seeds.

Bloom Period: Spring, summer.

Distribution: Cameron, Hidalgo, Willacy, and Starr counties.

Comments: This is a widespread species, often cultivated. It also grows in South America and was possibly introduced from there. It is often found growing along water courses, spreading easily by seeds. It is sometimes called palo verde, along with *P. texana*. Comparison of the pinna lengths distinguishes the two species. Those of *P. aculeata* are much longer.

Retama *Parkinsonia aculeata*

Retama *P. aculeata* flowers

Fabaceae

Palo Verde

Family: Fabaceae (Leguminosae: Caesalpinoideae)
Scientific Name: *Parkinsonia texana* var. *macra* [*Cercidium macrum*]
Habit: Small compact trees up to 13' tall, with paired spines. Branches green.
Leaves: Alternate, twice compound; pinnae 1 or 2 pairs up to 1 ⅝" long; leaflets about ¼" long.
Flowers: Slightly bilateral; petals yellow, one of them with red marks, up to ½" long.
Fruit: Legumes brown, up to 2 ⅜" long.
Bloom Period: Spring, summer.
Distribution: Cameron, Hidalgo, Willacy, and Starr counties.
Comments: Retama, *Parkinsonia aculeata,* is sometimes also called palo verde, since it also has green branches (*palo verde* translates as "green stick"). It is easy to distinguish the two species by comparing the pinna lengths; those of *P. texana* are much shorter. *P. texana* var. *texana* occurs in Starr County. The base of its seed pod is hairy, and the leaves produce usually 1 or 2 pairs of pinnae. *P. texana* var. *macra* has a hairless seed pod, and the leaves usually produce 2 or 3 pairs of pinnae.

Palo Verde *Parkinsonia texana* var. *macra*

South Texas Rushpea

Family: Fabaceae (Leguminosae: Caesalpinoideae)
Scientific Name: *Pomaria austrotexana*
Habit: Small rounded, bad-smelling shrubs 6–24" tall; lower stems falling over, erect at the ends. Old stems reddish, roughened or with small lines, hairless; young stems ribbed, hairy, with scattered glandular hairs.
Leaves: Alternate, twice compound, hairy and glandular, with 5–7 pinnae (the first leaf division), up to 2 ¾" long; leaflets (the second division 7–11 per pinna, about ¼" long.
Flowers: Inflorescences at the stem ends, with up to 15 flowers; flowers about ¾" long and ⅜" broad, yellow, with red spots on one petal.
Fruit: Legumes up to 1 ⅜" long and ⅜" broad, with scattered glands.
Bloom Period: Spring, summer.
Distribution: Hidalgo and Starr counties.
Comments: This species is fairly rare, restricted to the deep sandy areas of South Texas and adjacent Mexico. The red glandular dots on the leaves and legumes are a major characteristic of this rushpea. Crushed leaves stain the hands yellow and leave a scent resembling peanut butter.

South Texas Rushpea *Pomaria austrotexana*

Two-Leaved Senna

Family: Fabaceae (Leguminosae: Caesalpinoideae)

Scientific Name: *Senna bauhinioides* [*Cassia bauhinioides*]

Habit: Herbs or shrublike plants with leaning stems, gray-hairy, up to 16" long.

Leaves: Alternate, compound with only 2 leaflets about 1 ¾" long.

Flowers: Inflorescences with 1 or 2 flowers; corollas slightly bilateral; petals yellow, about ⅝" long.

Fruit: Legumes hairy, up to 1 ⅝" long, at right angles to the stem or slanting upward.

Bloom Period: Spring, early summer.

Distribution: Hidalgo and Starr counties.

Comments: This *Senna* is easy to recognize with its 2 leaflets. Dwarf Senna, *S. pumilio,* also has 2 leaflets, but they are much narrower. Isely's Durango senna, *S. durangensis,* also has 2 leaflets, but they are shorter and broader, and its legumes are held more or less vertically. *S. bauhinioides* grows in sandy loam and also caliche or gypsum soils.

Two-Leaved Senna *Senna bauhinioides*

Isely's Durango Senna

Family: Fabaceae (Leguminosae: Caesalpinoideae)

Scientific Name: *Senna durangensis* var. *iselyi* [*Cassia durangensis* var. *iselyi*]

Habit: Erect or leaning shrubs up to 20" tall, with hairy stems.

Leaves: Alternate, compound with only 2 oblong or somewhat roundish leaflets up to 2" long.

Flowers: Inflorescences with up to 6 flowers; corollas slightly bilateral; petals pale yellow, up to ⅝" long.

Fruit: Legumes up to 1 ⅝" long, flattened, hairy, held erect, straight or slightly curved.

Bloom Period: Mostly spring and summer.

Distribution: Cameron, Hidalgo, and Starr counties.

Comments: The main features of this species are its 2 leaflets and its seed pods, which are held vertically or nearly so. The hairiness can vary, with the hairs from ¹⁄₁₆" to ⅛". *S. durangensis* grows in deep sand.

Isely's Durango Senna *Senna durangensis* var. *iselyi*

Fabaceae

Lindheimer's Senna

Family: Fabaceae (Leguminosae: Caesalpinoideae)

Scientific Name: *Senna lindheimeriana* [*Cassia lindheimeriana*]

Habit: Perennial herbs up to 6' tall, arising from a woody root.

Leaves: Alternate, hairy, compound with 5–8 pairs of leaflets; leaflets silver in color around the margins, the tips angular rather than rounded.

Flowers: Inflorescences with 5–10 flowers; petals yellow, about ⅝" long.

Fruit: Legumes brown, up to 2 ⅜" long.

Bloom Period: Summer, fall.

Distribution: Cameron and Starr counties.

Comments: The silvery margins and the angular leaf tips are good indicators of this species. It prefers low places where water accumulates and stands for a time. *S. lindheimeriana* was named in honor of Ferdinand Lindheimer, who immigrated from Germany and became a prominent Texas botanist in the nineteenth century.

Lindheimer's Senna *Senna lindheimeriana*

Coffee Senna

Family: Fabaceae (Leguminosae: Caesalpinoideae)

Scientific Name: *Senna occidentalis* [*Cassia occidentalis*]

Habit: Bad-smelling, somewhat woody annuals up to 6½' tall.

Leaves: Alternate, compound with 4–6 pairs of leaflets; leaflets up to 2 ¼" long, the ends forming an acute angle.

Flowers: Inflorescences with few flowers on short pedicels; corollas slightly bilateral; petals yellow, about ¾" long.

Fruit: Legumes brown, linear, up to 4 ¾" long. The sides are compressed around the seeds, leaving the outlines of the seeds.

Bloom Period: Summer, fall, winter.

Distribution: Cameron and Hidalgo counties.

Comments: The angular leaflet tips and the offensive smell are noticeable characteristics of this species. It is usually found in disturbed places.

Coffee Senna *Senna occidentalis*

Smooth Senna, Christmas Senna, Climbing Senna

Family: Fabaceae (Leguminosae: Caesalpinoideae)
Scientific Name: *Senna pendula* [*Cassia colunteoides*]
Habit: Shrubs up to 15' or taller.
Leaves: Alternate, compound, with 3–5 pairs of leaflets; leaflets elliptic to oval, about 1½" long.
Flowers: Inflorescences from the upper leaf axils, bearing 3–12 flowers; corollas bilateral, yellow with red veins, about ½" long
Fruit: Legumes more or less round in cross section, about 6" or longer, with many small beans.
Bloom Period: Fall, early winter.
Distribution: Cameron and Hidalgo counties.
Comments: Christmas senna is widely cultivated and has escaped and become established in our area. It has the ability to clamber to the tops of tall trees, 20' or more, and has become a pest in some low places that retain moisture. It is listed by the Florida Exotic Pest Plant Council as a Category I species that is invading and disrupting natural communities in Florida. Caterpillars of Sulphur butterflies feed on the leaves, and adults are attracted to the nectar. A variety of *S. pendula* that has flattened legumes is also cultivated in this area, and also sometimes escapes. According to Dr. Tom Wendt at the University of Texas at Austin, the species is highly variable, with 17 named varieties and several more that have not been named. Our plants were at one time misidentified as *Cassia splendida,* and that was the accepted name until recently.

Smooth Senna, Christmas Senna, Climbing Senna *Senna pendula*

Dwarf Senna

Family: Fabaceae (Leguminosae: Caesalpinoideae)
Scientific Name: *Senna pumilio* [*Cassia pumilio*]
Habit: Perennials up to 8" tall, usually less, the stems arising from small tubers.
Leaves: Alternate, compound; leaflets 2, up to 2" long and ⅛" broad.
Flowers: Inflorescences bearing only one flower; corollas slightly bilateral; petals yellow, about ½" long.
Fruit: Legumes inflated, about ⅝" long.

Dwarf Senna *Senna pumilio*

Bloom Period: Spring, early summer.
Distribution: Hidalgo and Starr counties.
Comments: The narrow leaflets of this species distinguish it from two-leaved senna, *S. bauhinioides,* which also has 2 leaflets. The specific epithet comes from a Latin word for "dwarf."

Fabaceae

White Bristly Acacia, Prairie Acacia

Family: Fabaceae (Leguminosae: Mimosoideae)
Scientific Name: *Acacia angustissima*
Habit: Small spineless shrubs up to 3' or taller, sometimes forming new plants from rhizomes.
Leaves: Alternate, twice compound; leaflets linear, close together.
Flowers: Radial, white, in globose clusters.
Fruit: A brown, flattened legume up to 3" long.
Bloom Period: Spring.
Distribution: Hidalgo and Starr counties.
Comments: This acacia is fairly rare in our area. It grows in seasonally wet, sandy soils. *A. angustissima* is noticeable for its lack of spines and its fernlike foliage.

White Bristly Acacia, Prairie Acacia *Acacia angustissima*

Guajillo

Family: Fabaceae (Leguminosae: Mimosoideae)
Scientific Name: *Acacia berlandieri*
Habit: Shrubs up to 10' tall, with small, slightly curved prickles (occasionally unarmed).
Leaves: Alternate, twice compound, up to 6" long; pinnae, 10–12 pairs; leaflets very narrow, about 1/16" broad.
Flowers: Tiny, radial, white, in globose clusters abut 3/8" broad.
Fruit: A reddish, woody, flattened legume up to 6" long.
Bloom Period: Spring.
Distribution: Hidalgo and Starr counties.
Comments: This is one of our earliest-blooming shrubs. It was named in honor of Jean Louis Berlandier, an early collector of plants from Mexico and Texas. Honey derived from guajillo is prized for its superior flavor. The leaves are toxic to livestock. They can cause what is called "guajillo wobbles" and death. The plant is recognized by the prickles, which are scattered around the stem rather than arranged in rows or pairs.

Guajillo *Acacia berlandieri*

Guajillo *A. berlandieri* fruit

Huisache

Family: Fabaceae (Leguminosae: Mimosoideae)
Scientific Name: *Acacia farnesiana* [*A. smallii*]
Habit: Small trees up to 20' tall, with paired straight spines up to ⅜" long.
Leaves: Alternate, twice compound, up to 3 ⅛" long; pinnae, 2–6 pairs; leaflets very narrow, about 1/16" broad.
Flowers: Tiny, radial, orange, very fragrant, in globose clusters about ⅜" broad.
Fruit: A black, nearly cylindrical legume up to 3 ⅛" long.
Bloom Period: Spring.
Distribution: Cameron, Hidalgo, Willacy, and Starr counties.
Comments: This is a good honey plant. For several hundred years, Huisache flowers have been used to produce fine perfume in Europe. An interesting small population of huisaches grows at Boca Chica Beach in Cameron County. The stems are prostrate, and the spines are 1" or longer.

Huisache *Acacia farnesiana*

Huisache *A. farnesiana* prostrate form

Wright's Catclaw

Family: Fabaceae (Leguminosae: Mimosoideae)
Scientific Name: *Acacia greggii* var. *wrightii* [*A. wrightii*]
Habit: Shrubs or trees up to 20' tall with curved prickles resembling a cat's claws.
Leaves: Alternate, twice compound, up to 2 ½" long; pinnae 1 or 2 pairs; leaflets up to ¼" broad.
Flowers: Tiny, radial, whitish, fragrant, in cylindrical clusters up to ⅜" broad.
Fruit: Legumes brown or reddish, up to 3 ⅛" long.
Bloom Period: Spring, summer.
Distribution: Cameron, Hidalgo, and Starr counties.
Comments: This shrub or small tree produces a noticeable aroma and is very attractive when in bloom. The wood is tough and hard. The *Atlas of the Vascular Plants of Texas* (Turner et al. 2003) places *A. greggii* var. *greggii* in Trans-Pecos Texas, but we have plants corresponding to that variety in the western LRGV. *A. greggii* var. *wrightii* has leaflets ¼" to ⅝" long, and legumes 1" to 1 ¼" wide. *A. greggii* var. *greggii* has leaflets ¼" or less in length, and legumes ¾" broad or less, distinctly constricted between the seeds. This variety was named in honor of Charles Wright, one of the best known botanists in the United States in the nineteenth century.

Wright's Catclaw *Acacia greggii* var. *wrightii*

Fabaceae

Black Brush

Family: Fabaceae (Leguminosae: Mimosoideae)
Scientific Name: *Acacia rigidula*
Habit: Shrubs up to 10' tall with paired straight spines up to 1 ⅛" long.
Leaves: Alternate, twice compound, up to 1 ⅛" long; pinnae mostly 1 pair; leaflets up to ¼" broad.
Flowers: Tiny, radial, whitish, in cylindrical clusters about ⅜" broad.
Fruit: Legumes black or reddish, up to 3 ⅛" long.
Bloom Period: Spring.
Distribution: Cameron, Hidalgo, Willacy, and Starr counties.
Comments: This is probably the most common shrub in Starr County. The flowers are an excellent source of honey.

Black Brush *Acacia rigidula*

Huisachillo

Family: Fabaceae (Leguminosae: Mimosoideae)
Scientific Name: *Acacia schaffneri*
Habit: Shrubs or small trees up to 7' tall with paired straight spines up to ½" long.
Leaves: Alternate, twice compound, up to 1 ⅜" long; pinnae, 2–6 pairs; leaflets very narrow, less than 1/16" broad.
Flowers: Tiny, radial, fragrant, orange colored, in globose clusters about ⅜" broad.
Fruit: Legumes black or reddish, velvety-hairy, up to 6" long.
Bloom Period: Spring.
Distribution: Cameron, Hidalgo, and Starr counties.
Comments: The genus name is derived from a Greek word meaning "sharp point." Fruit and flowers are often found together on huisachillo. It is easy to confuse with huisache. The two species can be distinguished by examination of the legumes. Huisachillo's legumes are flattened, hairy, and longer; huisache's are terete (round in cross section), hairless, and shorter. *Huisachillo* is a diminutive form of *huisache*.

Huisachillo *Acacia schaffneri*

False Mesquite Calliandra, Feather Duster

Family: Fabaceae (Leguminosae: Mimosoideae)
Scientific Name: *Calliandra conferta*
Habit: Small shrubs up to 12" tall, without spines.
Leaves: Alternate, twice compound; pinnae, 1 pair; leaflets about ⅛" long.
Flowers: Tiny, radial, pinkish or whitish in clusters up to ¾" broad.
Fruit: Flattened legumes 1 ¼" or sometimes longer.
Bloom Period: Spring, summer, occasionally fall.
Distribution: Hidalgo and Starr counties.
Comments: Feather duster is uncommon, mostly seen on caliche and gypsum hills of Starr and western Hidalgo County. It is heavily browsed by deer, producing low, compact growth. The genus name comes from two Greek words meaning "beautiful" and "male," referring to the attractive stamens.

False Mesquite Calliandra, Feather Duster *Calliandra conferta* with flowers

False Mesquite Calliandra, Feather Duster *C. conferta* with dehisced (opened) fruit

Texas Ebony

Family: Fabaceae (Leguminosae: Mimosoideae)
Scientific Name: *Chloroleucon ebano*
  [*Pithecellobium ebano, P. flexicaule*]
Habit: Trees up to 50' tall, with paired spines. Young branches often in a zigzag shape.
Leaves: Alternate, twice compound, about 1 ⅝" long; pinnae, one pair; leaflets, 3–6 pairs per pinna.
Flowers: Tiny, radial, whitish, fragrant, in cylindrical clusters up to 1 ⅛" long.
Fruit: Legumes woody, curved, up to 8" long.
Bloom Period: Spring, summer, fall; typically just before or after rains.
Distribution: Cameron, Hidalgo, Willacy, and Starr counties.
Comments: This is an attractive shade tree. The blooms usually appear after rains and perfume the entire vicinity. Texas ebony branches are the preferred growing site for the epiphytic Bailey's ball moss, *Tillandsia baileyi*. The dense branches and spines make Texas ebony a desirable nesting site for doves and other birds. Immature seeds are cooked and eaten.

Texas Ebony *Chloroleucon ebano*

Fabaceae

The older seeds have been used as a substitute for coffee. The dark, dense heartwood has long been prized for making furniture, sculpture, and bowls and has even been used as flooring. Dark, rich humus is formed below undisturbed old colonies and has long been collected and used as a soil conditioner in gardens. This species is a host plant for the Large Orange Sulphur butterfly.

Bee Wild Bundleflower
Family: Fabaceae (Leguminosae: Mimosoideae)
Scientific Name: *Desmanthus bicornutus*
Habit: Shrubs 4–8' tall, with weeping branches. Stems turn red in the fall when temperatures cool.
Leaves: Alternate, twice compound, up to 7" long, sagging downward at night; pinnae (the first division) usually 7 pairs, with a gland between the lowest pair of pinnae.
Flowers: Tiny, radial, white, with 10 stamens, in clusters about ⅝" broad.
Fruit: Legumes turning red in the fall when temperatures cool; up to 3½" long, straight, in small clusters, each usually bearing 10–11 seeds; mature seeds brown.
Bloom Period: Spring, summer, fall.
Distribution: Hidalgo County.
Comments: *Desmanthus bicornutus* is native to Arizona and western Mexico. A small branch of this species would look almost exactly like a branch of *D. virgatus* var. *depressus* (see following account). A seed mix sold as Bee Wild Bundle Flower is a combination of seeds gathered from four selected *Desmanthus bicornutus* cultivars. The mixture is used on ranches to provide cover and food for wildlife. It is used in agriculture for its nitrogen-fixing properties to improve soil quality and is sometimes seen in fallow fields, growing in rows. Since it is a mixture of four selected strains, variation in some characteristics from the wild species can be expected.

Bee Wild Bundleflower *Desmanthus bicornutus*

Depressed Wand-Like Bundle Flower

Family: Fabaceae (Leguminosae: Mimosoideae)
Scientific Name: *Desmanthus virgatus* var. *depressus*
Habit: Unarmed sprawling or leaning perennials.
Leaves: Alternate, twice compound, up to 2 ⅜" long; pinnae, up to 7 pairs; leaflets up to 14" long.
Flowers: Tiny, radial, white, in small clusters up to ⅜" broad.
Fruit: Legumes flattened, sometimes curved, up to 2 ⅜" long, usually in clusters.
Bloom Period: Spring, summer, fall.
Distribution: Cameron, Hidalgo, Willacy, and Starr counties.
Comments: The genus name comes from two Greek words meaning "bundle" and "flower," referring to the bundling (or clustering) together of the flowers. Unlike *D. bicornutus*, *D. virgatus* var. *depressus* is a native, known in Texas and Florida. It usually occurs in disturbed places such as lawns and roadsides. The shrubby *D. bicornutus* is easily distinguished from the much smaller, prostrate-growing *D. virgatus* var. *depressus*.

Depressed Wand-Like Bundle Flower *Desmanthus virgatus* var. *depressus*

Tenaza

Family: Fabaceae (Leguminosae: Mimosoideae)
Scientific Name: *Havardia pallens* [*Pithecellobium pallens*]
Habit: Trees up to 16' tall, with smooth bark, and paired spines on trunk and branches.
Leaves: Alternate, twice compound, up to 4" long; pinnae up to 6 pairs; leaflets up to 20 per pinna, about ¼" long.
Flowers: Tiny, radial, white, in globose clusters about ⅜" broad.
Fruit: Legumes papery, up to 4" long.
Bloom Period: Spring, summer, fall, often after rains.
Distribution: Cameron, Hidalgo, Willacy, and Starr counties.
Comments: The attractive, fragrant blooms make tenaza a desirable ornamental and an important butterfly nectar source. The smooth bark and the paired spines emerging from the old wood are distinctive characteristics of this tree. The Spanish common name translates as "pliers." Possibly the paired spines emerging from the trunk appeared to resemble an open pair of pliers.

Tenaza *Havardia pallens*

Fabaceae

Popinac

Family: Fabaceae (Leguminosae: Mimosoideae)
Scientific Name: *Leucaena leucocephala*
Habit: Trees up to 26' tall, with brittle branches without spines.
Leaves: Alternate, twice compound, up to 10" long; pinnae up to 25 pairs; leaflets up to 3/16" broad.
Flowers: Tiny, radial, white, in globose clusters up to 3/4" broad.
Fruit: Legumes reddish, papery, up to 6" long or longer.
Bloom Period: Spring, summer.
Distribution: Cameron, Hidalgo, and Starr counties.
Comments: This invasive fast-growing species has been introduced from tropical America. It is distinguished from the closely related tepeguaje, *L. pulverulenta*, by its broader leaflets, which have noticeable spaces between them.

Popinac *Leucaena leucocephala*

Tepeguaje

Family: Fabaceae (Leguminosae: Mimosoideae)
Scientific Name: *Leucaena pulverulenta*
Habit: Trees up to 33' or taller, with brittle branches without spines.
Leaves: Alternate, twice compound, up to 10" long; pinnae up to 25 pairs; leaflets about 1/16" broad.
Flowers: Tiny, radial, white, in globose clusters up to 3/4" broad.
Fruit: Legumes reddish, papery, up to 6" or longer.
Bloom Period: Spring, summer.
Distribution: Cameron, Hidalgo, Starr, and Willacy counties.
Comments: This species is distinguished from the closely related popinac, *L. leucocephala*, by its narrow leaflets having little or no spaces between them. Tepeguaje is used extensively as a shade tree in the LRGV. With its fast growth, it is important in the restoration of disturbed areas, providing shelter for various seedlings and nitrifying the soil. It is short-lived, allowing more long-lived species to replace it.

Tepeguaje *Leucaena pulverulenta*

Comparison of leaves of Tepeguaje *L. pulverulenta* (left) and Popinac *L. leucocephala* (right)

Zarza, Black Mimosa

Family: Fabaceae (Leguminosae: Mimosoideae)

Scientific Name: *Mimosa asperata* [*M. pigra* var. *berlandieri*]

Habit: Shrubs up to 6 ½' tall, with prominent straight (sometimes curved), white prickles.

Leaves: Sensitive to touch; alternate, twice compound, up to 2 ⅜" long; pinnae up to 6 pairs; leaflets narrow, 20 or more pairs per pinna.

Flowers: Tiny, radial, pink, in globose clusters about ⅜" broad.

Fruit: Legumes, called loments, jointed and breaking between the seeds, brown or reddish, up to 2" long.

Bloom Period: Spring, summer, fall.

Distribution: Cameron, Hidalgo, Willacy, and Starr counties.

Comments: This shrub is usually found near wet areas. It is native to the LRGV, but in South Florida and Australia it has become a serious pest. It crowds the edges of wet areas, creating a dense hedge with large, sharp prickles. *Zarza* is the Spanish word for "blackberry bush."

Zarza, Black Mimosa *Mimosa asperata*

Zarza, Black Mimosa *M. asperata* branch with prickles

Fabaceae

Sensitive Brier

Family: Fabaceae (Leguminosae: Mimosoideae)
Scientific Name: *Mimosa latidens* [*Schrankia latidens*]
Habit: Sprawling perennials with many small prickles (rarely without them).
Leaves: Alternate, twice compound, sensitive to touch, up to 2 ¾" long; pinnae up to 3 pairs; leaflets up to 12 pairs per pinna, up to ¼" long.
Flowers: Tiny, radial, pink, in globose clusters up to ¾" broad.
Fruit: Legumes up to 2 ⅜" long, usually with prickles.
Bloom Period: Spring, summer, fall.
Distribution: Cameron, Hidalgo, Willacy, and Starr counties.
Comments: This is distinguished from the other *Mimosa* species by its sprawling habit and the small prickles on the stems and legumes. It is more common in sandy areas.

Sensitive Brier *Mimosa latidens*

Sensitive Brier *M. latidens* fruit

Raspilla

Family: Fabaceae (Leguminosae: Mimosoideae)
Scientific Name: *Mimosa malacophylla*
Habit: Vinelike plants with woody, clambering stems. Stems and leaves armed with curved prickles.
Leaves: Alternate, compound, up to 8" long; pinnae 3–5 pairs.
Flowers: Tiny, radial, white, in globose clusters up to ¾" broad, fragrant.
Fruit: Legumes up to 4" long, curved, breaking between the seeds.
Bloom Period: All seasons following rains.
Distribution: Cameron, Hidalgo, and Starr counties.
Comments: When this species blooms, all the flowers open and are spent in a few days' time. A large plant is spectacular in bloom. This is a "look, don't touch" plant. The curved prickles tend to grab and hold on, ripping clothing and skin. The Spanish common name, translating as "little rasp," is very appropriate. This is a good plant for hiding an unsightly fence. It is a host plant to the Gray Ministreak butterfly.

Raspilla *Mimosa malacophylla*

Raspilla *M. malacophylla* flower cluster

Fabaceae

Powderpuff, Sensitive Plant

Family: Fabaceae (Leguminosae: Mimosoideae)
Scientific Name: *Mimosa strigillosa*
Habit: Sprawling perennials rarely with prickles.
Leaves: Alternate, twice compound, sensitive to touch, up to 4 ¾" long; pinnae up to 6 pairs; leaflets linear, up to 15 pairs per pinna, about ¼" long.
Flowers: Tiny, radial, pink (occasionally white), in globose clusters up to ¾" broad.
Fruit: Legumes hairy, jointed, up to ¾" long.
Bloom Period: Spring, summer.
Distribution: Cameron, Hidalgo, and Willacy counties.
Comments: This is a very attractive ground cover and a source of entertainment for children (as well as adults) with its sensitive leaves which fold up upon being touched.

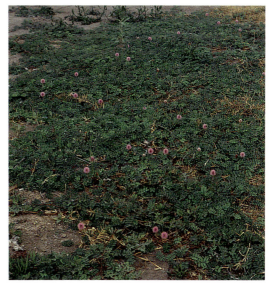

Powderpuff, Sensitive Plant *Mimosa strigillosa*

Powderpuff, Sensitive Plant *M. strigillosa* fruit

Texas Mimosa, Wherry's Mimosa

Family: Fabaceae (Leguminosae: Mimosoideae)
Scientific Name: *Mimosa texana* [*M. wherryana*]
Habit: Shrubs up to 5' or taller, with curved prickles.
Leaves: Alternate, twice compound; pinnae up to 3 pairs; leaflets up to 6 pairs per pinna, about ⅛" long.
Flowers: Tiny, radial, white, in globose clusters.
Fruit: Legumes dark brown, about 1 ⅛" long, with curved prickles on the margins.
Bloom Period: Spring, summer, fall.
Distribution: Starr County
Comments: This species is commonly found growing on hills and bluffs overlooking the Rio Grande in Starr County. The genus name comes from the Greek word *mimos,* meaning "mimic," a reference to the way the leaf closes or collapses when touched. Perhaps the author was reminded of animals like the opossum, which mimic death.

Texas Mimosa, Wherry's Mimosa *Mimosa texana*

Fabaceae

Water Sensitive

Family: Fabaceae (Leguminosae: Mimosoideae)
Scientific Name: *Neptunia plena*
Habit: Terrestrial or semiaquatic with sprawling prostrate stems; in aquatic environment, older stems in the water sometimes covered with a spongy fungal growth; younger stems in water producing adventitious roots.
Leaves: Alternate, twice compound, up to 2½" long; petioles with a gland just below the lowest pair of pinnae (the first divisions of the compound leaf); leaflets without raised veins.
Flowers: Two types in ovoid heads; one section of the head has tiny, radial whitish flowers; the other section has flowers with tiny petals and large, prominent yellow staminodes (sterile petal-like stamens).
Fruit: Legumes about ¾" long and ¼" broad.
Bloom Period: Fall.
Distribution: Kenedy and Cameron counties.
Comments: This species is rare for the flora of Texas. The previous collection was made in October 1938 by Robert Runyon from a temporary pond south of Armstrong, Kenedy County. Ours is the first report of this species since that time (Richardson & King, 2008). Our collection was made on October 21, 2007, in an ephemeral pond 4.8 miles north of the Willacy County border. Another population was later found in Cameron County. It was thought that the species must not be a persistent member of our flora. Probably it persists as seed in dry times, then emerges during our periodic wet seasons, which produce ephemeral ponds in that area. Tropical puff, *N. pubescens*, is also reported for our area. It has flower heads like those of *N. plena*, but its leaves lack the gland below the lowest pair of pinnae.

Water Sensitive *Neptunia plena*

Water Sensitive *N. plena* flower

Water Sensitive *N. plena* leaf gland

Tropical Puff

Family: Fabaceae (Leguminosae: Mimosoideae)
Scientific Name: *Neptunia pubescens*
Habit: Sprawling perennials without prickles or spines. Older stems corky.
Leaves: Alternate, twice compound, sensitive to touch; pinnae 2–6 pairs.
Flowers: Radial, in more or less globose heads with two types of flowers. In the yellow section of the head, the flower petals are tiny but with large yellow staminodes (sterile stamens); in the other whitish section, the petals are tiny and the large staminodes are lacking.
Fruit: Legumes flattened,
Bloom Period: Spring, summer, fall.
Distribution: Cameron, Hidalgo, and Willacy counties.
Comments: This plant seems to prefer moist places. The flower heads are similar to those of the rare *N. plena,* but the staminodes (sterile petal-like stamens) are smaller, and the leaves lack a gland below the lowest pair of pinnae.

Tropical Puff *Neptunia pubescens*

Honey Mesquite

Family: Fabaceae (Leguminosae: Mimosoideae)
Scientific Name: *Prosopis glandulosa*
Habit: Trees or shrubs up to 36' tall, with sharp spines on the young growth; more mature trees often almost without spines.
Leaves: Alternate, twice compound, up to 6" or longer; pinnae usually 1 pair; leaflets 15 or more pairs per pinna, usually about 1 ⅛" long.
Flowers: Tiny, radial, whitish, in cylindrical clusters up to 2" long.
Fruit: Legumes brown or orange, often curved, up to 7" long.
Bloom Period: Spring, summer.
Distribution: Cameron, Hidalgo, Willacy, and Starr counties.
Comments: Honey mesquite makes a beautiful shade tree. The legumes ("beans") are sweet and pleasant tasting. In the past, children would often be seen with a handful of them, chewing them as they would candy. The seeds are not eaten and can be toxic. The legumes are also a source of animal fodder. However, if excessive amounts are eaten by cattle, the high sugar content of the legumes upsets the rumen microbes, and death can result. The seeds are not digested and pass undamaged

Honey Mesquite *Prosopis glandulosa*

Honey Mesquite *P. glandulosa* fruit

Fabaceae

through the digestive tracts of the cattle; seedlings are often seen growing out of a cow patty. Mesquite wood is desirable for lumber, flooring, fence posts, furniture making, and barbecuing.

Tornillo, Screw Bean Mesquite
Family: Fabaceae (Leguminosae: Mimosoideae)
Scientific Name: *Prosopis reptans* var. *cinerascens*
Habit: Little shrubs up to 16" tall, with sharp spines.
Leaves: Alternate, twice compound, up to 5/8" long; pinnae 1 pair; leaflets up to 13 pairs, about 1/8" or slightly longer.
Flowers: Tiny, radial, orange in color, reddish basally, in globose clusters up to 5/8" broad.
Fruit: Legumes orange in color, tightly coiled to resemble the threads of a screw or bolt, up to 1 5/8" long.
Bloom Period: Spring, summer, fall.
Distribution: Cameron, Hidalgo, Willacy, and Starr counties.
Comments: This species is commonly seen growing in saline soils and is reported to be the host plant to the Blue Metalmark butterfly. The Spanish common name comes from the resemblance of the legume to a screw or bolt.

Tornillo, Screw Bean Mesquite *Prosopis reptans* var. *cinerascens*

Joint Vetch
Family: Fabaceae (Leguminosae: Papilionoideae)
Scientific Name: *Aeschynomene indica*
Habit: Herbaceous perennials with erect or leaning stems.
Leaves: Alternate, compound; leaflets 19–63, up to 1/4" long.
Flowers: Strongly bilateral, pea shaped, yellow with reddish stripes, about 3/8" long.
Fruit: Legumes up to 2" long, constricted and breaking between the seeds.
Bloom Period: Spring, summer, fall.
Distribution: Cameron, Hidalgo, and Willacy counties.
Comments: This species usually grows in sand near the coast. One of its major characteristics is the fruit, which breaks into one-seeded segments. This type of fruit is called a loment. We observed the fork-tailed Legume Caterpillar feeding on this plant. Its forked tail flipped up and down like a snake's tongue.

Joint Vetch *Aeschynomene indica*

Small Flower Milk Vetch

Family: Fabaceae (Leguminosae: Papilionoideae)
Scientific Name: *Astragalus nuttallianus* var. *austrinus*
Habit: Annuals, stems running along the ground.
Leaves: Alternate, compound; leaflets 7–19, elliptic, notched apically, up to ½" long.
Flowers: Strongly bilateral, pea shaped; corollas purplish, about ⅜" long.
Fruit: Legumes straight but curved near the base, up to ¾" long.
Bloom Period: Spring.
Distribution: Cameron, Hidalgo, and Starr counties.
Comments: We have five species of *Astragalus* in this area: *A. brazoensis, A. emoryanus, A. leptocarpus, A. nuttallianus,* and *A. reflexus.* Keys must be used to distinguish them. This is one legume that is definitely considered inedible. Almost all species of *Astragalus* contain toxic compounds that are very dangerous for humans if eaten. A well-known common name for the genus is loco weed. Some species contain toxic compounds causing "locoism" in livestock.

Small Flower Milk Vetch *Astragalus nuttallianus* var. *austrinus*

Smooth Stem Wild Indigo

Family: Fabaceae (Leguminosae: Papilionoideae)
Scientific Name: *Baptisia leucophaea* [*B. bracteata*]
Habit: Erect perennials up to 32" tall.
Leaves: Alternate, palmately compound, with 3 obovate or spatulate leaflets up to 2 ⅜" long. Petioles are absent, the 3 leaflets attached directly to the stem.
Flowers: Strongly bilateral, fragrant, pea shaped; corollas light yellow, up to ⅞" long.
Fruit: Legumes brown or black, inflated, up to 2" long.
Bloom Period: Spring.
Distribution: Cameron and Willacy counties.
Comments: This species is found growing in the beach sand dunes. The leaves remain attached after the plant dies, turning whitish or silvery.

Smooth Stem Wild Indigo *Baptisia leucophaea*

Fabaceae

Swordbean
Family: Fabaceae (Leguminosae: Papilionoideae)
Scientific Name: *Canavalia rosea* [*C. maritima*]
Habit: Trailing herbs, occasionally twining.
Leaves: Alternate, compound; leaflets 3, leathery, ovate to roundish, up to 1 ½–4 ¾" long.
Flowers: Strongly bilateral, pea shaped; corollas purplish or lavender; corollas up to 1 ⅛" long.
Fruit: Legumes leathery, up to 6" long.
Bloom Period: Spring.
Distribution: Cameron and Willacy counties.
Comments: Swordbean is widespread in sand on the Gulf Coast and in other warm areas. It makes an excellent ground cover in coastal areas.

Swordbean *Canavalia rosea*

Butterfly Pea
Family: Fabaceae (Leguminosae: Papilionoideae)
Scientific Name: *Centrosema virginianum*
Habit: Perennial twining or trailing vines.
Leaves: Alternate, compound; leaflets 3, variously shaped, up to 2 ⅜" long.
Flowers: Strongly bilateral, pea shaped; corollas purple to lavender, up to 1 ⅛" long.
Fruit: Legumes linear, up to 4 ¾" long.
Bloom Period: Spring, summer, fall.
Distribution: Cameron County.
Comments: The flowers on this vine are bigger than most in the Papilionoideae, and they are very showy.

Butterfly Pea *Centrosema virginianum*

Baby Bonnets
Family: Fabaceae (Leguminosae: Papilionoideae)
Scientific Name: *Coursetia axillaris*
Habit: Shrubs up to 5' tall.
Leaves: Alternate, compound; leaflets up to 10, obovate or spatulate, up to ½" or longer.
Flowers: Strongly bilateral, pea shaped; corollas pink or white, about ⅜" long.
Fruit: Legumes constricted around each seed, up to 1 ⅝" long.
Bloom Period: Spring, summer, fall.
Distribution: Hidalgo County.
Comments: This rare shrub is often cultivated for its beauty. It is a host plant to the Southern Dogface butterfly.

Baby Bonnets *Coursetia axillaris*

Hoary Rattlebox, Shakeshake

Family: Fabaceae (Leguminosae: Papilionoideae)
Scientific Name: *Crotalaria incana*
Habit: Erect annuals 12–40" tall, the younger stems hairy.
Leaves: Alternate, compound; leaflets 3, obovate, up to 1 ⅛" long.
Flowers: Strongly bilateral, pea shaped; corollas yellow, about ⅜" long.
Fruit: Legumes hairy, inflated, up to 1 ⅛" long.
Bloom Period: Spring, summer, fall.
Distribution: Cameron, Hidalgo, and Willacy counties.
Comments: This is a species of sandy soils. The name of the genus is derived from a Greek word meaning "rattle."

Hoary Rattlebox, Shakeshake *Crotalaria incana*

Notched Rattleweed

Family: Fabaceae (Leguminosae: Papilionoideae)
Scientific Name: *Crotalaria retusa*
Habit: Erect annuals up to 36" tall.
Leaves: Alternate, simple, obovate, about 3 ⅛" long.
Flowers: Strongly bilateral, pea shaped; corollas yellow streaked with reddish marks, about 1" long.
Fruit: Legumes round in cross section, about 1 ⅝" long.
Bloom Period: Late summer, fall.
Distribution: Cameron County
Comments: A population of this species is located on South Padre Island. Native to the Asian tropics, it is known in Africa, Australia, Central America, and the Caribbean Islands. Possibly it escaped from cultivation here and became naturalized. The plants, especially the seeds, are poisonous.

Notched Rattleweed *Crotalaria retusa*

Fabaceae

Showy Crotalaria

Family: Fabaceae (Leguminosae: Papilionoideae)
Scientific Name: *Crotalaria spectabilis*
Habit: Erect annuals up to 40" or taller
Leaves: Alternate, simple, obovate, up to 6" or longer.
Flowers: Strongly bilateral, pea shaped; corollas yellow, about 1" long.
Fruit: Legumes up to 2" long.
Bloom Period: Spring, summer, or (mostly) fall.
Distribution: Cameron County.
Comments: This legume was introduced into the United States in 1921 to improve the soil for agriculture. Because the vegetative parts and seeds are poisonous, the use was discontinued. It is cultivated in gardens and sometimes escapes but would not usually be encountered in a natural setting.

Showy Crotalaria *Crotalaria spectabilis*

Golden Dalea

Family: Fabaceae (Leguminosae: Papilionoideae)
Scientific Name: *Dalea aurea*
Habit: Erect perennials with dense reddish hairs.
Leaves: Alternate, compound; leaflets 5–7, up to ⅝" long; glands are sometimes seen on the lower leaf surfaces but are often obscured by the hairs.
Flowers: In tight, terminal inflorescences; corollas strongly bilateral, pea shaped, yellow, about ⅜" long.
Fruit: Legumes hairy, up to ⅜" long.
Bloom Period: Spring, summer, fall.
Distribution: Hidalgo, Willacy, and Starr counties.
Comments: This species is typically found in sandy soils. The genus is named for the English botanist and physician Samuel Dale.

Golden Dalea *Dalea aurea*

Woolly Dalea

Family: Fabaceae (Leguminosae: Papilionoideae)
Scientific Name: *Dalea austrotexana* [*D. lanata*]
Habit: Prostrate white-hairy perennials.
Leaves: Woolly; alternate, compound; leaflets 9–13, up to 3/16" long.
Flowers: In tight, low inflorescences; corollas strongly bilateral, pea shaped, purple, about 3/16" long.
Fruit: Legumes hairy, 1/4" or less in length.
Bloom Period: Summer, fall.
Distribution: Cameron and Starr counties.
Comments: This species is usually found in sandy soils. The glands on the calyxes and lower leaf surfaces are often obscured by the hairs. *D. austrotexana* in our area has for many years been confused with *D. lanata*. In December of 2006, it was described as a new species by B. L. Turner.

Woolly Dalea *Dalea austrotexana*

Prairie Dalea

Family: Fabaceae (Leguminosae: Papilionoideae)
Scientific Name: *Dalea compacta*
Habit: Leaning or erect perennials with large fleshy roots.
Leaves: Hairless, alternate, up to 1" long, compound with 5–7 leaflets, the undersurface dotted with glands.
Flowers: Light purple, in dense cylindrical spikes; longest petals 1/8" or longer.
Fruit: Legumes less than 1/8" long, sometimes hairy.
Bloom Period: Summer.
Distribution: Hidalgo County.
Comments: This species is widespread in north-central Texas but has not previously been reported for our area. It was found growing in western Hidalgo County, in roadside sandy soil and caliche/gravel hillsides. Daleas characteristically have small glands dotted on the lower leaf surfaces. Prairie dalea is distinguished from our other daleas by its broader inflorescence and larger purple flowers. Some daleas, including this species, were previously known as *Petalostemon* or *Petalostemum*.

Prairie Dalea *Dalea compacta*

Fabaceae

Prairie Clover

Family: Fabaceae (Leguminosae: Papilionoideae)
Scientific Name: *Dalea emarginata* [*Petalostemum emarginatum*]
Habit: Hairless weak-stemmed annuals.
Leaves: Alternate, compound; leaflets 13–17, up to ⅜" long.
Flowers: In tight, terminal inflorescences, pleasantly aromatic when handled; corollas strongly bilateral, pea shaped, purple, about 3⁄16" long.
Fruit: Legumes ¼" or less in length.
Bloom Period: Spring, summer, fall.
Distribution: Cameron, Hidalgo, Willacy, and Starr counties.
Comments: This species is widespread, from sand dunes at the coast to clay soils inland. Glands may be easily seen on the calyxes and the lower leaf surfaces.

Prairie Clover *Dalea emarginata*

Dwarf Dalea

Family: Fabaceae (Leguminosae: Papilionoideae)
Scientific Name: *Dalea nana*
Habit: Perennial herbs with reddish hairs; stems leaning.
Leaves: Alternate, compound; leaflets 5–9, about ⅝" long, gland-dotted on the lower surfaces. Sometimes the glands are obscured by hairs.
Flowers: Strongly bilateral, pea shaped; corollas yellow, sometimes fading to rose, about 3⁄16" long.
Fruit: Legumes hairy, about ¼" long.
Bloom Period: Spring.
Distribution: Cameron, Hidalgo, and Starr counties.
Comments: This species is similar to golden dalea, *D. aurea,* but smaller and much lower to the ground.

Dwarf Dalea *Dalea nana*

Pussyfoot

Family: Fabaceae (Leguminosae: Papilionoideae)
Scientific Name: *Dalea obovata* [*Petalostemum obovata*]
Habit: Leaning or sprawling perennials with reddish hairs; glandular but glands often hidden by the hairs.
Leaves: Alternate, compound, leaflets usually 5, up to ¾" long.
Flowers: Strongly bilateral, pea shaped; in dense inflorescences; petals pale yellow or whitish, about 3/16" long.
Fruit: Legumes small, nearly roundish, with only 1 or 2 seeds.
Bloom Period: Spring.
Distribution: Cameron, Hidalgo, and Willacy counties.
Comments: This species is endemic to Texas, growing in sandy soils.

Pussyfoot *Dalea obovata*

Bearded Dalea

Family: Fabaceae (Leguminosae: Papilionoideae)
Scientific Name: *Dalea pogonathera* var. *walkerae*
Habit: Hairless glandular perennial herbs; stems leaning or often flat on the ground.
Leaves: Alternate, compound, glandular on the lower surfaces; leaflets 5–7, up to 5/16" long.
Flowers: Strongly bilateral, pea shaped; corollas purple, about 3/16" long.
Fruit: Legumes small, hairy, hidden by the calyxes.
Bloom Period: Winter, spring, summer.
Distribution: Cameron, Hidalgo, Willacy, and Starr counties.
Comments: This species has 5–7 leaflets per leaf, distinguishing it from prairie clover, *D. emarginata*, which has 13–17 leaflets per leaf.

Bearded Dalea *Dalea pogonathera* var. *walkerae*

Fabaceae

Few-Flowered Climbing Dalea

Family: Fabaceae (Leguminosae: Papilionoideae)
Scientific Name: *Dalea scandens* var. *pauciflora* [*D. thyrsiflora*]
Habit: Erect hairy shrubs up to 48" tall.
Leaves: Alternate, compound, glandular on the lower surfaces; leaflets 5–9, up to ⅝" long.
Flowers: Strongly bilateral, pea shaped; corollas yellow, turning purple, about ⅜" long.
Fruit: Small, hairy, about as long as the sepals.
Bloom Period: All seasons.
Distribution: Cameron and Willacy counties
Comments: This is our only shrubby *Dalea.* It is a host plant to the Southern Dogface butterfly. The crushed leaves are aromatic and leave yellow stains, reminiscent of the leaves of *Pomaria.* We have never seen this plant climbing.

Few-Flowered Climbing Dalea *Dalea scandens* var. *pauciflora*

Few-Flowered Climbing Dalea *D. scandens* var. *pauciflora* lower leaf surface showing the dotted glands

Coral Bean

Family: Fabaceae (Leguminosae: Papilionoideae)
Scientific Name: *Erythrina herbacea*
Habit: Thorny shrubs swollen basally, with many low branches, up to 6' or taller.
Leaves: Alternate, compound; leaflets 3, up to 5 ⅛" long.
Flowers: Strongly bilateral; corollas red to pink, up to 2" long.
Fruit: Legumes up to 9" long, black when mature, constricted between the seeds; seeds red.
Bloom Period: Spring.
Distribution: Cameron County
Comments: Although all parts of the plant are poisonous, it is frequently planted as an ornamental. The red seeds were used as necklace beads by the Indians. The seeds have also been used as rat poison. Hummingbirds utilize this species as a nectar plant.

Coral Bean *Erythrina herbacea*

Texas Kidneywood

Family: Fabaceae (Leguminosae: Papilionoideae)
Scientific Name: *Eysenhardtia texana*
Habit: Shrubs up to 13' tall with airy, upright growth.
Leaves: Alternate, compound; leaflets 15–47, glandular on the lower surfaces, up to ⅜" long.
Flowers: Fragrant, strongly bilateral; corollas white or yellowish, about 3/16" long.
Fruit: Legumes curved, up to ⅜" long, with only 1 seed.
Bloom Period: Spring, summer, fall.
Distribution: Cameron, Hidalgo, and Starr counties.
Comments: Dyes have been made from the wood. The flowers provide nectar for honey production. The crushed leaves are aromatic. This is an excellent butterfly nectar plant. It is a host plant to the Southern Dogface butterfly.

Texas Kidneywood *Eysenhardtia texana*

Hoary Milkpea

Family: Fabaceae (Leguminosae: Papilionoideae)
Scientific Name: *Galactia canescens*
Habit: Perennials with trailing (not twining) stems.
Leaves: Alternate, compound, with 3 ovate to elliptic leaflets up to 2 ⅜" long.
Flowers: Strongly bilateral, pea shaped; corollas pink or purple, about ⅜" long. Cleistogamous (self-pollinating) underground flowers are also produced.
Fruit: Legumes up to 1 ⅝" long. Small underground legumes (like peanuts) also produced.
Bloom Period: Spring, summer, fall.
Distribution: Cameron, Hidalgo, Willacy, and Starr counties.
Comments: This species is endemic to Texas. The trailing stems distinguish it from Texas milkpea, *G. texana,* which has twining stems. The above-ground flowers provide cross pollination. It also produces underground flowers, which are self-pollinated and do not open (cleistogamous). The underground legumes are smaller than those above ground.

Hoary Milkpea *Galactia canescens*

Fabaceae

Margined Milkpea

Family: Fabaceae (Leguminosae: Papilionoideae)
Scientific Name: *Galactia marginalis*
Habit: Weak-stemmed plants producing tubers on the roots.
Leaves: Alternate, simple; blades lanceolate, up to 2 ¾" long.
Flowers: Few-flowered inflorescences, usually one flower opening at a time; corollas purple, strongly bilateral, pea shaped, up to ⅝" long.
Fruit: Legumes straight.
Bloom Period: Spring, summer, fall.
Distribution: Hidalgo County
Comments: This is a new report for the LRGV. This is one of the few legumes with simple rather than compound leaves. The simple leaves plus the single or few purple flowers, combined with the underground tubers, are identifying characteristics of this species. This species likely produces self-pollinating flowers that do not open (cleistogamous) in addition to the showy flowers. On a cultivated plant, we have observed fruit production but no open flowers.

Margined Milkpea *Galactia marginalis*

Texas Milkpea

Family: Fabaceae (Leguminosae: Papilionoideae)
Scientific Name: *Galactia texana*
Habit: Twining perennials.
Leaves: Alternate, compound, with 3 ovate to elliptic leaflets up to 2 ⅜" long.
Flowers: Strongly bilateral, pea shaped; corollas lavender, up to ⅝" long
Fruit: Legumes up to 2 ⅛" long.
Bloom Period: Spring, summer.
Distribution: Cameron County
Comments: This species is endemic to Texas. Our population is disjunct from the Hill Country populations.

Texas Milkpea *Galactia texana*

Scarlet Pea

Family: Fabaceae (Leguminosae: Papilionoideae)
Scientific Name: *Indigofera miniata*
Habit: Prostrate, hairy perennials.
Leaves: Alternate, compound; leaflets up to 9, about 1" long.
Flowers: Strongly bilateral, pea shaped, salmon color, up to ¾" long.
Fruit: Legumes up to 1 ⅛" long.
Bloom Period: Spring, summer, fall.
Distribution: Cameron, Hidalgo, Willacy, and Starr counties.
Comments: This is a host plant to the Cassius Blue, Reakirt's Blue, and Funereal Dusky Wing butterflies. An indigo dye is extracted from some species of *Indigofera;* hence the genus name. Unfortunately, many people assume the flowers are indigo in color because of the name.

Scarlet Pea *Indigofera miniata*

Shrubby Indigo

Family: Fabaceae (Leguminosae: Papilionoideae)
Scientific Name: *Indigofera suffruticosa*
Habit: Erect shrubs up to 6' or taller, woody basally.
Leaves: Alternate, compound, leaflets 9–15, about ⅝" long.
Flowers: Strongly bilateral, pea shaped, salmon color, about ¼" long.
Fruit: Legumes curved, about ⅜" long.
Bloom Period: Summer, fall.
Distribution: Cameron and Brooks counties.
Comments: This species was introduced from Asia because of its importance in producing indigo dye. In Cameron County this species is reported to grow on coastal sands. In Brooks County, just north of Hidalgo County, it grows in inland sands. The flowers are much like those of scarlet pea, *I. miniata,* but they are smaller, and the plants have different growth habits.

Shrubby Indigo *Indigofera suffruticosa*

Fabaceae

Sandy Land Bluebonnet

Family: Fabaceae (Leguminosae: Papilionoideae)
Scientific Name: *Lupinus subcarnosus*
Habit: Annuals with older stems prostrate, the younger portions erect, up to 16" tall.
Leaves: Alternate, hairy, palmately compound; leaflets usually 5, up to 1" long.
Flowers: Strongly bilateral, pea shaped, about ½" long; petals blue, one of them with a white center.
Fruit: Legumes hairy, up to 1 ⅜" long.
Bloom Period: Spring.
Distribution: Brooks County.
Comments: At one time this species was designated as the state flower of Texas. Now all the *Lupinus* species native to Texas are designated as state flower. This species has inflorescences with bluish, rounded tips; our other species, Texas bluebonnet, *L. texensis,* has inflorescences with pointed, white tips. Since *L. subcarnosus* was found growing in sandy soil in Brooks County, it is likely also to occur in Hidalgo County sands. It is endemic to Texas. The plants are poisonous to livestock.

Sandy Land Bluebonnet *Lupinus subcarnosus*

Texas Bluebonnet

Family: Fabaceae (Leguminosae: Papilionoideae)
Scientific Name: *Lupinus texensis*
Habit: Hairy annuals or biennials; older stems prostrate, the younger portions erect, up to 16" tall.
Leaves: Alternate, hairy, compound; leaflets usually 5, up to 1" long.
Flowers: Strongly bilateral, pea shaped, about ½" long; petals blue, one of them with a white center.
Fruit: Legumes hairy, up to 1 ⅜" long.
Bloom Period: Spring.
Distribution: Cameron, Hidalgo, Willacy, and Starr counties.
Comments: This species is endemic to Texas. The inflorescences have pointed, white tips, compared with the sandy land bluebonnet, *L. subcarnosus,* in which inflorescences have rounded, bluish tips. *L. texensis* is the species most often pictured as the state flower, although all the lupines of Texas are designated thus. When the bluebonnet was selected as the state flower, the other two candidates were cotton (because of its commercial value) and prickly pear, which does produce beautiful flowers. Probably most people are happy with the choice of the legislature. Bluebonnets are poisonous to livestock.

Texas Bluebonnet *Lupinus texensis*

Purple Bean

Family: Fabaceae (Leguminosae: Papilionoideae)
Scientific Name: *Macroptilium atropurpureum*
Habit: Perennial twining vines, sometimes hairy.
Leaves: Alternate, compound; leaflets 3, up to 2" long.
Flowers: Strongly bilateral, velvety reddish purple to almost black, up to ¾" long.
Fruit: Legumes up to 3 ½" long.
Bloom Period: Spring.
Distribution: Cameron and Willacy counties.
Comments: The distribution of this species in Texas is unusual. Besides Cameron and Willacy counties, it is found in Presidio County in West Texas, and in Kenedy and Kleberg counties in the Coastal Bend.

Purple Bean *Macroptilium atropurpureum*

Bur Clover

Family: Fabaceae (Leguminosae: Papilionoideae)
Scientific Name: *Medicago polymorpha*
Habit: Erect or sprawling annual herbs.
Leaves: Alternate, compound; leaflets 3, up to ¾" long.
Flowers: Strongly bilateral, pea shaped, yellow, up to 3⁄16" long.
Fruit: Legumes curled up, prickly.
Bloom Period: Spring.
Distribution: Cameron, Hidalgo, and Willacy counties.
Comments: Although the burlike legume makes it easy to recognize this species, it is not easily recognized as a legume because of the legume's coiled appearance. It is an introduced species, native of Eurasia. Handling the plant and then touching the eyes can cause eye irritation.

Bur Clover *Medicago polymorpha*

Fabaceae

White Sweet Clover
Family: Fabaceae (Leguminosae: Papilionoideae)
Scientific Name: *Melilotus albus*
Habit: Erect annual herbs.
Leaves: Alternate, compound; leaflets 3, up to ¾" long.
Flowers: Strongly bilateral, pea shaped, white, up to 3/16" long.
Fruit: Legumes more or less globose, usually with one seed.
Bloom Period: Spring, summer.
Distribution: Cameron, Hidalgo, and Willacy counties.
Comments: This species was introduced from Eurasia and has now become naturalized.

White Sweet Clover *Melilotus albus*

Sour Clover
Family: Fabaceae (Leguminosae: Papilionoideae)
Scientific Name: *Melilotus indicus*
Habit: Erect annuals.
Leaves: Alternate, compound; leaflets 3, up to ¾" long.
Flowers: Strongly bilateral, pea shaped, yellow, about ⅛" long.
Fruit: Legumes more or less globose, usually with one seed.
Bloom Period: Spring, summer.
Distribution: Cameron and Hidalgo counties.
Comments: This species was introduced from the Mediterranean area and has now become naturalized.

Sour Clove *Melilotus indicus*

Round Leaf Scurf Pea
Family: Fabaceae (Leguminosae: Papilionoideae)
Scientific Name: *Pediomelum rhombifolium* [*Psoralea rhombifolia*]
Habit: Trailing perennial vines, somewhat hairy.
Leaves: Alternate, compound; leaflets 3, up to 1 ¼" long, the upper surfaces with tiny glands.
Flowers: Strongly bilateral, pea shaped, reddish orange, up to 5/16" long.
Fruit: Legumes up to ⅜" or longer, enclosed by the calyxes.
Bloom Period: Spring, summer, fall.
Distribution: Cameron, Hidalgo, Willacy, and Starr counties.
Comments: This species is usually found in sandy soils. The specific epithet refers to the more or less rhombic shape of the leaf.

Round Leaf Scurf Pea *Pediomelum rhombifolium*

Snoutbean

Family: Fabaceae (Leguminosae: Papilionoideae)
Scientific Name: *Rhynchosia americana*
Habit: Twining or trailing perennial vines.
Leaves: Alternate, simple; blades kidney shaped or heart shaped, up to 1 ⅛" long.
Flowers: Strongly bilateral, pea shaped, yellow, about ⅜" long.
Fruit: Legumes up to ¾" long.
Bloom Period: Spring, summer, fall.
Distribution: Cameron, Hidalgo, Willacy, and Starr counties.
Comments: This is one of the rare members of the family with simple leaves.

Snoutbean *Rhynchosia americana*

Least Snout Bean

Family: Fabaceae (Leguminosae: Papilionoideae)
Scientific Name: *Rhynchosia minima*
Habit: Twining or trailing perennial vines.
Leaves: Alternate, compound; leaflets 3, up to 1" or slightly longer.
Flowers: Strongly bilateral, pea shaped, yellow, ¼" or slightly longer.
Fruit: Legumes curved, up to ⅝" long.
Bloom Period: All seasons.
Distribution: Cameron and Hidalgo counties.
Comments: The seeds are eaten by birds and other animals.

Least Snout Bean *Rhynchosia minima*

Drummond's Rattlebush

Family: Fabaceae (Leguminosae: Papilionoideae)
Scientific Name: *Sesbania drummondii*
Habit: Shrubs up to 9' or taller.
Leaves: Alternate, compound; leaflets as many as 50, up to 1" long.
Flowers: Strongly bilateral, pea shaped, yellow, often with red lines, up to ⅝" long.
Fruit: Legumes up to 3" long, slightly constricted between the seeds, squarish in cross section, with a narrow "wing" at each corner.
Bloom Period: Spring, summer.
Distribution: Cameron, Hidalgo, Willacy, and Starr counties.
Comments: *S. drummondii* was named in honor of Thomas Drummond, who immigrated from Scotland and made significant contributions to the knowledge of Texas plants. This species is especially found near water. The winged legumes distinguish this species from bequilla,

Drummond's Rattlebush *Sesbania drummondii*

S. herbacea, with its linear legumes. The seeds are toxic to livestock.

Fabaceae

Bequilla

Family: Fabaceae (Leguminosae: Papilionoideae)
Scientific Name: *Sesbania herbacea*
  [*S. exaltata, S. macrocarpa, Emerus herbacea*]
Habit: Annuals, sometimes woody, up to 13' tall.
Leaves: Alternate, compound; leaflets as many as 70, up to 1" long.
Flowers: Strongly bilateral, pea shaped, yellow, often with red lines, up to ⅝" long.
Fruit: Legumes linear, up to 7 ½" long.
Bloom Period: Summer, fall.
Distribution: Cameron, Hidalgo, and Starr counties.
Comments: The plants usually grow near water, sometimes standing in water. A water fungus sometimes grows on the plant parts that are immersed in the water. The linear legumes of this species distinguish it from Drummond's rattlebush, *S. drummondii,* with its winged legumes.

Bequilla *Sesbania herbacea*

Bequilla *S. herbacea* plant base covered with fungal growth

Texas Mountain Laurel, Mescal Bean

Family: Fabaceae (Leguminosae: Papilionoideae)
Scientific Name: *Sophora secundiflora*
Habit: Evergreen shrubs up to 11' or slightly taller.
Leaves: Alternate, compound; leaflets about 11, stiff, about 2 ⅜" long.
Flowers: Strongly bilateral, pea shaped, bluish violet, about ¾" long, in clusters.
Fruit: Legumes up to 4 ¾" long, slightly constricted between the seeds. Seeds red.
Bloom Period: Spring.
Distribution: Cameron and Hidalgo counties.
Comments: The shrub in bloom is very fragrant. The odor, resembling that of grape juice, can be detected some distance from the shrub. However, a spray of flowers brought into the house will soon have to be removed because the fragrance, although pleasant, is too strong for indoors. Children used to trick friends by rubbing the seeds on concrete until they would get very hot, and then touching the seed to an arm or leg. This shrub or small tree is often cultivated for its beauty and fragrance. The leaves and seeds are toxic. The seeds were used as trade articles by Native Americans. A powder from the seeds, in a very small amount, was added to mescal to produce intoxication, delirium, and sleep. The original populations of this species may have been introduced into our area through trade and other traditional uses.

Texas Mountain Laurel, Mescal Bean *Sophora secundiflora*

Yellow Sophora, Necklace Pod

Family: Fabaceae (Leguminosae: Papilionoideae)
Scientific Name: *Sophora tomentosa* var. *occidentalis*
Habit: Hairy shrubs up to 6 ½' tall.
Leaves: Alternate, compound, hairy; leaflets about 19, up to 1 ⅝" long.
Flowers: Strongly bilateral, pea shaped, yellow, up to ⅞" long.
Fruit: Legumes up to 5 ½" long, constricted between the seeds, resembling a short string of beads.
Bloom Period: Spring, summer, fall.
Distribution: Cameron and Willacy counties.
Comments: In its natural habitat in coastal sands, this shrub dies back severely during drought. When rains come, it quickly begins to grow.

Yellow Sophora, Necklace Pod *Sophora tomentosa* var. *occidentalis*

When cultivated and watered regularly, it remains attractive, with the hairy leaves and the yellow flowers. In the garden it responds well to a severe pruning every two or three years.

Fabaceae

Sticky Pencilflower

Family: Fabaceae (Leguminosae: Papilionoideae)
Scientific Name: *Stylosanthes viscosa*
Habit: Perennial, prostrate herbs.
Leaves: Alternate, compound, with glandular hairs; leaflets 3, up to 1" or longer.
Flowers: Strongly bilateral, pea shaped, orange-yellow, up to ⅜" long.
Fruit: Legumes flattened, about ¼" long, usually one-seeded.
Bloom Period: Spring, summer, early fall.
Distribution: Willacy County.
Comments: This species grows along the coast but is also found growing in inland sandy areas.

Sticky Pencilflower *Stylosanthes viscosa*

Hoary Pea

Family: Fabaceae (Leguminosae: Papilionoideae)
Scientific Name: *Tephrosia lindheimeri*
Habit: Sprawling hairy perennials, sometimes woody basally.
Leaves: Alternate, compound, very hairy; leaflets as many as 15, up to 1 ¾" long, the margins silver colored.
Flowers: Strongly bilateral, pea shaped, purple, up to ⅝" long.
Fruit: Legumes flattened, up to 2" long.
Bloom Period: Spring, summer.
Distribution: Hidalgo and Willacy counties.
Comments: This showy species is endemic to Texas. It grows in sandy areas. The crushed leaves, stems, and roots (containing rotenone) were used as fish poison. The genus name comes from a Greek word meaning "ash-colored," referring to the gray-hairy leaves. This species was named in honor of Ferdinand Lindheimer, a German-born botanist who contributed greatly to the knowledge of Texas plants.

Hoary Pea *Tephrosia lindheimeri*

Slim Vetch

Family: Fabaceae (Leguminosae: Papilionoideae)
Scientific Name: *Vicia ludoviciana*
Habit: Herbaceous annual vines.
Leaves: Alternate, compound, with a terminal tendril; leaflets as many as 12, up to 5/8" long.
Flowers: Strongly bilateral, pea shaped, blue, up to 5/16" long.
Fruit: Legumes flattened, up to 3/4" long.
Bloom Period: Spring.
Distribution: Cameron, Hidalgo, and Willacy counties.
Comments: This vigorous little vine is an untidy nuisance in lawns or growing on fences, but it disappears with warm weather. The tendril at the leaf tip is an identifying characteristic.

Slim Vetch *Vicia ludoviciana*

Snail Vine

Family: Fabaceae (Leguminosae: Papilionoideae)
Scientific Name: *Vigna caracalla* [*Phaseolus caracalla*]
Habit: Twining perennial vines.
Leaves: Alternate, compound; leaflets 3, ovate to roundish, up to 2½" long.
Flowers: Strongly bilateral, violet color, sometimes yellow, about 1 ¼" long
Fruit: Legumes narrow, 4" or longer.
Bloom Period: All seasons.
Distribution: Cameron County.
Comments: This species is native to tropical America and is often cultivated. It has escaped and become naturalized in our area. A population was found in Brownsville growing along a resaca bank. The specific epithet and the common name come from the flower's resemblance to a snail.

Snail Vine *Vigna caracalla*

Fabaceae

Yellow Cowpea

Family: Fabaceae (Leguminosae: Papilionoideae)
Scientific Name: *Vigna luteola*
Habit: Perennial twining vines.
Leaves: Alternate, compound; leaflets 3, up to 2 ¾" long.
Flowers: Strongly bilateral, pea shaped, yellow, up to ¾" long.
Fruit: Legumes flattened, up to 2" long.
Bloom Period: All seasons.
Distribution: Cameron, Hidalgo, and Starr counties.
Comments: Yellow cowpea is frequently encountered near water. The specific epithet *luteola* comes from a Latin word meaning "yellow."

Yellow Cowpea *Vigna luteola*

Yellow Cowpea *V. luteola* flower

Bracted Zornia

Family: Fabaceae (Leguminosae: Papilionoideae)
Scientific Name: *Zornia bracteata*
Habit: Sprawling or prostrate herbs.
Leaves: Alternate, compound; leaflets 4, up to 1 ⅛" long.
Flowers: Strongly bilateral, pea shaped, yellow, up to ½" or longer.
Fruit: Legumes spiny, about ⅝" long, constricted between the seeds and breaking apart at maturity.
Bloom Period: Spring, fall.
Distribution: Hidalgo, Willacy, and Starr counties.
Comments: This species prefers sandy soils. The 4 leaflets are a major characteristic, since few legumes have 4 leaflets.

Bracted Zornia *Zornia bracteata*

Net Leaf Rabbit's Ears

Family: Fabaceae (Leguminosae: Papilionoideae)
Scientific Name: *Zornia reticulata*
Habit: Prostrate or sprawling herbs.
Leaves: Alternate, compound; leaflets 2, up to 1 ⅛" long.
Flowers: Strongly bilateral, pea shaped, yellow, up to ½" or longer.
Fruit: Legumes not spiny, about ⅝" long, constricted between the seeds and breaking apart at maturity.
Bloom Period: Spring, fall.
Distribution: Willacy County.
Comments: This is a plant of sandy soils. The 2 leaflets are a major characteristic. Very few legumes have 2 leaflets.

Net Leaf Rabbit's Ears *Zornia reticulata*

Live Oak, Encino

Family: Fagaceae
Scientific Name: *Quercus virginiana*
Habit: Evergreen trees up to 40' or taller.
Leaves: Alternate, simple; blades stiff, sometimes toothed, up to 3 ½" long.
Flowers: Tiny, unisexual, on separate spikes. The male spikes are called catkins.
Fruit: Acorns (a nut in a cup of scales).
Bloom Period: Spring.
Distribution: Cameron, Hidalgo, Willacy, and Starr counties.
Comments: Live oak is more common in sandy soils from Kenedy County and northward. Although the trees drop their old leaves annually, they remain evergreen by dropping them in the spring as the new leaves are emerging. Mature trees produce young trees from their roots, often forming colonies called mottes. The trees are commonly cultivated, making it difficult to determine the original native range. The acorns are a food source for white-tailed deer, javelinas, fox squirrels, and turkeys.

Live Oak, Encino *Quercus virginiana*

Live Oak, Encino *Q. virginiana* inflorescences (catkins)

Brush Holly

Family: Flacourtiaceae
Scientific Name: *Xylosma flexuosa*
Habit: Thorny shrubs up to 6' or taller.
Leaves: Alternate or in clusters, simple; blades ovate and toothed, up to 2 ½" long.
Flowers: Very small, in clusters in the leaf axils. Sometimes male and female flowers are on separate plants.
Fruit: Drupes, red to black when ripe, about 3/16" broad.
Bloom Period: Spring, summer, fall.
Distribution: Cameron, Hidalgo, and Starr counties.
Comments: The plants resemble holly, but they only occasionally produce the large quantities of fruit that are seen on holly bushes. The fruit are a food source for birds.

Brush Holly *Xylosma flexuosa*

Fumariaceae

Brush Holly *X. flexuosa* plant in flower

Johnston's Frankenia

Family: Frankeniaceae
Scientific Name: *Frankenia johnstonii*
Habit: Gray-hairy shrubs 12" to 24" tall.
Leaves: Opposite, hairy; blades more or less oblanceolate, about ½" long, the margins curled under.
Flowers: Solitary and sessile, radial; petals 5 or 6, white, about 5⁄16" long.
Fruit: A small capsule, retained inside the calyx.
Bloom Period: Fall, winter, spring.
Distribution: Starr County.
Comments: This species is listed as endangered by the state and federal governments. More populations have been found since that designation (Poole et al. 2007), and population sizes can number in the thousands. The future of Johnston's frankenia appears to be more secure than it was formerly thought to be.

Johnston's Frankenia *Frankenia johnstonii*

Scrambled Eggs

Family: Fumariaceae
Scientific Name: *Corydalis micrantha* subsp. *texensis*
Habit: Erect annual herbs up to 12" tall.
Leaves: Alternate; blades deeply divided, up to 4 3⁄8" long.
Flowers: Bilateral, with a spur, yellow, up to 5⁄8" long.
Fruit: Capsules up to 1 1⁄8" long.
Bloom Period: Spring.
Distribution: Hidalgo, Willacy, and Starr counties.
Comments: This is a fairly widespread cool-weather species. It is endemic to Texas. The subspecies *texensis* is known from Bexar County to the LRGV.

Scrambled Eggs *Corydalis micrantha* subsp. *texensis*

Fine Leaf Fumitory

Family: Fumariaceae
Scientific Name: *Fumaria parviflora*
Habit: Annuals with stems leaning or sprawling, older plants somewhat rounded.
Leaves: Alternate; blades deeply divided, the segments very narrow, the sides curving upward producing tiny channels.
Flowers: Bilateral, with a spur; white with tiny dark tips, up to ¼" long.
Fruit: Capsules roundish, often with a tiny point at the apex, about ¹⁄₁₆" or slightly longer.
Bloom Period: Spring.
Distribution: Cameron and Hidalgo counties.
Comments: This species was introduced from Europe. Although the flowers are tiny, the plants are attractive when in full bloom. The name of the species comes from the combination of two Latin words for "small" and "flower." Another species, *F. officinalis,* occurs in our area. Its flowers are purple.

Fine Leaf Fumitory *Fumaria parviflora*

Rosita

Family: Gentianaceae
Scientific Name: *Centaurium breviflorum*
Habit: Annuals up to 20" tall.
Leaves: Opposite, simple, without petioles; blades thick, lanceolate, up to 2 ⅜" long.
Flowers: Radial; petals 5, pink, fused basally; lobes about ⁵⁄₁₆" long.
Fruit: Capsules about ⁵⁄₁₆" long.
Bloom Period: Spring, summer.
Distribution: Cameron and Hidalgo counties.
Comments: This family is easy to recognize in our area. All our species have opposite leaves without petioles, and radial flowers with 5 pink or bluish lobes. *Rosita* is a diminutive of the Spanish word *rosa,* which translates as "pink."

Rosita *Centaurium breviflorum*

Gentianaceae

Bluebell Gentian

Family: Gentianaceae
Scientific Name: *Eustoma exaltatum*
Habit: Annuals up to 30" tall.
Leaves: Opposite, simple, without petioles; blades thick, more or less ovate, up to 3 ½" long.
Flowers: Radial, various bluish shades; petals 5, fused basally; lobes about ⅝" long.
Fruit: Capsules about ⅝" long.
Bloom Period: Spring, summer.
Distribution: Cameron, Hidalgo, Willacy, and Starr counties.
Comments: This showy species appears in profusion in low places with spring rains. In drier years, the plants are somewhat rare.

Bluebell Gentian *Eustoma exaltatum*

Saltmarsh Pink

Family: Gentianaceae
Scientific Name: *Sabatia arenicola*
Habit: Erect annuals, usually of sandy soils, up to 8" tall.
Leaves: Opposite, simple, without petioles, not clasping; blades thick, elliptic or ovate, up to ⅞" long.
Flowers: Radial, pink or sometimes white; petals 5, fused basally; lobes about ½" long, shorter than the calyx lobes.
Fruit: Capsules up to ⅜" long.
Bloom Period: Spring, summer.
Distribution: Cameron, Hidalgo, and Willacy counties.
Comments: This species is very similar to the meadow pink, *S. campestris,* which has corolla lobes equal to or longer than the calyx lobes and leaves that are broadest at the base and clasp the stem.

Saltmarsh Pink *Sabatia arenicola*

Meadow Pink

Family: Gentianaceae

Scientific Name: *Sabatia campestris*

Habit: Erect annuals, usually of sandy soils, up to 16" tall.

Leaves: Opposite, simple, without petioles, clasping the stem; blades usually broadest at or near the base, up to 1 ¾" long.

Flowers: Radial, pink or sometimes white, with yellow centers; lobes up to ⅞" long, equal in length to or longer than the calyx lobes.

Fruit: Capsules up to ⅜" long.

Bloom Period: Spring, summer, fall.

Distribution: Cameron, Hidalgo, and Willacy counties.

Comments: This species is very similar to the saltmarsh pink, *S. arenicola,* which has corolla lobes shorter than the calyx lobes and leaves that are broadest above the base and do not clasp the stem.

Meadow Pink *Sabatia campestris*

Red Stem Stork's Bill

Family: Geraniaceae

Scientific Name: *Erodium cicutarium*

Habit: Mostly prostrate-growing annuals or biennials.

Leaves: Alternate; blades compound and variously lobed, ovate in outline, up to 1 ⅜" long.

Flowers: Radial, about ¼" broad; petals 5, lavender.

Fruit: Dry, the seeds about ⅛" long; styles remain attached, growing to about 2" long, giving the "stork's bill" appearance.

Bloom Period: Mostly spring, also summer.

Distribution: Hidalgo County.

Comments: *Erodium,* which usually has 5 stamens, is very similar to *Geranium,* which usually has 10 stamens. Our species are not showy like the garden pelargoniums that are often called geraniums. The name of the genus comes from the Greek *erodios,* meaning "heron."

Red Stem Stork's Bill *Erodium cicutarium*

Geraniaceae

Carolina Crane's Bill
Family: Geraniaceae
Scientific Name: *Geranium carolinianum*
Habit: Herbaceous hairy annuals; stems leaning or falling over; stem hairs, at least some of them, pointing outward.
Leaves: Alternate; blades circular or kidney shaped in outline, up to 2 ¾" broad, lobed.
Flowers: Radial; petals 5, pink or whitish, about ¼" long.
Fruit: Separating into 5 one-seeded segments, the long style persisting and hardening, forming a long "bill."
Bloom Period: Spring.
Distribution: Cameron County.
Comments: This species, with stem hairs pointing outward, is similar to Texas Geranium, *G. texanum*, which has stem hairs flattened against the stem, pointing downward. The genus name comes from the Greek *geranios*, meaning "crane."

Carolina Crane's Bill *Geranium carolinianum*

Texas Geranium
Family: Geraniaceae
Scientific Name: *Geranium texanum*
Habit: Herbaceous hairy annuals; stems leaning or falling over; hairs more or less flattened against the stem and pointing downward.
Leaves: Alternate; blades circular in outline, up to 1 ⅝" broad, lobed.
Flowers: Radial; petals 5, white tinged with purple, about 3⁄16" long.
Fruit: Separating into 5 one-seeded segments, the long style persisting and hardening, forming a long "bill."
Bloom Period: Spring.
Distribution: Cameron and Hidalgo counties.
Comments: This species is distinguished from Carolina crane's bill, *G. carolinianum,* by its stem hairs, which are more or less flattened against the stem and pointing downward.

Texas Geranium *Geranium texanum*

Sandbell

Family: Hydrophyllaceae
Scientific Name: *Nama hispidum*
Habit: Hairy annuals up to 12" tall, the stems often falling over.
Leaves: Alternate (rarely opposite), various shapes from spatulate to linear, varying from ¼" to 2" long.
Flowers: Radial, funnel shaped; petals united basally, pink to purple, up to ⅝" long.
Fruit: Capsules roundish, about ¼" or shorter.
Bloom Period: Spring.
Distribution: Cameron, Hidalgo, and Starr counties.
Comments: We have five species of *Nama* (*N. stenocarpum* is not illustrated). *N. hispidum* is recognized by its hairiness, alternate leaves with short pedicels, purple or pink flowers, and leaves with smooth margins. When rains are abundant during the growing season, the ground can be covered by beautiful purple mounds of *N. hispidum*.

Sandbell *Nama hispidum*

Fiddleleaf, Jamaican Weed

Family: Hydrophyllaceae
Scientific Name: *Nama jamaicense*
Habit: Annuals; stems prostrate or leaning upward.
Leaves: Alternate, hairy, more or less obovate, up to 2" long.
Flowers: Radial, tubelike; petals white, united basally, up to ⁵⁄₁₆" long.
Fruit: Capsules roundish, about ¼" or shorter.
Bloom Period: Spring, summer.
Distribution: Cameron, Hidalgo, Willacy, and Starr counties.
Comments: *N. jamaicense* can be distinguished from our other four species of *Nama* by its white flowers. (*N. stenocarpum* is not illustrated.)

Fiddleleaf, Jamaican Weed *Nama jamaicense*

Hydrophyllaceae

Small-Leaf Nama

Family: Hydrophyllaceae
Scientific Name: *Nama parvifolium*
Habit: Prostrate branching perennials; stems slender, minutely hairy
Leaves: Alternate, sometimes opposite apically; blades hairy, ovate to elliptic, up to 1 ⅛" long, the margins smooth.
Flowers: Single or in pairs from the leaf axils; pedicels up to ⅜" long or absent; corollas purple to pink, radial, funnel shaped, less than ½" long.
Fruit: Capsules elliptical, up to 3/16" long.
Bloom Period: Spring, summer, fall.
Distribution: Cameron, Hidalgo, and Starr counties.
Comments: This can be distinguished from our other four *Nama* species (*N. stenocarpum* is not shown) by its free sepals and its longest pedicels, more than 3/16" long. In Texas it occurs mostly in counties along the Rio Grande from Cameron County to Val Verde County. It is abundant in Mexico.

Small-Leaf Nama *Nama parvifolium*

Wavy Leaf Nama

Family: Hydrophyllaceae
Scientific Name: *Nama undulatum*
Habit: Annuals; stems prostrate or leaning upward.
Leaves: Alternate; blades more or less spatulate, up to 1 ½" long, the margins undulate (wavy, up and down).
Flowers: Radial, funnel shaped; petals pink, united basally, up to ⅜" long.
Fruit: Capsules roundish, about ¼" or shorter.
Bloom Period: Spring, summer.
Distribution: Cameron and Hidalgo counties.
Comments: *N. undulatum* can be distinguished from our other four *Nama* species by its wavy leaf margins, as its name suggests, and by its pink flowers. (*N. stenocarpum* is not shown.)

Wavy Leaf Nama *Nama undulatum*

Blue Curls

Family: Hydrophyllaceae
Scientific Name: *Phacelia congesta*
Habit: Annuals 14" or taller; stems glandular and sticky.
Leaves: Alternate; blades toothed and lobed, occasionally compound, up to 2 ⅜" long.
Flowers: Inflorescences branching; flowers in one-sided coils, bluish to white, up to ⁵⁄₁₆" long.
Fruit: Capsules roundish, about ⅛" tall.
Bloom Period: Spring, summer.
Distribution: Cameron, Hidalgo, and Starr counties.
Comments: This species can be distinguished from South Texas sand scorpionweed, *P. patuliflora,* by its glandular stems and its branching inflorescences. It grows in sandy soils.

Blue Curls *Phacelia congesta*

Blue Curls *P. congesta* white form

South Texas Sand Scorpionweed

Family: Hydrophyllaceae
Scientific Name: *Phacelia patuliflora* var. *austrotexana*
Habit: Annuals up to 12" tall; stems erect or leaning.
Leaves: Alternate; blades ovate, toothed and usually lobed, up to 3 ⅛" long.
Flowers: Inflorescences not branching; flowers in one-sided coils, pinkish to purple with white centers, about 1" broad.
Fruit: Capsules roundish, about ¼" broad.
Bloom Period: Spring.
Distribution: Cameron, Hidalgo, and Willacy counties.
Comments: This species can be distinguished from blue curls, *P. congesta,* by its unbranched inflorescences and its lack of sticky glands on the stems.

South Texas Sand Scorpionweed *Phacelia patuliflora* var. *austrotexana*

Krameriaceae

Prairie Bur

Family: Krameriaceae
Scientific Name: *Krameria lanceolata*
Habit: Trailing or prostrate, hairy perennials.
Leaves: Alternate, linear, up to ¾" long.
Flowers: Bilateral, borne singly from the leaf axils; petals 5, purple, unequal, up to 5/16" long.
Fruit: Roundish, spiny, up to 5/16" broad.
Bloom Period: Spring.
Distribution: Hidalgo and Starr counties.
Comments: This species is fairly widespread in Texas but uncommon in the LRGV.

Prairie Bur *Krameria lanceolata*

Prairie Bur *K. lanceolata* spiny fruit

Calderona

Family: Krameriaceae
Scientific Name: *Krameria ramosissima*
Habit: Shrubs up to 40" tall.
Leaves: Alternate or in clusters, gray-hairy, linear, about ¼" long.
Flowers: Bilateral; petals 5, unequal, 3 purplish and 2 greenish, about 5/16" long.
Fruit: Globose with prickles, less than ⅛" long.
Bloom Period: Spring, summer.
Distribution: Hidalgo and Starr counties.
Comments: In Texas this species is found in the counties bordering Mexico, from western Hidalgo County to Val Verde County. It frequently provides a protected "nursery" area for seedling cacti.

Calderona *Krameria ramosissima*

Sand Brazoria

Family: Lamiaceae (Labiatae)
Scientific Name: *Brazoria arenaria*
Habit: Mostly erect herbs up to 16" tall.
Leaves: Opposite, basal; blades usually ovate, up to 3 1/8" long.
Flowers: Bilateral; calyxes 2-lipped; corollas purplish, 2-lipped, about 3/4" long.
Fruit: 4 nutlets.
Bloom Period: Spring.
Distribution: Hidalgo County
Comments: This is among the most attractive LRGV wildflowers. It is endemic to Texas, growing in sandy soils. The flowers sometimes resemble those of *Scutellaria,* but the calyxes are different. In *Brazoria,* the upper calyxes are smooth; in *Scutellaria* the upper calyxes have a noticeable lip about halfway to the end.

Sand Brazoria *Brazoria arenaria*

Mock Pennyroyal, Desert Mint

Family: Lamiaceae (Labiatae)
Scientific Name: *Hedeoma drummondii*
Habit: Erect or leaning annual herbs up to 24" tall.
Leaves: Opposite; blades usually elliptic, 5–7" or longer
Flowers: Bilateral; corollas white to bluish, 2-lipped, up to 5/8" long, usually less.
Fruit: 4 nutlets.
Bloom Period: Spring, summer, fall.
Distribution: Hidalgo and Starr counties.
Comments: The genus name comes from two Greek words meaning "sweet" and "scent," referring to the aromatic leaves. The species was named in honor of Thomas Drummond, a well-known botanist in the nineteenth century. It grows in well-drained gypseous, gravelly, sandy, or caliche soils.

Mock Pennyroyal, Desert Mint *Hedeoma drummondii*

Lamiaceae

Henbit

Family: Lamiaceae (Labiatae)
Scientific Name: *Lamium amplexicaule*
Habit: Annuals with scattered hairs, branching from the bases and lower stems; stems erect, leaning, or lying down.
Leaves: Opposite; blades roundish and crinkled, up to 1" broad, the lower ones with petioles, and the upper ones with very short petioles or none.
Flowers: Bilateral; corollas purple or red, 2-lipped, up to ⅝" long. In late fall and winter, tiny self-pollinating flowers are produced.
Fruit: 4 nutlets
Bloom Period: Mostly spring.
Distribution: Cameron and Hidalgo counties.
Comments: Henbit was introduced from Europe and has become naturalized throughout much of North America. It usually is found in lawns and other cultivated areas, disturbed places, and along roadsides.

Henbit *Lamium amplexicaule*

Browne's Savory, Lorna's Savory

Family: Lamiaceae (Labiatae)
Scientific Name: *Micromeria brownei* var. *pilosiuscula*
Habit: Weak, sprawling herbs forming mats.
Leaves: Opposite; blades ovate to elliptic, up to ⅝" long; margins minutely toothed.
Flowers: Bilateral; corollas 2-lipped, lavender (sometimes very pale), about ⁵⁄₁₆" long.
Fruit: 4 nutlets.
Bloom Period: Spring, summer.
Distribution: Cameron and Hidalgo counties.
Comments: This species usually grows near water, or in low places that periodically retain water and slowly dry out. It is being used as an ornamental in wet areas. The common name, Lorna's Savory, was given for one of the daughters of Mike Heep, a local native plant nurseryman who introduced *M. brownei* as an ornamental.

Browne's Savory, Lorna's Savory *Micromeria brownei* var. *pilosiuscula*

Horsemint

Family: Lamiaceae (Labiatae)
Scientific Name: *Monarda citriodora*
Habit: Erect annual herbs up to 32" tall, with a lemon odor.
Leaves: Opposite; blades usually lanceolate, up to 2 1/8" long; margins usually with fine teeth.
Flowers: In dense axillary clusters; corollas bilateral, 2-lipped, white to pink, usually dotted purple, up to 3/4" long.
Fruit: 4 nutlets.
Bloom Period: Spring, summer, fall.
Distribution: Cameron and Hidalgo counties.
Comments: We have two other species of *Monarda*. Shrubby beebalm, *M. fruticulosa*, is a shrub, whereas *M. punctata* is herbaceous. Plume tooth beebalm, *M. citriodora*, has needlelike calyx teeth; *M. punctata* has narrowly triangular calyx teeth. The leaves are sometimes dried and used as an insect repellant. Their effectiveness is not yet established.

Horsemint *Monarda citriodora*

Shrubby Beebalm

Family: Lamiaceae (Labiatae)
Scientific Name: *Monarda fruticulosa*
Habit: Shrubs up to 26" tall.
Leaves: Opposite; blades linear, up to 1 3/8" long.
Flowers: In dense axillary clusters; corollas bilateral, 2-lipped, white or pinkish, up to 1/2" or longer.
Fruit: 4 nutlets.
Bloom Period: Spring, summer, fall.
Distribution: Hidalgo and Starr counties.
Comments: This species grows in sandy soils. The woodiness and the linear leaves separate this species from our other two species of *Monarda*.

Shrubby Beebalm *Monarda fruticulosa*

Lamiaceae

Plume Tooth Beebalm

Family: Lamiaceae (Labiatae)
Scientific Name: *Monarda punctata* var. *lasiodonta*
Habit: Erect annuals or short-lived perennials up to 39" tall.
Leaves: Opposite; blades lanceolate, up to 1 ¾" long.
Flowers: In dense axillary clusters; corollas bilateral, 2-lipped, white or cream color, up to ¾" long.
Fruit: 4 nutlets.
Bloom Period: Spring, summer, fall.
Distribution: Cameron and Willacy counties.
Comments: The herbaceous nature and the narrowly triangular calyx teeth separate this species from our other two species. Two varieties are recognized, var. *punctata* and var. *lasiodonta,* based on corolla color (var. *punctata* has pale yellow corollas) and hairiness of the leaves.

Plume Tooth Beebalm *Monarda punctata* var. *lasiodonta*

Shrubby Blue Sage, Mejorana

Family: Lamiaceae (Labiatae)
Scientific Name: *Salvia ballotiflora*
Habit: Shrubs up to 8' tall, usually less, with brittle stems.
Leaves: Opposite, aromatic; blades ovate, up to 1" long, the margins toothed.
Flowers: Bilateral; corollas blue, rarely white, 2-lipped, up to ½" long.
Fruit: 4 nutlets.
Bloom Period: Spring, summer, fall.
Distribution: Cameron, Hidalgo, Willacy, and Starr counties.
Comments: This would make an excellent shrub for the garden. Like many members of the family, this species has aromatic leaves. The Spanish common name translates as "sweet marjoram."

Shrubby Blue Sage, Mejorana *Salvia ballotiflora*

Shrubby Blue Sage, Mejorana *S. ballotiflora* white-flowering form

Red Sage, Scarlet Sage

Family: Lamiaceae (Labiatae)
Scientific Name: *Salvia coccinea*
Habit: Herbs up to 39" tall, usually less.
Leaves: Opposite, aromatic; blades ovate, up to 2 ⅜" long, the margins toothed.
Flowers: Bilateral; corollas red, 2-lipped, up to 1" long.
Fruit: 4 nutlets.
Bloom Period: All seasons.
Distribution: Cameron, Hidalgo, Willacy, and Starr counties.
Comments: Other color forms, pink or white, have been reported. This species reseeds itself and would make an excellent addition to the natural garden. It grows in full sun and also partial shade and tolerates clay soils. This is a good nectar plant for hummingbirds and butterflies.

Red Sage, Scarlet Sage *Salvia coccinea*

Blue Creeping Sage, River Sage

Family: Lamiaceae (Labiatae)
Scientific Name: *Salvia misella*
Habit: Prostrate, creeping herbs, sometimes rooting at the nodes.
Leaves: Opposite; blades more or less ovate, 1" or longer, the margins toothed.
Flowers: Bilateral, 2-lipped; corollas blue, occasionally white, about ¼" long.
Fruit: 4 nutlets.
Bloom Period: All seasons.
Distribution: Cameron and Hidalgo counties.
Comments: This is a Mexican species that was first discovered in this area by Mike Heep. It was growing in eastern Hidalgo County between Weslaco and Mercedes. Whether it has been growing here naturally or was introduced is not known. Because of the native plant trade, it is becoming more widespread in the LRGV. It spreads rapidly in well-watered gardens and can become a pest. Judicious watering is needed to keep it under control. The common name selena is sometimes used, but it is easy to confuse that name with *Silene* (Caryophyllaceae), or *Selenia* (Brassicaceae).

Blue Creeping Sage, River Sage *Salvia misella*

Lamiaceae

Runyon's Skullcap

Family: Lamiaceae (Labiatae)
Scientific Name: *Scutellaria drummondii* var. *runyonii*
Habit: Annuals up to 12" tall; stem hairs mostly pointing straight outward.
Leaves: Opposite; blades ovate, up to ¾" long, the margins of the mid and upper leaves usually entire, the lower ones sometimes minutely toothed, the teeth rounded.
Flowers: Bilateral, 2-lipped; calyxes with a transverse lip or ridge; corollas blue to violet, up to ⅝" long but variable.
Fruit: 4 nutlets, under the microscope resembling tiny pine cones.
Bloom Period: Late winter and spring.
Distribution: Cameron and Willacy counties.
Comments: The transverse ridge or lip on the calyx is an identifying characteristic of the genus. Another species, Rio Grande skullcap, *S. muriculata,* also grows here; its leaf margins usually have rounded teeth, the teeth smaller on the mid and upper leaves. Also, its stem hairs are mostly curved and pressed against the stems. The stem hairs on Runyon's skullcap are mostly pointed straight outward. In Texas this variety is limited to the LRGV. The populations we have observed were localized, growing at the edges of brush in clay loam. *S. drummondii* was named in honor of Thomas Drummond, who immigrated from Scotland and did extensive work with plants of Central Texas. The variety was named in honor of Robert Runyon, a former mayor of Brownsville and self-taught botanist who made many contributions to the understanding of plants of the Lower Rio Grande Valley.

Runyon's Skullcap *Scutellaria drummondii* var. *runyonii*

Rio Grande Skullcap

Family: Lamiaceae (Labiatae)
Scientific Name: *Scutellaria muriculata*
Habit: Annuals up to 12" tall; stem hairs mostly curved and pressed against the stems.
Leaves: Opposite; blades ovate, up to ¾" long, the margins of the lower leaves usually with rounded teeth; mid and upper leaves usually with smaller teeth.
Flowers: Bilateral, 2-lipped; calyxes with a transverse lip or ridge; corollas bluish, up to ⅝" long but variable.
Fruit: 4 nutlets with tiny pointed outgrowths (seen under high magnification).
Bloom Period: Spring, summer, fall.
Distribution: Cameron, Hidalgo, and Willacy counties.
Comments: This species is widespread in deep sands. Runyon's skullcap, *S. drummondii* var. *runyonii*, also grows here; the margins on its upper leaves are usually entire, and its stems have hairs that point straight outward. The transverse ridge or lip on the calyx is an identifying characteristic of the genus.

Rio Grande Skullcap *Scutellaria muriculata*

Shade Betony

Family: Lamiaceae (Labiatae)
Scientific Name: *Stachys crenata*
Habit: Erect annuals up to 18" tall.
Leaves: Opposite; blades ovate or heart shaped, up to 1 ⅜" long, the margins toothed.
Flowers: Bilateral; corollas pink, about ¼" long, barely extending beyond the calyxes.
Fruit: 4 nutlets.
Bloom Period: Spring.
Distribution: Cameron and Hidalgo counties.
Comments: This species is easily distinguished from pink mint, *S. drummondii*, by its much smaller corollas.

Shade Betony *Stachys crenata*

Lamiaceae

Pink Mint

Family: Lamiaceae (Labiatae)
Scientific Name: *Stachys drummondii*
Habit: Erect annuals up to 36" tall.
Leaves: Opposite; ovate to heart shaped, up to 3 1/8" long, the margins toothed.
Flowers: Bilateral; corollas pink, rarely white, about 3/8" long.
Fruit: 4 nutlets.
Bloom Period: Fall, winter, spring.
Distribution: Cameron, Hidalgo, Willacy, and Starr counties.
Comments: In this species the corolla extends well beyond the calyx, easily distinguishing it from shade betony, *S. crenata*. Pink mint is a showy species and is one of our earliest spring wildflowers, disappearing quickly as soon as the weather gets hot. It was named in honor of Thomas Drummond, an immigrant from Scotland who contributed much to the knowledge of Texas flora, especially from Central Texas.

Pink Mint *Stachys drummondii*

American Germander

Family: Lamiaceae (Labiatae)
Scientific Name: *Teucrium canadense* var. *occidentale*
Habit: Erect perennial herbs up to 39" tall, usually shorter, growing from underground rhizomes.
Leaves: Opposite; blades ovate, up to 4 3/8" long, the margins toothed.
Flowers: Bilateral; corollas pink, up to 3/4" long.
Fruit: 4 nutlets.
Bloom Period: Winter, spring, summer.
Distribution: Cameron and Hidalgo counties.
Comments: The pink flowers of this species distinguish it from coastal germander, *T. cubense*, which has white flowers. American germander usually prefers moist places, forming colonies in well-watered gardens. The plants spread by underground rhizomes.

American Germander *Teucrium canadense* var. *occidentale*

Coastal Germander

Family: Lamiaceae (Labiatae)
Scientific Name: *Teucrium cubense*
Habit: Erect perennial herbs up to 28" tall.
Leaves: Opposite; blades ovate, up to 1 ¾" long, the margins toothed and lobed.
Flowers: Bilateral; corollas white, up to ⅝" long.
Fruit: 4 nutlets
Bloom Period: All seasons.
Distribution: Cameron, Hidalgo, Willacy, and Starr counties.
Comments: The white flowers of this species distinguish it from American germander, *T. canadense*. This is among the few wildflowers that continue to bloom even in the hot dry summer of the LRGV.

Coastal Germander *Teucrium cubense*

Love Vine

Family: Lauraceae
Scientific Name: *Cassytha filiformis*
Habit: Parasitic vines resembling dodder, with yellow or orange stems, with a spicy odor.
Leaves: Absent or reduced to scales.
Flowers: Radial, about ⅛" broad; petals absent; sepals 6.
Fruit: A globose drupe, about ¼" broad, black when mature.
Bloom Period: Spring, summer.
Distribution: Cameron and Willacy counties.
Comments: This species is rare. It occurs along tropical seashores. In our area, it was found on South Padre Island by Tom Patterson.

Love Vine *Cassytha filiformis*

Love Vine *C. filiformis* fruit

Linaceae

Bladderwort

Family: Lentibulariaceae
Scientific Name: *Utricularia gibba*
Habit: Small perennial herbs, usually submerged in water.
Leaves: Alternate, threadlike, up to ⅜" long, bearing tiny bladders that trap aquatic microbes.
Flowers: Inflorescences raised above the water, with up to 4 flowers; corollas yellow, bilateral, 4-lobed, up to ⅜" broad.
Fruit: A tiny capsule.
Bloom Period: Summer.
Distribution: Cameron and Willacy counties.
Comments: This species is rare in our area. It is unusual in its lifestyle of trapping and digesting microorganisms in its submerged bladders.

Bladderwort *Utricularia gibba*

Winged Flax

Family: Linaceae
Scientific Name: *Linum alatum*
Habit: Annuals up to 16" tall.
Leaves: Opposite on lower stems, alternate above; blades linear, up to ¾" long.
Flowers: Petals yellow with reddish lines at the base, about ¾" long, falling away early.
Fruit: Capsules about 3/16" tall.
Bloom Period: Spring, summer, fall.
Distribution: Cameron, Hidalgo, and Willacy counties.
Comments: This is our most abundant *Linum*. It is especially abundant at the coast.

Winged Flax *Linum alatum*

Laredo Flax

Family: Linaceae
Scientific Name: *Linum elongatum*
Habit: Annuals up to 20" tall.
Leaves: Opposite on lower stems, alternate above; blades linear, up to 1 1/8" long.
Flowers: Petals yellow with purplish bands toward the center, about 5/8" long.
Fruit: Capsules about 3/16" tall.
Bloom Period: Spring, summer, fall.
Distribution: Hidalgo and Starr counties.
Comments: The *Linum* species have many characteristics in common and are difficult to distinguish. The corollas of all species fall away easily when the plants are disturbed.

Laredo Flax *Linum elongatum*

Tufted Flax

Family: Linaceae
Scientific Name: *Linum imbricatum*
Habit: Annuals up to 12" tall.
Leaves: Opposite on lower stems, alternate above; blades linear, up to 3/8" long.
Flowers: Petals yellow, sometimes with reddish bases, up to 5/16" long.
Fruit: Capsules about 1/8" tall.
Bloom Period: Spring, summer.
Distribution: Hidalgo County.
Comments: This species is more common north of our area.

Tufted Flax *Linum imbricatum*

Lundell's Flax

Family: Linaceae
Scientific Name: *Linum lundellii*
Habit: Annuals up to 16" tall.
Leaves: Opposite on lower stems, alternate above; blades linear, up to 1 1/8" long.
Flowers: Petals yellow to orange, with faint bands near the base, about 1/2" long or less.
Fruit: Capsules about 1/8" tall.
Bloom Period: Spring, summer.
Distribution: Hidalgo and Starr counties.
Comments: In Texas this species has been reported only in the counties along the Rio Grande from Hidalgo County to Dimmit County.

Lundell's Flax *Linum lundellii*

Loasaceae

Stinging Cevallia

Family: Loasaceae
Scientific Name: *Cevallia sinuata*
Habit: Somewhat woody plants up to 24" tall, with stinging hairs.
Leaves: Alternate; blades with sinuate (wavy, in and out) or lobed margins, up to 2 ½" long.
Flowers: In tight, rounded clusters; sepals and petals yellow, up to 5/16" long.
Fruit: A small achene.
Bloom Period: Spring, summer, fall.
Distribution: Cameron, Hidalgo, Willacy, and Starr counties.
Comments: Stinging cevallia prefers rocky soils. The common name is appropriate. One touch is a memorable event, not to be willingly repeated. It is a good nectar plant for butterflies. Hairs on the seeds tend to coil, digging the seeds into the ground.

Stinging Cevallia *Cevallia sinuata*

Yellow Rock Nettle

Family: Loasaceae
Scientific Name: *Eucnide bartonioides*
Habit: Hairy annuals, sometimes perennials; stems brittle, creeping.
Leaves: Alternate; blades roundish to heart shaped with rounded or sharp teeth. The petioles are as long as or longer than the blades.
Flowers: Radial, petals 5, yellow, up to 1 ½" long.
Fruit: Roundish capsules up to ½" long. The seeds are tiny, like reddish powder.
Bloom Period: Spring, summer, occasionally fall.
Distribution: Starr County.
Comments: This species is found more often in Central and West Texas. It has not previously been reported for our area. It grows on limestone cliffs in protected places. There are two flower forms. Ours is the smaller one. The flowers open by midmorning. It is sometimes incorrectly assumed that this species was named for Barton Warnock, a well-known botanist of West Texas. But the specific epithet *bartonioides* refers to the genus *Bartonia*, named for Benjamin Smith Barton, a Philadelphia botanist.

Yellow Rock Nettle *Eucnide bartonioides*

Cut Leaf Stickleaf

Family: Loasaceae

Scientific Name: *Mentzelia incisa*

Habit: Sprawling, sometimes leaning plants up to 24" tall, with barbed hairs.

Leaves: Alternate, more or less ovate, sometimes with a pair of basal lobes, up to 2 3/8" long, with many barbed hairs.

Flowers: Solitary or in terminal inflorescences; petals orange-yellow, up to 5/8" long. The stamens are of two different lengths.

Fruit: Capsules up to 5/8" long.

Bloom Period: Spring, summer, fall.

Distribution: Cameron and Hidalgo counties.

Comments: This is definitely a stickleaf. Once attached to clothing, the leaves will not let go, but tear into smaller and smaller pieces as removal is attempted. The flowers open early in the morning and close around noon.

Cut Leaf Stickleaf *Mentzelia incisa*

Lindheimer's Stickleaf

Family: Loasaceae

Scientific Name: *Mentzelia lindheimeri*

Habit: Perennials up to 24" tall from a fleshy tuber, with barbed hairs, the stems often falling over.

Leaves: Alternate; blades ovate, often 3-lobed, up to 3 1/8" long; margins toothed.

Flowers: Opening in the morning, closing about 10:30 in the morning; petals orange-yellow, up to 1/2" or slightly longer. The stamens are all more or less the same length.

Fruit: Capsules up to 9/16" long; seeds up to 8 per capsule.

Bloom Period: Spring, summer, fall.

Distribution: Cameron, Hidalgo, and Starr counties.

Comments: The flower opening time contrasts with that of the closely related sand lily, *M. nuda,* which opens late afternoon. The leaves of Lindheimer's stickleaf, like those of cut leaf stickleaf, attach firmly to clothing and are almost impossible to remove. This species was named in honor of Ferdinand Lindheimer, a German immigrant who made great contributions to Texas botany.

Lindheimer's Stickleaf *Mentzelia lindheimeri*

Lythraceae

Sand Lily

Family: Loasaceae
Scientific Name: *Mentzelia nuda* var. *stricta*
Habit: Erect perennials up to 40" tall, with barbed hairs.
Leaves: Alternate; blades linear to lanceolate, up to 2" or longer, the margins toothed.
Flowers: Opening late afternoon; petals white, up to 1 ⅝" long.
Fruit: Capsules up to 1 ⅛" long; seeds almost ⅛" long, winged.
Bloom Period: Summer, fall.
Distribution: Brooks County.
Comments: The flowers open late in the afternoon. Those we observed did not open until 6:30 P.M. on July 4, 2005. The white flower color combined with the late opening time of this species suggests an evening or night pollinator.

Sand Lily *Mentzelia nuda* var. *stricta*

Tooth Cup

Family: Lythraceae
Scientific Name: *Ammannia coccinea* [*A. teres*]
Habit: Annual herbs up to 24" or taller.
Leaves: Opposite, sessile, linear to narrowly lanceolate, broadened at the base, up to 2 ¾" long.
Flowers: Inflorescences usually with a very short peduncle from the leaf axils, with usually 2–5 flowers; sepals 4; petals 4, purple, about ⅛" long.
Fruit: Roundish capsules about 3/16" broad.
Bloom Period: Spring, summer, fall.
Distribution: Cameron, Hidalgo, and Willacy counties.
Comments: This species prefers wet or damp places.

Tooth Cup *Ammannia coccinea*

Large Redstem

Family: Lythraceae
Scientific Name: *Ammannia robusta*
Habit: Robust annuals up to 40" tall.
Leaves: Opposite, linear to occasionally elliptic, up to 3 1/8" long.
Flowers: Inflorescences sessile from the leaf axils with up to 3 or more flowers; sepals 4; petals 4, pinkish, sometimes with darker midveins or darker bases, about 1/8" long.
Fruit: Roundish capsules about 1/4" broad.
Bloom Period: Spring, summer, fall.
Distribution: Cameron and Willacy counties.
Comments: Like tooth cup, *A. coccinea,* this species prefers wet or damp places. It is distinguished from *A. coccinea* by its sessile inflorescences and by the pinkish flower petals, which sometimes have darker veins.

Large Redstem *Ammannia robusta*

Colombian Waxweed

Family: Lythraceae
Scientific Name: *Cuphea carthagenensis*
Habit: Erect annuals up to 36" tall, with bristly hairs and glandular hairs.
Leaves: Opposite; blades elliptic to ovate or obovate, up to 2 3/8" long.
Flowers: Solitary or in small groups from the leaf axils; calyxes about 1/4" long, the lobes minute; petals pinkish, narrow, about 1/16" long.
Fruit: Capsules about 1/4" long.
Bloom Period: Summer, fall.
Distribution: Cameron County.
Comments: This species has not previously been reported for our area. It is better known, but not abundant, in East Texas. This is an introduced species, native to South America, Central America, and Mexico. It prefers moist places and thrives where other plants are cultivated, such as gardens or nurseries.

Colombian Waxweed *Cuphea carthagenensis*

Lythraceae

Hachinal, Willow Leaved Heimia

Family: Lythraceae
Scientific Name: *Heimia salicifolia*
Habit: Shrubs up to 5' or occasionally taller.
Leaves: Opposite, sessile, narrowly elliptic to lanceolate, up to 2 ⅜" long.
Flowers: Borne singly; petals 5–7, yellow, up to ¾" long.
Fruit: Capsules roundish, about 3/16" tall.
Bloom Period: Spring, summer, fall.
Distribution: Cameron, Hidalgo, Willacy, and Starr counties.
Comments: In Texas this species is reported only from Cameron, Hidalgo, Willacy, and Starr counties. It seems to prefer, but is not limited to, moist places. Hachinal is often used as a landscape plant.

Hachinal, Willow Leaved Heimia *Heimia salicifolia*

Lance Leaf Loosestrife

Family: Lythraceae
Scientific Name: *Lythrum alatum* var. *lanceolatum* [*L. lanceolatum*]
Habit: Perennials up to 40" or taller, with square, ribbed stems.
Leaves: Opposite, sessile, mostly elliptic, up to 1 ⅜" long.
Flowers: Usually single; petals purple to reddish or bluish, up to 3/16" long.
Fruit: Capsules cylindrical, about 5/16" long.
Bloom Period: Spring, summer.
Distribution: Cameron and Willacy counties.
Comments: This species is distinguished from California loosestrife, *L. californicum,* by its smaller, less showy flowers, the petals being ⅛–3/16" long.

Lance Leaf Loosestrife *Lythrum alatum* var. *lanceolatum*

California Loosestrife

Family: Lythraceae
Scientific Name: *Lythrum californicum*
Habit: Perennials up to 24" or taller, with square, ribbed stems.
Leaves: Opposite, sessile, lanceolate to linear, about ¾" long.
Flowers: Usually single; petals purple, up to 5⁄16" long.
Fruit: Capsules cylindrical, about 5⁄16" long.
Bloom Period: Spring, summer.
Distribution: Cameron, Hidalgo, and Starr counties.
Comments: This species is distinguished from lance leaf loosestrife, *L. alatum*, by its larger, showier, usually darker-colored flowers, the petals 3⁄16–5⁄16" long.

California Loosestrife *Lythrum californicum*

Branched Tooth Cup

Family: Lythraceae
Scientific Name: *Rotala ramosior*
Habit: Sprawling annuals, the stems often rooting at the nodes.
Leaves: Opposite; petioles very short or absent; blades elliptic to oblanceolate, up to 1 ¼" or longer.
Flowers: Solitary; sepals 4, petals 4, white or pink, attached to the edge of the calyx tube, less than 1⁄16" long.
Fruit: Capsules roundish, about 1⁄16" or slightly broader.
Bloom Period: Spring, summer, fall.
Distribution: Cameron and Willacy counties.
Comments: This species prefers wet or damp places. It is distinguished from the two species of *Ammannia* by its flowers being borne singly from the leaf axils.

Branched Tooth Cup *Rotala ramosior*

Narrow Leaf Goldshower

Family: Malpighiaceae
Scientific Name: *Galphimia angustifolia* [*Thryallis angustifolia*]
Habit: Low, slender, hairless shrubs up to 16" tall.
Leaves: Opposite; linear to lanceolate, up to 2 ⅛" long.
Flowers: Inflorescences up to 6" long; petals yellow, turning reddish with age, up to ¼" or longer, attached to the receptacle by a narrow stalk.
Fruit: Capsules roundish, about 3⁄16" tall.
Bloom Period: Spring, summer, fall.

Narrow Leaf Goldshower *Galphimia angustifolia*

Distribution: Hidalgo, Willacy, and Starr counties.
Comments: This species grows in hard soils, gypsum, caliche, and gravel. Petals attached to the receptacle by a narrow stalk are referred to as "clawed." This is a characteristic of the family Malpighiaceae.

Malvaceae

Barbados Cherry, Manzanita

Family: Malpighiaceae
Scientific Name: *Malpighia glabra*
Habit: Shrubs up to 6' or taller.
Leaves: Opposite; blades ovate, up to 1 ⅝" or longer, the margins crinkled.
Flowers: Slightly bilateral; petals pinkish, clawed (attached to the receptacle by a narrow stalk), about ¼" long.
Fruit: A bright red roundish drupe about ³⁄₁₆" broad.
Bloom Period: Spring, summer, fall.
Distribution: Cameron, Hidalgo, Willacy, and Starr counties.
Comments: This shrub is excellent for the garden. It requires little care other than occasional pruning. In full bloom it is covered with the pink flowers, which are followed by the very attractive red drupes. When not in bloom, it is still attractive with its waxy green leaves. It is a host plant for White-patched Skipper, Cassius Blue, and Brown-banded Skipper butterflies. Barbados cherry has two recognizable growth forms, one prostrate, the other erect.

Barbados Cherry, Manzanita *Malpighia glabra*

Shrubby Indian Mallow

Family: Malvaceae
Scientific Name: *Abutilon abutiloides*
Habit: Densely hairy shrubs up to 5' tall.
Leaves: Alternate; blades hairy, ovate to roundish, heart shaped basally, up to 4" long.
Flowers: Solitary from the leaf axils; corollas yellow to orange; petals about ½" long.
Fruit: A smooth capsule about ⅝" tall, the individual compartments opening, each with 2 or 3 seeds.
Bloom Period: Summer.
Distribution: Cameron, Hidalgo, Willacy, and Starr counties.
Comments: The dense hairiness and the yellow to orange petals about ½" long are important diagnostic characteristics of this species.

Shrubby Indian Mallow *Abutilon abutiloides*

Berlandier's Indian Mallow

Family: Malvaceae
Scientific Name: *Abutilon berlandieri*
Habit: Erect, woody plants up to 5' tall, usually less, with dense star-shaped hairs and glandular hairs.
Leaves: Alternate, blades hairy, ovate to broadly ovate, heart shaped basally, up to 4" long.
Flowers: Usually single from the leaf axils; corollas radial, the petals up to ⅝" long.
Fruit: Capsules about ⅝" tall, with 2 or more seeds in each compartment.
Bloom Period: Spring, summer.
Distribution: Cameron, Hidalgo, Willacy, and Starr counties.
Comments: *A. berlandieri* was named in honor of Jean Louis Berlandier, who made major collections of plants from Mexico and Texas. This species has been mistakenly called *A. abutiloides*. It has also been called *A. lignosum*, which is now considered a synonym of *A. abutiloides*. The two species are very similar but are easily distinguished by presence or absence of sticky glands. *A. berlandieri* is glandular and sticky; *A. abutiloides* is not.

Berlandier's Indian Mallow *Abutilon berlandieri*

Texas Indian Mallow

Family: Malvaceae
Scientific Name: *Abutilon fruticosum*
Habit: Herbs up to 24" tall
Leaves: Alternate; blades heart shaped, up to 2" long.
Flowers: Radial, solitary from the leaf axils; petals orange-yellow, up to ⅜" long.
Fruit: A smooth capsule taller than it is broad, the individual compartments opening, each with 2–3 seeds.
Bloom Period: All seasons.
Distribution: Cameron, Hidalgo, and Starr counties.
Comments: A well-grown plant can produce many flowers at a time, making this an attractive plant for the garden. This species is fairly widespread in Central and South Texas.

Texas Indian Mallow *Abutilon fruticosum*

Malvaceae

Mauve Indian Mallow, Jann's Indian Mallow

Family: Malvaceae
Scientific Name: *Abutilon hulseanum*
Habit: Shrubs up to 5' or taller, densely hairy.
Leaves: Alternate; blades soft and velvety with hairs, more or less heart shaped, up to 4 3/8" long, the margins toothed.
Flowers: Radial; petals 5, pinkish with yellowish centers, up to 5/8" long; stamen filaments grown together around the style, the stamens seeming to emerge from the style.
Fruit: A capsule, the individual compartments opening, each with 2 seeds.
Bloom Period: Spring, summer, fall.
Distribution: Cameron and Willacy counties.
Comments: This species was known only in Cameron County from old historic records. Recently a population in southern Willacy County was discovered by Jann Miller, a local native plant enthusiast. The flowers open in the afternoon. Preferring moist places, this would make an excellent landscape plant. The flowers are edible, making a pleasing garnish or addition to a salad. This is a host plant for Laviana White-skipper butterflies.

Mauve Indian Mallow, Jann's Indian Mallow *Abutilon hulseanum*

Rio Grande Abutilon

Family: Malvaceae
Scientific Name: *Abutilon hypoleucum*
Habit: Shrubs up to 4' or taller.
Leaves: Alternate; petioles reddish; blades heart shaped, up to 4 1/8" long, whitish on the lower surfaces; margins toothed.
Flowers: Radial; calyx white-hairy; petals 5, yellow, about 3 1/8" long; stamen filaments grown together around the style, the stamens seeming to emerge from the style.
Fruit: A capsule, the individual compartments opening, each with 2 seeds.
Bloom Period: Spring.
Distribution: Cameron County.
Comments: The sepals appear to be heart shaped. This characteristic and the whitish lower surfaces of the leaves make the Rio Grande abutilon easy to recognize. This species would be an attractive addition to a native plant landscape. Abutilons produce a stem fiber that has been used in making cords.

Rio Grande Abutilon *Abutilon hypoleucum*

Dwarf Indian Mallow

Family: Malvaceae
Scientific Name: *Abutilon parvulum*
Habit: Perennials with weak, erect to trailing stems arising from a woody root.
Leaves: Small, up to 2" long, blades broadly ovate to rounded, heart shaped basally, the margins toothed.
Flowers: Arising singly from the leaf axils; petals ¼" long, pinkish to orange or yellow.
Fruit: Capsules with 5 or 6 compartments opening, each with 3 seeds.
Bloom Period: Spring, summer, fall.
Distribution: Hidalgo County.
Comments: Lindley Lentz and John Pemelton reported this species growing naturally in Hidalgo County in an area where soil from Starr County had been deposited. It has previously been reported to be limited to West Texas, but since it is known to grow in northern Tamaulipas, Mexico, finding it in our area is not surprising. *A. parvulum* closely resembles Texas Indian mallow, *A. fruticosum,* which has larger flowers.

Dwarf Indian Mallow *Abutilon parvulum*

Velvet Leaf Indian Mallow, Pie Print

Family: Malvaceae
Scientific Name: *Abutilon theophrasti*
Habit: Erect hairy annuals 2' to 4' tall.
Leaves: Alternate; blades heart shaped, up to 6" long.
Flowers: Radial, from the leaf axils; petals yellow, up to ⅝" long; stamen filaments grown together around the style, the stamens seeming to emerge from the style.
Fruit: Capsules about ⅝" tall, topped by long awns; each compartment opening with 2 seeds.
Blooming Period: Mostly summer and fall; also winter and spring.
Distribution: Hidalgo and Willacy counties.
Comments: This species is a native of Asia. Several *Abutilon* species have flowers of similar size. This one is easily differentiated from the others by the long awns at the top of the fruit. The common name pie print comes from the use of the fruit to pierce pie crusts. This species is named for the Greek naturalist Theophrastus, considered the father of botany. He lived from 372 to 287 B.C.

Velvet Leaf Indian Mallow, Pie Print *Abutilon theophrasti*

Velvet Leaf Indian Mallow, Pie Print *A. theophrasti* fruit

Malvaceae

Three Furrowed Indian Mallow

Family: Malvaceae

Scientific Name: *Abutilon trisulcatum*

Habit: Shrubs up to 5' tall; stems 3-cornered, especially the younger ones.

Leaves: Alternate; blades heart shaped, hairy, up to 3 ¾" long, the margins with rounded teeth.

Flowers: Radial; solitary from the leaf axils; petals yellow with reddish bases, up to ¼" long; stamen filaments grown together around the style, the stamens seeming to emerge from the style.

Fruit: A capsule about ⁵⁄₁₆" long, the individual compartments opening, each with 2 seeds.

Bloom Period: Spring, summer, fall.

Distribution: Cameron, Hidalgo, and Starr counties.

Comments: This is our most common *Abutilon*, usually growing in clay soils. Some of the *Abutilon* species are difficult to distinguish from each other, but this is our only *Abutilon* with flowers that are reddish toward the center, according to Dr. Paul Fryxell (personal communication)—although members of other genera may, too. That characteristic, combined with the 3-cornered stems, makes this species one of the easier ones to recognize. The plants multiply rapidly in cultivation, becoming difficult to control. *A. trisulcatum* competes successfully with exotic invasive grasses.

Three Furrowed Indian Mallow *Abutilon trisulcatum*

Umbrella Indian Mallow

Family: Malvaceae

Scientific Name: *Abutilon umbellatum*

Habit: Perennials, sometimes woody, up to 40" tall.

Leaves: Alternate; blades heart shaped, up to 4" long, the margins toothed.

Flowers: Clusters arising from a single point (in umbels); corollas radial; petals yellow, about ⅜" long; stamen filaments grown together around the style, the stamens seeming to emerge from the style.

Fruit: A capsule about ⁵⁄₁₆" long, the individual compartments opening, each with usually 3 seeds.

Bloom Period: Fall.

Distribution: Cameron and Hidalgo counties.

Comments: This is a fairly rare species in our area. Thirteen species of *Abutilon* are reported for Texas. We have nine of those in our area. Keys must be used to distinguish most of them.

Umbrella Indian Mallow *Abutilon umbellatum*

Umbrella Indian Mallow *A. umbellatum* fruit, produced in an umbel

Wright's Indian Mallow

Family: Malvaceae

Scientific Name: *Abutilon wrightii*

Habit: Herbs with woody bases, up to about 24" tall, often trailing on the ground.

Leaves: Alternate; blades ovate to heart shaped, about 2" long, the margins toothed.

Flowers: Radial; petals yellow, about ½" long; stamen filaments grown together around the style, the stamens seeming to emerge from the style.

Fruit: A capsule about ½" long, the individual compartments opening, each with 2 or more seeds.

Bloom Period: Spring, summer, fall.

Distribution: Cameron, Hidalgo, and Starr counties.

Comments: *A. wrightii* was named in honor of Charles Wright, one of the best known American botanists in the nineteenth century. This species has a rather broad distribution from southern Texas to western Texas. Evidently the stems are somewhat weak, since none of the plants we observed grew strongly erect. The flowers usually open late in the evening.

Wright's Indian Mallow *Abutilon wrightii*

Lozano's False Indian Mallow

Family: Malvaceae

Scientific Name: *Allowissadula lozanii* [*Pseudabutilon lozanii*]

Habit: More or less shrubby plants up to 5' tall.

Leaves: Alternate; blades dark green, ovate or rounded, up to 5 ¼" long; margins toothed.

Flowers: Radial; petals yellow, up to ¾" long; stamen filaments grown together around the style, the stamens seeming to emerge from the style.

Fruit: A capsule about 5/16" long, the individual compartments opening, each one divided into 2 one-seeded compartments.

Bloom Period: Spring, summer, fall.

Distribution: Cameron, Hidalgo, and Willacy, and Starr counties.

Comments: This species grows in heavy clay soils. The flowers are edible and are reported to be quite tasty if eaten raw. They should be used in salads or added to food after it has been cooked. This is a host plant for the Texas Powdered Skipper butterfly.

Lozano's False Indian Mallow *Allowissadula lozanii*

Malvaceae

Field Anoda

Family: Malvaceae
Scientific Name: *Anoda pentaschista*
Habit: Slender erect herbs up to 24" tall.
Leaves: Alternate; blades more or less ovate, with a pair of basal lobes, up to 2 ⅛" long.
Flowers: Solitary from the leaf axils; radial; petals yellow to orange, about ⅜" long; stamen filaments grown together around the style, the stamens seeming to emerge from the style.
Fruit: Capsules more or less hemispheric, the compartments usually 5, breaking apart, each with one seed.
Bloom Period: Summer, fall.
Distribution: Cameron and Hidalgo counties.
Comments: The basal lobes on the leaves are a noticeable characteristic of this species. The yellow flowers distinguish it from *A. cristata* (not pictured), which has purple flowers.

Field Anoda *Anoda pentaschista*

Mexican Bastardia

Family: Malvaceae
Scientific Name: *Bastardia viscosa*
Habit: Perennial herbs up to 40" tall, bad-smelling when bruised.
Leaves: Alternate, heart shaped, up to 3 ¾" long, the margins toothed.
Flowers: Radial, usually borne singly in the leaf axils; petals yellow, up to ¼" long; stamen filaments grown together around the style, the stamens seeming to emerge from the style.
Fruit: Capsules somewhat globose, but broader than they are tall, about 5⁄16" tall; individual compartments usually one-seeded; seeds with tiny hairs.
Bloom Period: Spring, summer, fall.
Distribution: Cameron, Hidalgo, Willacy, and Starr counties.
Comments: The bad smell when the plant is bruised is a fairly good field test for this species. The flowers usually open in the afternoon, a characteristic of many members of the Malvaceae family.

Mexican Bastardia *Bastardia viscosa*

Coppery False Fanpetals, Copper Mallow

Family: Malvaceae

Scientific Name: *Billieturnera helleri* [*Sida helleri*]

Habit: Woody, hairy plants with prostrate or leaning stems.

Leaves: Alternate; blades roundish or fan shaped, up to 5/8" long, the margins toothed.

Flowers: Radial, borne singly from the leaf axils; petals yellow, up to 5/16" long; stamen filaments grown together around the style, the stamens seeming to emerge from the style.

Fruit: Capsules more or less globose, about 3/16" tall, enclosed in the calyx; compartments usually 5, each with one seed.

Bloom Period: All seasons.

Distribution: Cameron, Hidalgo, and Starr counties.

Comments: This species was named for Dr. Billie L. Turner, a well-known professor of botany at the University of Texas at Austin. It grows in coastal and inland saline clay soils. The flowers open in the afternoon. They are edible.

Coppery False Fanpetals, Copper Mallow *Billieturnera helleri*

Winecup

Family: Malvaceae

Scientific Name: *Callirhoe involucrata* var. *lineariloba*

Habit: Perennial herbs, the stems mostly creeping.

Leaves: Alternate; more or less ovate in outline, the margins deeply cut, up to 1" or longer.

Flowers: Radial, borne singly from the leaf axils; petals reddish purple, up to 1" long; stamen filaments grown together around the style, the stamens seeming to emerge from the style.

Fruit: Capsules with many compartments, which break apart at maturity.

Bloom Period: Spring.

Distribution: Cameron, Hidalgo, Willacy, and Starr counties.

Comments: This species is quite widespread in Texas, being absent only in West Texas and parts of Northeast Texas. The roots are edible and are considered a delicacy. The petals have been used to make an orange dye. Some wildflowers have common names that are somewhat obscure, but this attractive plant has an apt common name. The flowers are more or less wine colored and remain cup shaped rather than opening fully as most flowers do. Even after seeing them only once, the name winecup comes readily to mind when they are next encountered.

Winecup *Callirhoe involucrata* var. *lineariloba*

Winecup *C. involucratus* var. *lineariloba* flower

Malvaceae

Sulphur Mallow

Family: Malvaceae
Scientific Name: *Cienfuegosia drummondii*
Habit: Perennial herbs, occasionally woody, the stems leaning or sprawling.
Leaves: Alternate; blades ovate, up to 2 ½" long, the margins with broad teeth.
Flowers: Radial; petals yellow, up to 1 ⅛" long; stamen filaments grown together around the style, the stamens seeming to emerge from the style.
Fruit: Capsules globose; compartments separating and breaking open, each with 2 seeds.
Bloom Period: Summer.
Distribution: Cameron and Willacy counties.
Comments: *C. drummondii* was named in honor of Thomas Drummond, who immigrated from Scotland and made significant contributions to the knowledge of the flora of Texas. This attractive plant grows along the Gulf Coast, preferring low places that retain water. Seldom seen, it is said to be an alternate host for the cotton boll weevil.

Sulphur Mallow *Cienfuegosia drummondii*

Sulphur Mallow *C. drummondii* fruit

Bladder Mallow

Family: Malvaceae
Scientific Name: *Herissantia crispa* [*Bogenhardia crispa*]
Habit: Perennial herbs with creeping stems.
Leaves: Alternate, hairy; blades heart shaped, up to 2" long, the margins toothed.
Flowers: Radial; pale yellow to white, up to ½" long; stamen filaments grown together around the style, the stamens seeming to emerge from the style.
Fruit: Capsules separating into inflated papery compartments with 2 seeds, sometimes more.
Bloom Period: All seasons.
Distribution: Cameron, Hidalgo, Willacy, and Starr counties.
Comments: Together, the creeping habit, pale yellow to white flowers, and inflated papery compartments of the fruit are good indicators of this species. It would make a good native ornamental.

Bladder Mallow *Herrissantia crispa*

Tulipán Del Monte, Heart Leaf Hibiscus

Family: Malvaceae
Scientific Name: *Hibiscus martianus*
   [*H. cardiophyllus*]
Habit: Short-lived perennials with woody bases, up to 24" tall.
Leaves: Alternate; blades heart shaped, up to 3 ½" long, the margins toothed.
Flowers: Radial; petals red, up to 1 ⅛" long; stamen filaments grown together around the style, the stamens seeming to emerge from the style.
Fruit: Capsules about ⅝" tall; compartments opening but not separating from each other; seeds minutely hairy.
Bloom Period: Spring, summer, fall, after rains.
Distribution: Cameron, Hidalgo, Willacy, and Starr counties.
Comments: This attractive species makes a startling splash of color in the natural setting, especially when rains are abundant. With a little care, it is a beautiful plant in the garden. Cultivated plants are often killed from frequent watering. It is difficult to grow this species from seed. The plants live only about three years. The Spanish common name translates as "wild hibiscus."

Tulipán Del Monte, Heart Leaf Hibiscus *Hibiscus martianus*

Saltmarsh Mallow

Family: Malvaceae
Scientific Name: *Kosteletzkya virginica*
Habit: Perennial herbs up to 6' tall, with prickly hairs.
Leaves: Alternate; blades heart shaped to ovate, up to 6" long, sometimes with basal lobes, the margins coarsely toothed.
Flowers: More or less radial (except for the column); petals 5, pink, up to 1 ¾" long; the column (the style surrounded by the anther filaments) arching downward.
Fruit: 5-angled, hairy, compressed.
Bloom Period: Summer, fall.
Distribution: Cameron County.
Comments: This plant prefers mildly salty wet soils, such as banks of ditches and brackish marshes. It is a beautiful plant and would be ideal in a wet place in the garden. It is easily distinguished from the other members of the family by its flower color combined with the habit of the column in the flower to arch downward. In most mallows it is straight.

Saltmarsh Mallow *Kosteletzkya virginica*

Malvaceae

K. depressa is reported to have been collected in the same area as the plants we photographed. However, we have been unable to find any of that species. In contrast to *K. virginica,* its flowers open white, and its petals are not more than ⅜" long. The genus name is sometimes spelled *Kosteletzyka.*

Malva De Caballo
Family: Malvaceae
Scientific Name: *Malachra capitata*
Habit: Perennials up to 5' tall, the younger parts often hairy; mature stems spiny.
Leaves: Alternate; blades broadly ovate, up to 4 ¾" long, the margins toothed, sometimes obscurely so.
Flowers: Radial, in clusters surrounded by large, whitish bracts; petals yellow-orange, up to ⅜" long; stamen filaments grown together around the style, the stamens seeming to emerge from the style.
Fruit: Capsules with 5 one-seeded compartments, about ⅛" tall.
Bloom Period: All seasons.
Distribution: Cameron and Hidalgo counties.
Comments: The large whitish bracts that surround the flowers are a good marker for this species. It usually grows at the edges of wetlands. The common name translates as "horse mallow."

Malva De Caballo *Malachra capitata*

Little Mallow
Family: Malvaceae
Scientific Name: *Malva parviflora*
Habit: Large annual herbs up to 6' or taller.
Leaves: Alternate; blades kidney shaped, sometimes rounded in outline, up to 2 ⅜" long; margins toothed and shallowly lobed.
Flowers: Radial, in clusters in the axils of the leaves; petals white or bluish, up to 1" long; stamen filaments grown together around the style, the stamens seeming to emerge from the style.
Fruit: Disk shaped, not splitting open.
Bloom Period: Winter, spring.
Distribution: Cameron, Hidalgo, and Starr counties.
Comments: This distinctive species can be recognized by its small white flowers, the lobed and toothed leaves, and the fruit, which does not split open. It is an introduction from the Old World.

Little Mallow *Malva parviflora*

Malva Loca

Family: Malvaceae
Scientific Name: *Malvastrum americanum*
Habit: Erect, densely hairy perennials up to 40" tall.
Leaves: Alternate; blades ovate, up to 2 ½" long, the margins toothed.
Flowers: Radial, in dense terminal, cylindrical clusters; petals orange-yellow, about ¼" long; stamen filaments grown together around the style, the stamens seeming to emerge from the style.
Fruit: Capsules ⅛" or less tall; the compartments do not split open.
Bloom Period: All seasons.
Distribution: Cameron, Hidalgo, and Willacy counties.
Comments: The tight, cylindrical inflorescences with the small orange-yellow flowers point to this species. It is common in our area but much less so in other parts of Texas. Rabbits feed on the leaves. The common name translates as "crazy mallow."

Malva Loca *Malvastrum americanum*

Wright's False Mallow

Family: Malvaceae
Scientific Name: *Malvastrum aurantiacum*
Habit: Perennials from a woody base; stems slender, erect or falling over.
Leaves: Alternate; blades more or less ovate, up to 1 ⅝" long, with toothed margins.
Flowers: Radial, usually solitary from the leaf axils; petals yellow-orange, up to ¾" long.
Fruit: Capsules in 12–20 segments, each with one seed, reddish at maturity. The calyx often persists below the capsule.
Bloom Period: Spring, summer, fall.
Distribution: Cameron County.
Comments: The larger flowers and the reddish capsules distinguish this species from our other two *Malvastrum* species. In our area, it seems to prefer an environment near the coast.

Wright's False Mallow *Malvastrum aurantiacum*

Yard Mallow

Family: Malvaceae
Scientific Name: *Malvastrum coromandelianum*
Habit: Erect herbs up to 40" tall, usually less, the stems often falling over.
Leaves: Alternate, ovate, up to 2 ⅛" long, the margins toothed.
Flowers: Radial, usually solitary from the axils of the leaves; petals yellow-orange, up to ⁵⁄₁₆" long; stamen filaments grown together around the style, the stamens seeming to emerge from the style.

Malvaceae

Fruit: of numerous sharp-pointed compartments, which do not split open.
Bloom Period: All seasons.
Distribution: Cameron, Hidalgo, Willacy, and Starr counties.
Comments: This species is common in the LRGV and is often seen in gardens and in various disturbed areas. The leaves are attractive to rabbits and are eaten by Texas tortoises.

Yard Mallow *Malvastrum coromandelianum*

Turk's Cap

Family: Malvaceae
Scientific Name: *Malvaviscus drummondii* [*M. arboreus* var. *drummondii*]
Habit: Shrubs up to 10' tall, usually about 5', the stems often falling over.
Leaves: Alternate, hairy, roundish to ovate, up to 3 ½" long, the bases more or less heart shaped, the margins toothed.
Flowers: Radial, usually solitary from the axils of the leaves; petals red, up to 1 ⅝" long, opening only partially.
Fruit: Red, somewhat juicy.
Bloom Period: Spring, summer, fall.
Distribution: Cameron, Hidalgo and Willacy counties.
Comments: *M. drummondii* was named in honor of Thomas Drummond, who did extensive plant collecting in Central Texas in the 1830s. Populations growing in sandy soils near mottes of Live Oak, *Quercus virginiana*, in Kenedy County are more densely hairy than the more commonly encountered populations. This species is a good choice for growing in partial shade in the landscape. It is a host plant for the Mallow Scrub-hairstreak and Turk's Cap White-skipper butterflies. The fruit is said to be edible but is not particularly tasty. It is eaten by wildlife. The flowers are an important nectar source for hummingbirds and butterflies. *Malvaviscus penduliflorus*, a cultivated species, is similar to *M. drummondii*. In the past, it was known as *M. arboreus* var. *penduliflorus*.

Turk's Cap *Malvaviscus drummondii*

Slender Stalked Mexican Mallow

Family: Malvaceae
Scientific Name: *Meximalva filipes* [*Sida filipes*]
Habit: Erect herbs, sometimes woody basally, up to 40" tall.
Leaves: Alternate; blades lanceolate, up to 1 ⅞" long, the margins toothed.
Flowers: Radial, solitary from the axils of the leaves, facing downward; pedicels long and slender; petals violet, up to 3/16" long.
Fruit: Capsules about ⅛" tall, opening at the apices.
Bloom Period: Spring, summer.
Distribution: Cameron, Hidalgo, and Starr counties.
Comments: This species frequents gravelly and rocky areas. The long pedicels and the deep violet-purple corollas are indicators of this species.

Slender Stalked Mexican Mallow *Meximalva filipes*

Wheel Mallow

Family: Malvaceae
Scientific Name: *Modiola caroliniana*
Habit: Creeping hairy herbs.
Leaves: Alternate; blades ovate in outline, deeply notched, up to 1 ¾" long.
Flowers: Radial, solitary from the axils of the leaves; petals pink to reddish, about ¼" long.
Fruit: Capsules compressed, about 5/16" broad, of many segments, falling apart at maturity.
Bloom Period: Spring.
Distribution: Cameron, Hidalgo, and Willacy counties.
Comments: Wheel mallow is especially found in disturbed places, lawns, and edges of marshes.

Wheel Mallow *Modiola caroliniana*

Bladderpod Sida

Family: Malvaceae
Scientific Name: *Rhynchosida physocalyx* [*Sida physocalyx*]
Habit: Sprawling herbs, sometimes woody basally, with taproots resembling parsnips.
Leaves: Alternate; blades somewhat succulent, ovate, up to 2" long, the margins toothed.
Flowers: Radial, solitary from the axils of the leaves; petals pale yellow, up to ⅜" long.
Fruit: Capsules enclosed by the swollen calyx, blackening when ripe.
Bloom Period: Spring, summer, fall.

Bladderpod Sida *Rhynchosida physocalyx*

Malvaceae

Distribution: Cameron, Hidalgo, Willacy, and Starr counties.

Comments: The mature calyx inflated around the capsule is an important characteristic in recognizing this species.

Spreading Sida

Family: Malvaceae
Scientific Name: *Sida abutifolia* [*S. filicaulis*]
Habit: Herbs, the stems weak and falling over.
Leaves: Alternate; blades ovate to lanceolate, up to ⅝" long, the margins toothed.
Flowers: Radial, mostly solitary from the axils of the leaves; petals yellow, about ¼" long.
Fruit: Capsules about ¼" broad.
Bloom Period: Summer.
Distribution: Cameron, Hidalgo, and Starr counties.
Comments: This plant can grow in various habitats. Its leaves are a source of food for wildlife.

Bracted Sida

Family: Malvaceae
Scientific Name: *Sida ciliaris*
Habit: Herbs with leaning or prostrate stems.
Leaves: Alternate; blades narrow, up to 1" long, the margins toothed apically.
Flowers: Radial, mostly in tight apical clusters, with narrow hairy bracts; petals purple to yellow, up to ⅝" long.
Fruit: Capsules about ¼" broad.
Bloom Period: Spring, summer, fall.
Distribution: Cameron, Hidalgo, and Willacy counties.
Comments: The flower stalks grow onto the petioles of the adjacent leaves or bracts. This characteristic is useful for identifying this species. It grows primarily on sandy soils.

Spreading Sida *Sida abutifolia*

Bracted Sida *Sida ciliaris*

Bracted Sida *S. ciliaris* pink-flowering form

Bracted Sida *S. ciliaris* white-flowering form

Heart Leaf Fanpetals

Family: Malvaceae
Scientific Name: *Sida cordifolia*
Habit: Erect hairy herbs of sandy soils, sometimes woody basally, up to 5' tall.
Leaves: Alternate; blades heart shaped (often barely so), up to 5 ¾" long, densely hairy on both sides, the margins toothed.
Flowers: Radial, in clusters or borne singly from the axils of the leaves; petals yellow to pink, about ⅝" long.
Fruit: Capsules about ⅜" broad.
Bloom Period: Summer, fall.
Distribution: Cameron, Hidalgo, Willacy, and Starr counties.
Comments: The specific epithet *cordifolia* comes from Latin words meaning "heart-shaped leaves." Unfortunately, the characteristic of heart-shaped leaves is common in the Malvaceae and is not much use in recognizing this species, especially since its leaves are usually only vaguely heart-shaped. The large flowers, the erect growth habit, and the ovate to cordate leaves of *S. cordifolia* distinguish it from the other *Sida* species in the LRGV.

Heart Leaf Fanpetals *Sida cordifolia*

Lindheimer's Sida

Family: Malvaceae
Scientific Name: *Sida lindheimeri*
Habit: Erect or sprawling herbs up to 36" tall.
Leaves: Alternate; blades linear to lanceolate, up to 2 ⅛" long, the margins toothed.
Flowers: Radial; petals yellow to pinkish, up to ⅝" long.
Fruit: Dry, opening partially into usually 10 parts, each usually with one seed.
Bloom Period: Spring, summer, fall.
Distribution: Cameron, Hidalgo, Willacy, and Starr counties.
Comments: *S. lindheimeri* was named in honor of Ferdinand Lindheimer, a prominent botanist in the nineteenth century. The linear to lanceolate leaves with toothed margins and the relatively large size of the flowers are noticeable characteristics of this species. Lindheimer's Sida is fairly widespread in Texas, from Mason County in Central Texas into East Texas and extreme South Texas. The flowers open late in the afternoon.

Lindheimer's Sida *Sida lindheimeri*

Malvaceae

Indian Hemp

Family: Malvaceae
Scientific Name: *Sida rhombifolia*
Habit: Herbaceous or woody plants up to 6' or taller, usually less.
Leaves: Alternate; blades mostly rhombic, slightly cordate basally, up to 3 ⅛" long, silvery-hairy on the lower surfaces; margins with tiny teeth.
Flowers: Radial; calyxes with thickened veins; petals yellow, up to ¼" long.
Fruit: Capsules small, less than ¼" tall.
Bloom Period: All seasons.
Distribution: Cameron and Willacy counties.
Comments: The rhombic leaves with silvery undersurfaces call attention to this species.

Indian Hemp *Sida rhombifolia*

Prickly Mallow

Family: Malvaceae
Scientific Name: *Sida spinosa*
Habit: Herbs up to 28" tall, usually much shorter.
Leaves: Alternate; blades lanceolate, up to 2 ¼" long, the margins toothed. Some of the larger leaves have a stiff point at the base.
Flowers: Radial, solitary from the axils of the leaves; petals pale yellow, rarely white, about ¼" long.
Fruit: Capsules about ¼" broad.
Bloom Period: Summer.
Distribution: Cameron, Hidalgo, and Willacy counties.
Comments: This species can be recognized by feeling the small points at the bases of the larger leaves. The points are usually small and difficult to see. The specific epithet can be misleading because the plant does not have long sharp spines.

Prickly Mallow *Sida spinosa*

Prickly Mallow *S. spinosa* white-flowering form

Panicled Sand Mallow

Family: Malvaceae

Scientific Name: *Sidastrum paniculatum* [*Sida paniculata*]

Habit: Erect herbs, sometimes woody.

Leaves: Alternate; blades ovate, up to 1 ¾" long, the margins toothed.

Flowers: Radial, facing downward, on slender, branching inflorescences; petals purple to reddish, about ³⁄₁₆" long, reflexed at maturity.

Fruit: Capsules about ¼" broad.

Bloom Period: Winter, spring, summer.

Distribution: Cameron and Willacy counties.

Comments: In Texas this species is reported to grow only in Cameron, Willacy, and Kenedy counties. It has sometimes been confused with *Ayenia pilosa* because of the reflexed purple to reddish petals (see dwarf green-flowered ayenia).

Panicled Sand Mallow *Sidastrum paniculatum*

Pink Narrowleaf Globe Mallow

Family: Malvaceae

Scientific Name: *Sphaeralcea angustifolia* var. *angustifolia*

Habit: Erect-growing perennials up to 5' tall, usually less. Young plants arise from rhizomes.

Leaves: Alternate; blades narrowly lanceolate, grayish often toothed, up to 4 ⅜" long.

Flowers: Inflorescences from the leaf axils, usually one flower opening at a time; corollas radial, pink or salmon colored, the petals up to ¾" long.

Fruit: Capsules roundish, about ¼" broad, the divisions partially splitting apart.

Bloom Period: All seasons.

Distribution: Hidalgo and Starr counties.

Comments: This is our only erect-growing *Sphaeralcea* species. It is abundant in West Texas and only occasionally seen in the western part of the LRGV. The plants with pink flowers are var. *angustifolia*. The ones with pinkish orange flowers are var. *cuspidata*. A third variety of this species grows in Texas, var. *oblongifolia* (synonym, var. *lobata*). It is characterized by broader leaves with length not more than three times the width.

Pink Narrowleaf Globe Mallow *Sphaeralcea angustifolia* var. *angustifolia*

Copper Narrowleaf Globe Mallow *S. angustifolia* var. *cuspidata*

Malvaceae

Wrinkled Globe Mallow

Family: Malvaceae
Scientific Name: *Sphaeralcea hastulata*
Habit: Perennial hairy herbs, the stems usually sprawling on the ground.
Leaves: Alternate; blades ovate but variable, the margins lobed and toothed, up to 2 ⅜" long.
Flowers: Radial; petals brilliant reddish yellow, up to ¾" long.
Fruit: Capsules about ¼" tall.
Bloom Period: Spring, summer, fall.
Distribution: Starr County.
Comments: This is a species of western Texas, occurring only as far east as Starr County.

Wrinkled Globe Mallow *Sphaeralcea hastulata*

South Texas Globe Mallow

Family: Malvaceae
Scientific Name: *Sphaeralcea lindheimeri*
Habit: Perennial hairy herbs, the stems usually sprawling on the ground.
Leaves: Alternate; ovate to heart shaped, up to 1 ¾" long, the margins shallowly cut and toothed.
Flowers: Radial; petals orange-pink, up to ¾" long.
Fruit: Capsules about ³⁄₁₆" tall.
Bloom Period: Spring, summer.
Distribution: Cameron and Hidalgo counties.
Comments: *S. lindheimeri* was named in honor of Ferdinand Lindheimer, an immigrant from Germany who made great contributions to the knowledge of Texas plants in the nineteenth century. This is a species of sandy soils. It is endemic to Texas.

South Texas Globe Mallow *Sphaeralcea lindheimeri*

Palm Leaf Globe Mallow

Family: Malvaceae
Scientific Name: *Sphaeralcea pedatifida*
Habit: Biennial hairy herbs, sometimes short-lived perennials, the stems sprawling on the ground.
Leaves: Alternate; blades more or less ovate, the margins deeply lobed, up to 1 ⅝" long.
Flowers: Radial; petals orange to pink, up to ¾" long.
Fruit: Capsules about ³⁄₁₆" tall.
Bloom Period: Spring, summer.
Distribution: Hidalgo, Willacy, and Starr counties.
Comments: The specific epithet is a Latin word meaning "cut like a bird's foot," referring to the leaf shape.

Palm Leaf Globe Mallow *Sphaeralcea pedatifida*

Velvet Leaf Mallow

Family: Malvaceae
Scientific Name: *Wissadula amplissima*
   [*W. hernandioides*]
Habit: Erect perennials up to 6' tall.
Leaves: Alternate; blades cordate, up to 4 ⅜" long, with several major veins from the base.
Flowers: Radial; petals yellow or orange, about 5/16" long.
Fruit: Capsules about 5/16" tall, dividing into 5 segments; each segment with 2 compartments, the lower one with 2 seeds, the upper one with 1 seed.
Bloom Period: Spring, summer, fall.
Distribution: Cameron and Hidalgo counties.
Comments: In Texas, this species is reported to grow only in Cameron, Hidalgo, and Kenedy counties.

Velvet Leaf Mallow *Wissadula amplissima*

Small-Leaved Wissadula

Family: Malvaceae
Scientific Name: *Wissadula parvifolia*
Habit: Tall understory herbs 2' to 8' tall.
Leaves: Alternate; blades ovate to slightly heart shaped, up to 3 ¼" long.
Flowers: Radial, yellow with reddish centers, about ½" broad.
Fruit: Capsules compressed, about ½" broad, with 5 compartments, each with 3 seeds.
Bloom Period: Summer, fall.
Distribution: Hidalgo County.
Comments: When we first encountered this plant, we were not able to identify it. We made photographs and sent a specimen to Dr. Paul Fryxell, who recognized it to be an undescribed species. Using our collection as the type specimen, he published a description of the new species (Fryxell 2007). This species is known only from a very restricted area. The plants grow in low places as an understory of huisache and retama. The flowers of this species have a reddish center; those of velvet leaf mallow, *W. amplissima*, do not.

Small-Leaved Wissadula *Wissadula parvifolia*

Meliaceae

White Velvet Leaf Mallow

Family: Malvaceae
Scientific Name: *Wissadula periplocifolia*
Habit: Erect hairy herbs, sometimes woody, up to 40" tall.
Leaves: Alternate; blades ovate to lanceolate, up to 3 ½" long.
Flowers: Radial; petals white, sometimes pink basally, up to ¼" long.
Fruit: Capsules about 5/16" tall.
Bloom Period: Fall.
Distribution: Cameron County.
Comments: This globally widespread species also occurs in Mexico, the West Indies, Central and South America, and the Old World tropics. Its white flowers set it apart from our other two *Wissadula* species, which have yellow flowers.

White Velvet Leaf Mallow *Wissadula periplocifolia*

Chinaberry Tree

Family: Meliaceae
Scientific Name: *Melia azedarach*
Habit: Trees up to 50' tall.
Leaves: Alternate, twice compound, up to 16" long; leaflets up to 2 ⅜" long.
Flowers: Radial, lavender with dark purple centers, fragrant; petals about ⅜" long.
Fruit: A round drupe about 9/16" or less broad.
Bloom Period: Spring.
Distribution: Cameron and Hidalgo counties.
Comments: This species is an introduction from Asia, naturalized in low or wet areas. The flowers are attractive but not noticeable at the top of the tree. Small boys sometimes carry the clusters of drupes for ammunition in a mock battle, picking missiles from the cluster as needed. Bark, leaves, flowers, and fruit are all poisonous to humans. The green and ripened fruit are eaten by Red-crowned Parrots and Plain Chachalacas.

Chinaberry Tree *Melia azedarach*

Variable Leaf Snailseed
Family: Menispermaceae
Scientific Name: *Cocculus diversifolius*
Habit: Twining woody vines.
Leaves: Alternate, variable shape, up to 2 ⅜" long, with 3 or more obvious veins arising from the base.
Flowers: Radial, unisexual, on separate plants; sepals and petals similar, whitish, about 1/16" long.
Fruit: Roundish drupes, purple when ripe, about ¼" broad. The seeds are roughly snail shaped.
Bloom Period: Summer, fall.
Distribution: Cameron, Hidalgo, Willacy, and Starr counties.
Comments: The specific epithet aptly describes the leaf, which is quite variable. These vines are often found in the garden or climbing on fences. Because of their very deep roots, it is difficult to eliminate them if they are not wanted. When growing freely in nature, they can develop lower stems 2" or more in diameter.

Variable Leaf Snailseed *Cocculus diversifolius*

Variable Leaf Snailseed *C. diversifolius* flowers

Spreading Sweetjuice
Family Molluginaceae
Scientific Name: *Glinus radiatus*
Habit: Prostrate annuals with star-shaped hairs.
Leaves: Whorled, of 2 different sizes; blades elliptic to spatulate, up to 1" long.
Flowers: Small, in axillary clusters; sepals green, about 3/16" long; petals none.
Fruit: Capsules 3-valved, about ⅛" tall; seeds dark brown, usually shiny.
Bloom Period: Spring, summer, fall.
Distribution: Hidalgo and Starr counties.
Comments: This species is usually found in muddy soils, especially at the edges of drying ponds or other temporary bodies of water. It is native to the tropical and subtropical regions of the Americas, but it is not certain whether it is native or introduced in the LRGV. *Glinus* has a strong superficial resemblance to *Tidestromia,* but the leaves are yellowish green rather than silvery.

Spreading Sweetjuice *Glinus radiatus*

Moraceae

Indian Chickweed

Family Molluginaceae
Scientific Name: *Mollugo verticillata*
Habit: Annuals with spindly stems, mostly prostrate.
Leaves: Basal leaves in rosettes; stem leaves in whorls, spatulate to linear, up to 1 3/16" long.
Flowers: Tiny; petals absent; sepals 5, white.
Fruit: Capsules roundish, tiny.
Bloom Period: Spring, summer, fall.
Distribution: Cameron, Hidalgo, Willacy, and Starr counties.
Comments: This species is often found around moist places but is not limited to them. It is introduced from tropical America.

Indian Chickweed *Mollugo verticillata*

Indian Chickweed *M. verticillata* flowers

Mulberry Weed

Family: Moraceae
Scientific Name: *Fatoua villosa*
Habit: Annuals up to 32" tall, usually less, with many soft, short hairs.
Leaves: Alternate; blades ovate with 3 major veins from the base, 1 ½" long or more, the margins toothed.
Flowers: Tiny, in clusters in the leaf axils, unisexual, both sexes on the same plant.
Fruit: Achenes small, 3-cornered.
Bloom Period: All seasons.
Distribution: Cameron and Hidalgo counties.
Comments: This is a species introduced from Asia. It is gradually making its way through Texas, probably aided through shipment of nursery plants.

Mulberry Weed *Fatoua villosa*

White Mulberry

Family: Moraceae
Scientific Name: *Morus alba*
Habit: Trees up to 49' tall.
Leaves: Alternate, hairless or mostly so; blades broadly ovate up to 8" long, usually less; margins toothed and often lobed.
Flowers: Tiny, separate male and female flowers on the same tree.
Fruit: Small, juicy, crowded and compacted together to form a single fruit called a syncarp. It may be white, pinkish, to purple or reddish when ripe, resembling a blackberry.
Bloom Period: Spring.
Distribution: Cameron and Hidalgo counties.
Comments: This tree was introduced from China and is widespread. It is similar to *M. rubra* (not illustrated), which has leaves that are hairy on the lower surfaces. The fruits are edible and very attractive to birds.

White Mulberry *Morus alba*

Yellow Lotus

Family: Nelumbonaceae
Scientific Name: *Nelumbo lutea*
Habit: Aquatic perennials with slender rhizomes rooting in the mud.
Leaves: Arising from the rhizomes; petioles inserted in the middle of the blade (called peltate); blades roundish, up to 24" broad.
Flowers: Pale yellow, up to 10" broad; sepals and petals numerous.
Fruit: Similar to nuts, embedded in a dried receptacle.
Bloom Period: Spring, summer.
Distribution: Cameron, Hidalgo, and Willacy counties.
Comments: The seed pods are sometimes used in floral arrangements. The lotus has been held sacred in parts of Asia for at least five thousand years. It resembles the water lily and until recently was classified in the same family. It is much more aggressive than the water lily. If introduced into a pond with mixed plant species, it quickly crowds them all out and becomes the major if not the only occupant. Native to North and South America, yellow lotus is believed to have been dispersed by Native Americans as a food resource (Diggs et al. 1999). Its status as a native to the LRGV is not clear. The common name yellow lotus is not to be confused with the genus *Lotus*, a legume (Fabaceae).

Yellow Lotus *Nelumbo lutea*

Yellow Lotus *N. lutea* fruit with receptacle

Nyctaginaceae

Amelia's Sand Verbena

Family: Nyctaginaceae
Scientific Name: *Abronia ameliae*
Habit: Perennials up to 24" tall, all parts glandular and sticky.
Leaves: Opposite; blades ovate to rounded, up to 3 1/8" long.
Flowers: Numerous, crowded in globose heads; sepals purple, up to 1" long; petals absent.
Fruit: Dry, one-seeded, with 5 wings.
Bloom Period: Spring, summer.
Distribution: Hidalgo and Starr counties.
Comments: This is a species of sandy soils. It is endemic to Texas. When rain is abundant, it presents a spectacular show, covering hillsides with a purple haze. In drier years, it is still a beautiful wildflower but in smaller numbers.

Amelia's Sand Verbena *Abronia ameliae*

Unequal-Leaf Trumpets

Family: Nyctaginaceae
Scientific Name: *Acleisanthes anisophylla*
Habit: Perennial herbs arising from thick taproots; plants covered with minute, white hairs. Stems sometimes erect, often weak and prostrate except for the erect tips.
Leaves: Opposite, one of each pair two to ten times as large as the other; blades green (or young leaves sometimes with a copper-color tint), with prominent purplish veins, somewhat succulent, mostly ovate, the apices rounded to pointed; smaller blades up to 1 1/8" long, larger blades up to 2 3/4" long.
Flowers: Radial, white, white with a pink tube (Poole, 1975), or pink, 1"– 1 5/8" broad, the tube very long; petals absent. Green, self-fertilizing flowers which do not open also occur, about 1/8" long.
Fruit: Capsules 10-ribbed, sparsely hairy, about 1/4" tall, constricted at the base.
Bloom Period: Spring, fall.
Distribution: Starr County.
Comments: Collections of this species in Texas are scattered, from Starr County to as far north as Val Verde County. It generally grows on rocky soil, or sand. The pink color of the flower was a surprise to us. In recent major treatments, the color is not listed, or else is reported either as white, or white with a pink tube. Bill Carr, botanist for the Texas Nature Conservancy, confirmed the identification as *Acleisanthes anisophylla,* and reported having made several collections in Texas, all with pink flowers. Time of flowers opening was not observed. They begin to close before noon. *A. longiflora* and *A. obtusa,* which grow in our area, have white flowers.

Unequal-Leaf Trumpet *Acleisanthes anisophylla*

Angel Trumpets

Family: Nyctaginaceae
Scientific Name: *Acleisanthes longiflora*
Habit: Perennial herbs, woody basally; stems sprawling.
Leaves: Opposite but unequal, succulent; blades variable, triangular to lanceolate or linear, up to 1 ¾" long.
Flowers: Solitary from the axils of the leaves; calyx white, up to 6 ¾" long, with a long tube; corolla absent.
Fruit: Dry, 5-angled, about ⅜" long.
Bloom Period: Spring, summer, fall.
Distribution: Hidalgo and Starr counties.
Comments: The flowers open in the evening and close in early morning. On cloudy days, the flowers remain open much longer. The plants can be covered with flowers and are spectacular at dusk or dawn, or especially at night with a full moon. The flowers are fragrant.

Angel Trumpets *Acleisanthes longiflora*

Vine Four O'clock

Family: Nyctaginaceae
Scientific Name: *Acleisanthes obtusa*
Habit: Perennial sprawling herbs, often climbing into shrubs or trees.
Leaves: Opposite but unequal; blades triangular to ovate, up to 2 ⅜" long.
Flowers: In large clusters, fragrant; calyx white, up to 1 ¾" long, with a long tube; corolla absent.
Fruit: Dry, 5-angled, about ¼" long.
Bloom Period: All seasons.
Distribution: Cameron, Hidalgo, Willacy, and Starr counties.
Comments: The flowers open in the evening and close in early morning. On cloudy days, the flowers remain open much longer. The flowers are at their best when seen on a moonlit night.

Vine Four O'clock *Acleisanthes obtusa*

Nyctaginaceae

Smooth Umbrella Wort, Garrapatilla

Family: Nyctaginaceae
Scientific Name: *Allionia choisyi*
Habit: Annuals, sometimes perennials, with prostrate stems.
Leaves: Opposite; blades ovate, up to 2" long.
Flowers: Radial, in clusters of 3, enclosed by 3 bracts; calyxes pink to rose in color, up to 3/16" long; corollas absent.
Fruit: Dry, somewhat flattened, about 3/16" long, with 2 rows of teeth and 2 rows of glands.
Bloom Period: All seasons.
Distribution: Cameron, Hidalgo, and Starr counties.
Comments: The Spanish name for this plant is a diminutive of *garrapata,* meaning "tick," because the fruit resembles a small tick. It is usually an annual here; in Mexico, many are perennials. This plant was named for Carlo Allioni, an Italian botanist, and Jacque Denis Choisy, a Swiss botanist.

Smooth Umbrella Wort, Garrapatilla *Allionia choisyi*

Hierba De La Hormiga

Family: Nyctaginaceae
Scientific Name: *Allionia incarnata*
Habit: Annuals or perennials with hairy prostrate stems.
Leaves: Opposite; blades ovate, up to 3 3/8" long.
Flowers: Radial, in clusters of 3, enclosed by 3 bracts; calyxes white to red but usually pink, about 3/16" long; corollas absent.
Fruit: Dry, somewhat flattened, about 3/16" long, with 2 rows of teeth and 2 rows of glands.
Bloom Period: All seasons.
Distribution: Cameron, Hidalgo, and Starr counties.
Comments: This could make an excellent landscape plant, especially in sandy or gravelly soils. As an added feature, doves feed on the seeds. The specific epithet *incarnata* is a Latin word for "flesh colored," the common color of the flower. The common name translates as "ant plant."

Hierba De La Hormiga *Allionia incarnata*

Scarlet Spiderling

Family: Nyctaginaceae
Scientific Name: *Boerhavia coccinea*
Habit: Perennial, glandular herbs with slender leaning or prostrate stems arising from a thickened taproot.
Leaves: Opposite; blades ovate or roundish, up to 2 1/8" long.
Flowers: Radial, in small clusters; calyxes red, 1/8" or less broad; corollas absent.
Fruit: Dry, glandular, up to 3/16" long.
Bloom Period: Spring, summer, fall.
Distribution: Cameron, Hidalgo, Willacy, and Starr counties.
Comments: This is an extremely widespread species, sometimes becoming a nuisance in the garden.

Scarlet Spiderling *Boerhavia coccinea*

Erect Spiderling

Family: Nyctaginaceae
Scientific Name: *Boerhavia erecta*
Habit: Annuals or perennials, sometimes glandular, the stems often reddish colored.
Leaves: Opposite; blades broadly ovate to triangular, up to 3 1/8" long.
Flowers: Radial, in small clusters; calyxes white or pinkish, about 1/16" long; corollas absent.
Fruit: Dry, up to 3/16" long.
Bloom Period: Spring, summer, fall.
Distribution: Cameron, Hidalgo, Willacy, and Starr counties.
Comments: The glandular, sticky stems of this species have been used as a fly catcher and also as a fly swatter.

Erect Spiderling *Boerhavia erecta*

Nyctaginaceae

Five Wing Spiderling
Family: Nyctaginaceae
Scientific Name: *Boerhavia intermedia*
Habit: Annuals; stems erect, leaning, or falling over.
Leaves: Opposite; blades ovate to lanceolate, sometimes linear, up to 1 ¾" long.
Flowers: Radial, in small clusters; calyxes pink or whitish, less than ⅛" long; corollas absent.
Fruit: Dry, about ⅛" long.
Bloom Period: Summer.
Distribution: Cameron, Hidalgo, and Starr counties.
Comments: This species is more commonly found in West Texas. It could make a pleasing landscape plant.

Five Wing Spiderling *Boerhavia intermedia*

Climbing Wortclub
Family: Nyctaginaceae
Scientific Name: *Commicarpus scandens* [*Boerhavia scandens*]
Habit: Perennials with stems sprawling or clambering over nearby erect plants.
Leaves: Opposite; blades ovate, up to 2 ⅜" long.
Flowers: Radial, in small umbrella-like clusters; calyx greenish yellow, up to 3/16" long.
Fruit: Dry, with scattered glands, up to ⅜" long.
Bloom Period: Summer, fall.
Distribution: Cameron, Hidalgo, and Willacy counties.
Comments: This species is found in southeast Texas and in West Texas. The young leaves have light-colored veins.

Climbing Wortclub *Commicarpus scandens*

South Texas Four O'clock

Family: Nyctaginaceae

Scientific Name: *Mirabilis austrotexana* [*M. albida*]

Habit: Perennials up to 5' tall; stems few, swollen at the nodes.

Leaves: Opposite; blades lanceolate to ovate, up to 4 ¾" long.

Flowers: Radial, in groups of 3, surrounded by 5 fused usually greenish bracts; calyxes white to pink or reddish purple, up to ⅜" long.

Fruit: Dry, ribbed, about 3/16" tall, surrounded by the enlarged, papery bracts, which often have purplish veining.

Bloom Period: All seasons.

Distribution: Cameron, Hidalgo, and Willacy counties.

Comments: The large papery bracts are often mistaken for five flower petals. This species was originally identified as *Mirabilis albida,* a fairly widespread species. The southern populations were subsequently recognized as a separate species, *M. austrotexana,* by Dr. Billie L. Turner, professor at the University of Texas at Austin. This plant typically grows in sandy areas, at the coast and inland.

South Texas Four O'clock *Mirabilis austrotexana*

South Texas Four O'clock *M. austrotexana* papery bracts remaining after the petal-like calyxes fall

Nyctaginaceae

Linear Leaf Four O'clock

Family: Nyctaginaceae

Scientific Name: *Mirabilis linearis*

Habit: Perennials up to 40" tall; stems often whitish, erect to leaning or prostrate.

Leaves: Opposite; few to many, mostly linear, up to 4" long.

Flowers: Radial, in groups of 3, surrounded by 5 fused bracts; calyxes pink to purplish, up to ⅜" long; petals absent.

Fruit: Dry, about 3/16" tall, surrounded by the enlarged, papery bracts, which are whitish, often with purplish veining.

Bloom Period: Spring, summer.

Distribution: Cameron County; probably also Willacy County.

Comments: The large, papery bracts are often mistaken for five flower petals. The inflorescences of this species resemble those of South Texas four o'clock, *M. austrotexana,* but the difference in leaf shape makes it easy to distinguish the two species.

Linear Leaf Four O'clock *Mirabilis linearis*

Scarlet Musk Flower, Devil's Bouquet

Family: Nyctaginaceae

Scientific Name: *Nyctaginea capitata*

Habit: Perennial, glandular and sticky herbs up to 16" tall, growing from a tuberous root, the branches often falling over.

Leaves: Opposite; blades ovate to triangular, up to ⅜" long.

Flowers: In clusters, subtended by many bracts; calyxes pink to red, rarely yellow, up to 1 ⅝" long; corollas absent.

Fruit: Dry, leathery, about ¼" tall.

Bloom Period: Spring, summer, fall.

Distribution: Cameron, Hidalgo, and Starr counties.

Comments: This species is very showy, with the large, brilliantly colored flowers attracting attention whenever it is in bloom. It is an excellent butterfly nectar plant. Some find the flower's odor offensive.

Scarlet Musk Flower, Devil's Bouquet *Nyctaginea capitata*

Climbing Devil's Claw

Family: Nyctaginaceae

Scientific Name: *Pisonia aculeata*

Habit: Shrubs up to 6' or taller; stems clambering onto nearby vegetation, aided by the curved spines, which can be up to ¾" long, occasionally absent.

Leaves: Opposite; ovate to obovate, up to 4" long.

Flowers: Radial, sticky, in clusters, male and female on separate plants; calyxes yellowish green, about ⅛" long.

Fruit: Glandular sticky, up to ⅜" or slightly longer.

Bloom Period: Spring, summer.

Distribution: Cameron and Hidalgo counties.

Comments: In Texas this species is limited to Cameron and Hidalgo counties. It also grows in adjacent Mexico and southern Florida and widely in the tropics. It grows well in cultivation. The flowers are not particularly showy, but the gracefully arching stems provide perches for birds.

Climbing Devil's Claw *Pisonia aculeata*

Blue Water Lily, Lampazos

Family: Nymphaeaceae

Scientific Name: *Nymphaea elegans*

Habit: Plants submerged in water with floating leaves.

Leaves: Blades more or less rounded, about 8" broad, with a narrow V cut to the petiole.

Flowers: Solitary, held above the water surface, blue to pale violet, up to 6" broad.

Fruit: Roundish, up to 1 ⅛" broad, maturing underwater.

Bloom Period: Spring, summer, fall.

Distribution: Cameron, Hidalgo, and Willacy counties.

Comments: These plants have an amazing capacity for survival. For example, some low places such as old resaca beds may remain dry and barren for several years, with the ground cracked. Then, after copious rains, small ponds form. Immediately, the water lilies that have been lying dormant become active and produce flowers, seemingly in a few days.

Blue Water Lily, Lampazos *Nymphaea elegans*

Oleaceae

Yellow Water Lily, Banana Water Lily, Audubon's Water Lily

Family: Nymphaeaceae
Scientific Name: *Nymphaea mexicana*
Habit: Plants submerged in water, the leaves floating on the surface.
Leaves: Blades more or less rounded, about 8" broad, with a narrow V cut to the petiole.
Flowers: Solitary, yellow, held above the water surface, about 4" broad.
Fruit: Egg shaped, about 1" long, maturing underwater.
Bloom Period: Spring, summer, fall.
Distribution: Cameron and Hidalgo counties; probably also Willacy County.
Comments: These beautiful water lilies were originally found in Mexico and named accordingly. Later, they were found in Florida, where they were painted by John James Audubon. Subsequently, they were found in Texas. The common name, Banana Water Lily, was coined because the small clusters of roots resemble bunches of bananas. It has nothing to do with the flower color.

Yellow Water Lily, Banana Water Lily, Audubon's Water Lily
Nymphaea mexicana

Narrow Leaf Elbow Bush

Family: Oleaceae
Scientific Name: *Forestiera angustifolia*
Habit: Shrubs up to 8' or taller.
Leaves: Opposite; blades spatulate to nearly linear, up to 1 ⅜" long.
Flowers: Male and female flowers on separate plants; sepals very small; petals none.
Fruit: Drupes, blackish and slightly elongated when mature, up to ¼" or longer.
Bloom Period: Spring, summer, fall.
Distribution: Cameron, Hidalgo, Willacy, and Starr counties.
Comments: The fruit and leaves provide food for wildlife. This is an attractive thornless native ornamental.

Narrow Leaf Elbow Bush *Forestiera angustifolia* plant with male flowers

Narrow Leaf Elbow Bush *F. angustifolia* plant with female flowers

Narrow Leaf Elbow Bush *F. angustifolia* plant in fruit

Mexican Ash, Fresno

Family: Oleaceae
Scientific Name: *Fraxinus berlandieriana*
Habit: Trees up to 42' tall.
Leaves: Opposite, compound; leaflets obovate to elliptic, up to 4 ¾" long.
Flowers: Male and female on separate plants; calyxes small, 4-lobed; petals absent.
Fruit: Dry, one-seeded with a wing about 1 ⅛" long.
Bloom Period: Spring.
Distribution: Cameron, Hidalgo, and Starr counties.
Comments: This native tree primarily grows naturally along river banks. It is now extensively used as a shade tree around homes and public buildings. A number of non-native species of *Fraxinus* have been introduced, and sometimes it is difficult to distinguish them from *F. berlandieri*. This species is named for Jean Louis Berlandier, who made important plant collections in Mexico and Texas.

Mexican Ash, Fresno *Fraxinus berlandieriana* with winged fruit

Redbud

Family: Oleaceae
Scientific Name: *Menodora heterophylla*
Habit: Low-growing perennial herbs, woody basally; stems slender, often falling over.
Leaves: Opposite; blades ovate to narrowly obovate, usually lobed, up to 1 ⅜" long.
Flowers: Solitary at the ends of the stems; buds red; corolla lobes yellow, up to ½" long.
Fruit: Capsules in two segments up to ⅜" long, the tops falling as lids.
Bloom Period: All seasons.
Distribution: Cameron, Hidalgo, Willacy, and Starr counties.
Comments: The red buds are a striking feature. The 2-parted fruit is another indicator of this species.

Redbud *Menodora heterophylla*

Redbud *M. heterophylla* fruit

Onagraceae

Square Bud Primrose
Family: Onagraceae
Scientific Name: *Calylophus serrulatus* [*C. australis*]
Habit: Herbs 12" tall or less, sometimes woody basally.
Leaves: Alternate; blades linear with toothed margins, up to 1 ¼" long.
Flowers: Radial; petals 4, yellow, up to ½" long; stigmas globose, ovaries inferior.
Fruit: Capsules cylindrical, about ¾" long.
Bloom Period: Spring, summer, fall.
Distribution: Cameron and Willacy counties.
Comments: Square bud primrose is abundant in sand, especially at the coast. It is fairly widespread in other parts of Texas. The 4 petals, inferior ovary, and globose stigmas are indicators of this species.

Square Bud Primrose *Calylophus serrulatus*

Plains Gaura
Family: Onagraceae
Scientific Name: *Gaura brachycarpa*
Habit: Hairy annuals up to 26" tall.
Leaves: Alternate; blades usually lanceolate, up to 2½" long.
Flowers: Bilateral; petals 4, white turning pink, up to ½" long; stigmas 4-lobed; ovaries inferior.
Fruit: Capsules sessile; somewhat pyramid shaped, about ¼" long.
Bloom Period: Spring, summer, fall.
Distribution: Cameron, Hidalgo, and Willacy counties.
Comments: The genus name comes from a Greek word meaning "superb," referring to the unusual flowers.

Plains Gaura *Gaura brachycarpa*

Sweet Gaura

Family: Onagraceae
Scientific Name: *Gaura drummondii* [*G. odorata*]
Habit: Hairy, sweet-scented perennials up to 24" or taller.
Leaves: Alternate; blades lanceolate to elliptic, up to 2 ¾" long; margins can be entire, wavy, or toothed.
Flowers: Bilateral; petals 4, pink to red, up to ⅜" long; stigmas 4-lobed; ovaries inferior.
Fruit: Capsules up to ½" long, with a thick stalk up to half the length.
Bloom Period: Spring, summer, fall.
Distribution: Cameron, Hidalgo, Willacy, and Starr counties.
Comments: This is one of our most attractive *Gaura* species, forming colonies by underground rhizomes. This species was named in honor of Thomas Drummond, who did important work in Texas, especially the central part.

Sweet Gaura *Gaura drummondii*

Mckelvey's Beeblossom

Family: Onagraceae
Scientific Name: *Gaura mckelveyae*
Habit: Hairy perennials, woody at the base, up to 20" tall.
Leaves: Alternate; blades oblanceolate to linear, tapering onto the petioles, up to 4 ¾" long; margins toothed, sometimes shallowly lobed.
Flowers: Bilateral; petals white or pink, turning pink or reddish, up to ⅜" or longer; stigmas 4-lobed; ovaries inferior.
Fruit: Capsules up to ¾" long, with a slender stalk up to ⅜" long.
Bloom Period: Spring, summer.
Distribution: Cameron, Hidalgo, Willacy, and Starr counties.
Comments: McKelvey's beeblossom is recognized by its slender-stalked fruit and its deeply cleft or strongly sinuate (wavy) leaf margins.

Mckelvey's Beeblossom *Gaura mckelveyae*. The whitish *G. mckelveyae* flowers are surrounded by reddish *Gaillardia pulchella* flowers.

Onagraceae

Lizard Tail, Velvety Leaf Gaura

Family: Onagraceae

Scientific Name: *Gaura parviflora*

Habit: Annuals with glandular hairs, up to 5' or taller.

Leaves: Alternate; blades lanceolate to narrowly elliptic, up to 4" long; margins entire or with scattered tiny teeth.

Flowers: Bilateral; petals 4, white turning pink or rose, up to 3/16" long; stigmas 4-lobed; ovaries inferior.

Fruit: Capsules up to 7/16" long, often narrowed basally.

Bloom Period: Spring, summer.

Distribution: Cameron and Hidalgo counties.

Comments: With its very small flowers and robust vegetative habit, this species is easy to identify, but it does not attract much attention for its beauty. However, it is much appreciated by various insects, such as butterflies. It grows well in poor soils in full sun.

Lizard Tail, Velvety Leaf Gaura *Gaura parviflora*

Wavy Leaved Gaura

Family: Onagraceae

Scientific Name: *Gaura sinuata*

Habit: Moderately hairy perennials up to 24" tall.

Leaves: Alternate; blades linear to oblanceolate, up to 3 1/8" long; margins wavy and toothed or moderately to deeply lobed.

Flowers: Bilateral; petals 4, white or pink, turning pink or red, up to 1/2" or longer; stigmas 4-lobed; ovaries inferior.

Fruit: Capsules up to 3/8" long, with a slender stalk up to 5/16" long.

Bloom Period: Spring, summer.

Distribution: Cameron, Hidalgo, and Willacy counties.

Comments: This plant family is fairly easy to recognize with the flowers usually having 4 petals and inferior ovaries. More commonly among flowering plants, the floral parts are in threes or fives.

Wavy Leaved Gaura *Gaura sinuata*. Its red flowers are at the upper left of the photo

Wing Leaf Primrose Willow

Family: Onagraceae
Scientific Name: *Ludwigia decurrens*
Habit: Herbs up to 6' tall, or stems falling over and appearing much shorter. Stems with 4 wings.
Leaves: Alternate; blades lanceolate to elliptical, up to 4 ¾" long, often shorter.
Flowers: Radial; petals 4, yellow, up to ½" long; ovaries inferior.
Fruit: Capsules with 4 angles, up to 1 ¼" long.
Bloom Period: Summer, fall.
Distribution: Kenedy County.
Comments: This species is easily distinguished from the other *Ludwigia* species by its winged stems. Like them, it prefers wet places. Although it was seen in Kenedy County, the site was less than five miles from the Willacy County border. Since similar habitats occur in Willacy County, it is highly probable that it grows there also. The species is better known in East Texas and has not previously been reported for our area or for Kenedy County.

Wing Leaf Primrose Willow *Ludwigia decurrens*

Wing Leaf Primrose Willow *L. decurrens* stem. Sometimes when the plant is standing in water, a water fungus attaches to the submerged parts, producing a spongy growth

Primrose Willow

Family: Onagraceae
Scientific Name: *Ludwigia octovalvis*
Habit: Erect herbs up to 40" or taller.
Leaves: Alternate, occasionally opposite; blades more or less lanceolate, up to 4 ¾" long.
Flowers: Radial; petals 4, yellow, up to ⅝" long; stigmas globose; ovaries inferior.
Fruit: Capsules cylindrical, up to 1 ¾" long.
Bloom Period: Summer, fall.
Distribution: Cameron and Hidalgo counties.
Comments: This species is found in wet places such as resaca and canal banks. It is a host plant for a diurnal Sphinx moth.

Primrose Willow *Ludwigia octovalvis*

Water Primrose

Family: Onagraceae
Scientific Name: *Ludwigia peploides*
Habit: Sprawling or floating herbs.
Leaves: Alternate; blades narrowly elliptic to obovate, up to 2 ⅜" long.
Flowers: Radial; petals 5, yellow, up to ½" or longer; stigmas 5 crowded globes; ovaries inferior.
Fruit: Capsules cylindrical, up to 1 ½" or longer.

Onagraceae

Bloom Period: Spring, summer.
Distribution: Cameron and Hidalgo counties.
Comments: This species usually occurs floating in water or on mud at the edge of a pond. It breaks from the general family characteristics in having 5 sepals and 5 petals rather than 4.

Water Primrose *Ludwigia peploides*

Beach Evening Primrose
Family: Onagraceae
Scientific Name: *Oenothera drummondii*
Habit: Hairy perennials with woody stems, up to 40" tall, usually less.
Leaves: Alternate; blades gray-hairy, spatulate, up to 3" long, the margins smooth.
Flowers: Radial; petals 4, yellow, fading to orange, up to 1 3/8" long; stigmas 4; ovaries inferior.
Fruit: Capsules cylindrical, up to 1 5/8" long.
Bloom Period: Spring, summer, fall.
Distribution: Cameron and Willacy counties.
Comments: This is a common plant in the coastal sand dunes and one of our showiest *Oenothera* species. It is distinguished from our other *Oenothera* species by its large yellow flowers and leaves with smooth margins. *O. drummondii* was named in honor of Thomas Drummond, who made important plant collections in the Edwards Plateau area and along the Brazos, Colorado, and Guadalupe rivers.

Beach Evening Primrose *Oenothera drummondii*

Showy Evening Primrose
Family: Onagraceae
Scientific Name: *Oenothera grandis*
Habit: Annuals up to 18" tall, from a basal rosette.
Leaves: Alternate; blades ovate to oblanceolate, up to 4" long, the margins toothed and often deeply lobed.
Flowers: Radial; petals 4, yellow, up to 1 3/8" long; stigmas 4; ovaries inferior.
Fruit: Capsules cylindrical, up to 1 3/8" long.
Bloom Period: Spring, summer.
Distribution: Cameron, Hidalgo, and Starr counties.
Comments: This species, like beach evening primrose, *O. drummondii*, prefers sandy soils; however, it is not limited to the coast. It is distinguished from our other *Oenothera* species by its large yellow flowers and leaves with toothed margins.

Showy Evening Primrose *Oenothera grandis*

Kunth's Evening Primrose

Family: Onagraceae
Scientific Name: *Oenothera kunthiana*
Habit: Annuals up to 16" tall from a basal rosette.
Leaves: Alternate; blades ovate to obovate, up to 3 ½" long.
Flowers: Radial; petals 4, white, turning to pink, up to ¾" long; stigmas 4; ovaries inferior.
Fruit: Capsules club shaped, about ⅜" long.
Bloom Period: Spring.
Distribution: Cameron, Hidalgo, and Willacy counties.
Comments: This species is very similar to *O. tetraptera* (not pictured). Its flowers are usually smaller (petals ¾" or less), and its stigmas are usually at the same level as the anthers.

Kunth's Evening Primrose *Oenothera kunthiana*

Cut Leaved Evening Primrose

Family: Onagraceae
Scientific Name: *Oenothera laciniata*
Habit: Annuals up to 18" tall from a basal rosette.
Leaves: Alternate; blades ovate to oblanceolate, up to 4" long; margins toothed and often deeply lobed.
Flowers: Radial; petals 4, yellow, 3/16" up to slightly less than ¾" long; stigmas 4; ovaries inferior.
Fruit: Capsules cylindrical, up to 1 ⅜" long.
Bloom Period: Spring, summer.
Distribution: Cameron, Hidalgo, Willacy, and Starr counties.
Comments: This species has the smallest flowers of our yellow *Oenotheras*. It prefers sandy soils.

Cut Leaved Evening Primrose *Oenothera laciniata*

Rose Sundrops

Family: Onagraceae
Scientific Name: *Oenothera rosea*
Habit: Perennials up to 16" tall.
Leaves: Alternate; blades narrowly ovate, up to 2 ⅜" long; margins toothed and occasionally deeply lobed basally.
Flowers: Radial; petals 4, pink, up to ½" long; stigmas 4; ovaries inferior.
Fruit: Capsules club shaped, up to ⅜" long, with a slim stalk about half the length.
Bloom Period: Spring, summer.
Distribution: Cameron and Hidalgo counties.

Rose Sundrops *Oenothera rosea*

Orobanchaceae

Comments: This species resembles *O. speciosa* (see following account), which has larger flowers and no flower pedicels. *O. rosea* has flower pedicels. Some sources have combined *O. rosea* with cut leaved evening primrose, *O. laciniata*. Since the two taxa are distinct in flower color, and in fruit shape and size, we decided it would be more useful to retain the older classification.

Pink Evening Primrose, Buttercup, Amapola Del Campo

Family: Onagraceae
Scientific Name: *Oenothera speciosa*
Habit: Perennials up to 20" tall.
Leaves: Alternate; blades obovate to lanceolate, up to 2" long; margins toothed and lobed.
Flowers: Radial; petals 4, pink (rarely white), up to 1 ⅝" long; stigmas 4; ovaries inferior.
Fruit: Capsules club shaped, up to ⅜" long, on a stalk of about equal length.
Bloom Period: Spring, summer.
Distribution: Cameron, Hidalgo, Willacy, and Starr counties.
Comments: This is our most abundant and widespread *Oenothera*. It is a very showy wildflower, at its best in early spring. The Spanish common name translates as "wild poppy."

Pink Evening Primrose, Buttercup, Amapola Del Campo
Oenothera speciosa

Large Flower Broomrape

Family: Orobanchaceae
Scientific Name: *Orobanche ludoviciana* subsp. *ludoviciana*
Habit: Root parasites 4–20" tall, somewhat sticky.
Leaves: Scalelike, about ½" long.
Flowers: Bilateral; corollas light purple to yellowish, up to 1 ⅜" long, emerging well beyond the calyx.
Fruit: Capsules about ⅝" tall.
Bloom Period: Spring, summer, fall.
Distribution: Starr County.
Comments: The two subspecies of *Orobanche* have in the past been considered two separate species. We have seen only one population of each subspecies. The most notable differences observed are the light purple to yellowish flower color and the emergence of the floral tube beyond the calyx in subsp. *ludoviciana*. Subsp. *multiflora* (see following account)

Large Flower Broomrape *Orobanche ludoviciana* subsp. *ludoviciana*

has dark purple flowers, and the floral tube remains within the calyx. The population of subsp. *ludoviciana* we observed appears to be parasitic on saladillo, *Varilla texana,* a member of the Asteraceae family. Both subspecies of *Orobanche* have been reported to parasitize various members of the Asteraceae.

Broomrape

Family: Orobanchaceae
Scientific Name: *Orobanche ludoviciana* subsp. *multiflora*
Habit: Root parasites up to 8" or taller.
Leaves: Whitish and scalelike, about ½" long.
Flowers: Bilateral; corollas purple to yellow, about 1 ⅜" long, barely emerging beyond the calyx.
Fruit: Capsules about ⅝" tall.
Bloom Period: Spring, summer.
Distribution: Cameron and Hidalgo counties.
Comments: The plants tend to prefer sand or sandy soil. They are not common. The only population we have observed is located at Boca Chica Beach. Camphor weed, *Heterotheca subaxillaris*, is reported to be a host plant. We have observed this subspecies often in close association with *Dalea emarginata*, suggesting another host plant possibility.

Broomrape *Orobanche ludoviciana* subsp. *multiflora*

Creeping Lady's-Sorrel

Family: Oxalidaceae
Scientific Name: *Oxalis corniculata* var. *wrightii*
Habit: Small creeping plants rooting at the nodes.
Leaves: Palmately compound; leaflets 3, obcordate, up to ¾" broad.
Flowers: Radial; petals yellow, up to ¼" long.
Fruit: Capsules hairy, cylindrical, about ¾" long.
Bloom Period: Spring, summer, fall, early winter.
Distribution: Cameron and Hidalgo counties.
Comments: This species reproduces so well that it can become a pest in the garden. When the capsules burst open, they throw the seeds some distance from the parent plant.

Creeping Lady's-Sorrel *Oxalis corniculata* var. *wrightii*

Penny Leaf Wood Sorrel

Family: Oxalidaceae
Scientific Name: *Oxalis dichondrifolia*
Habit: Erect or leaning perennials with woody bases.
Leaves: Simple; blades ovate to roundish, up to 1 ⅛" broad.
Flowers: Radial, yellow, about ⅝" long.
Fruit: Capsules cylindrical, about ⅜" long.
Bloom Period: All seasons.
Distribution: Cameron, Hidalgo, and Starr counties.
Comments: This is our only *Oxalis* with simple leaves.

Penny Leaf Wood Sorrel *Oxalis dichondrifolia*

Oxalidaceae

Drummond's Wood Sorrel

Family: Oxalidaceae
Scientific Name: *Oxalis drummondii*
Habit: Herbs arising from bulbs.
Leaves: Palmately compound; blades V or boomerang shaped, up to 2" broad.
Flowers: Radial, purple, up to ¾" long.
Fruit: Capsules cylindrical, up to ⅜" long.
Bloom Period: Spring, summer, fall.
Distribution: Cameron, Hidalgo, and Willacy counties.
Comments: This species is endemic to Texas. It can be a pest in flower beds and gardens. The leaves of *Oxalis* species have been eaten in salads; however, it is not usually recommended because they contain oxalic acid, which can be toxic if too much is consumed. Another bulb-producing species, *O. tuberosa,* is cultivated in the Andes of Colombia and Peru. Its bulbs are harvested like potatoes and treated to remove the oxalic acid before being eaten. *O. drummondii* was named in honor of Thomas Drummond, who did botanical work in Central Texas in the 1830s.

Drummond's Wood Sorrel *Oxalis drummondii*

Narrow Leaf Shrubby Wood Sorrel

Family: Oxalidaceae
Scientific Name: *Oxalis frutescens* subsp. *angustifolia* [*O. berlandieri*]
Habit: More or less erect perennials, often with woody bases.
Leaves: Pinnately compound; leaflets obovate, up to ⅝" or longer.
Flowers: Radial, yellow, about ⅝" long.
Fruit: Capsules cylindrical, about 5/16" long.
Bloom Period: Spring, summer, fall.
Distribution: Hidalgo and Willacy counties.
Comments: The yellow flowers and the pinnately compound leaves are identifying characteristics of this species. It commonly grows in sandy soils and can be very attractive.

Narrow Leaf Shrubby Wood Sorrel *Oxalis frutescens* subsp. *angustifolia*

Yellow Wood Sorrel

Family: Oxalidaceae
Scientific Name: *Oxalis stricta* [*O. dillenii*]
Habit: Creeping annuals.
Leaves: Palmately compound; leaflets obcordate, up to ¾" long.
Flowers: Radial, yellow, up to ⅜" long.
Fruit: Capsules often white-hairy, up to 1" long.
Bloom Period: Winter, spring, and summer.
Distribution: Cameron and Hidalgo counties.
Comments: Two important characteristics for recognizing this species are the palmately compound leaves and the yellow flowers.

Yellow Wood Sorrel *Oxalis stricta*

Golden Prickly Poppy

Family: Papaveraceae
Scientific Name: *Argemone aenea*
Habit: Erect prickly herbs up to 32" tall, with yellow sap.
Leaves: Alternate; blades deeply pinnately lobed, the lobes sharp-pointed.
Flowers: Radial, yellow to orange in color, 2 ¾" to 4 ¾" broad; stamens 150 or more; stamen filaments flattened, brown.
Fruit: Prickly capsules up to 1 ⅜" tall.
Bloom Period: Spring.
Distribution: Cameron, Hidalgo, and Starr counties.
Comments: This plant is beautiful when in flower but is a "look, don't touch" plant. It commonly grows in disturbed areas such as railroad tracks and roadsides. It resembles the yellow prickly poppy, *A. mexicana,* which has 50 or fewer stamens with threadlike, yellowish filaments and often has smaller flowers. Note the brown, flattened filaments and the spiny fruit.

Golden Prickly Poppy *Argemone aenea*

Yellow Prickly Poppy, Mexican Poppy

Family: Papaveraceae
Scientific Name: *Argemone mexicana*
Habit: Erect prickly herbs up to 32" tall, with yellow sap.
Leaves: Alternate; blades deeply pinnately lobed, the lobes sharp-pointed.
Flowers: Radial, yellow, up to 2 ¾" broad; stamens 50 or less; stamen filaments threadlike, yellowish.
Fruit: Prickly capsules up to 1 ¾" tall.
Bloom Period: Spring, summer.

Yellow Prickly Poppy, Mexican Poppy *Argemone mexicana*

Papaveraceae

Distribution: Cameron and Hidalgo counties.
Comments: This species commonly grows in disturbed soils. It resembles golden prickly poppy, *A. aenea,* which has 150 or more stamens with flattened, brown filaments, and usually has larger flowers. Sometimes there is a slight color difference. Oil from the seeds has been used for lamps in India.

Red Poppy
Family: Papaveraceae
Scientific Name: *Argemone sanguinea*
Habit: Erect prickly herbs up to 48" tall, with yellow sap.
Leaves: Alternate; blades deeply pinnately lobed, the lobes sharp-pointed.
Flowers: Radial, white to pink or purple, up to 3 ½" broad.
Fruit: Prickly capsules up to 2" tall.
Bloom Period: Spring, summer.
Distribution: Cameron, Hidalgo, Willacy, and Starr counties.
Comments: This species often grows in disturbed soils. The specific epithet *sanguinea,* suggesting a red color, can be misleading. Some plants have bright purple flowers. Others bear pure white flowers. Probably from hybridization between the two color forms, many shades of pink are also encountered. The various color forms make this species an interesting plant to examine (but not touch) in the natural setting.

Red Poppy *Argemone sanguinea* white-flowering form

Red Poppy *A. sanguinea* pink-flowering form

Slender Passion Flower, Palm Grove Passion Vine

Family: Passifloraceae
Scientific Name: *Passiflora filipes*
Habit: Herbaceous vines with tendrils.
Leaves: Alternate; blades shallowly 3-lobed, up to 1 ½" long and 2 ⅛" broad.
Flowers: Radial, yellowish or greenish white, up to ⅝" broad; sepals 5, petals 5, partially covered by a ring of threadlike appendages.
Fruit: Roundish berries up to ¼" or slightly broader, black when ripe, splitting open when dried.
Bloom Period: All seasons.
Distribution: Cameron County.
Comments: Cameron County is the only reported locality for this species in Texas. It grows abundantly in remnant groves of Texas sabal palm, *Sabal mexicana*. It is more common in Mexico. The leaves usually have light-colored blotches.

Slender Passion Flower, Palm Grove Passion Vine *Passiflora filipes*

Corona De Cristo

Family: Passifloraceae
Scientific Name: *Passiflora foetida* var. *gossypifolia*
Habit: Herbaceous, hairy vines usually sticky with glandular hairs, with tendrils.
Leaves: Alternate; blades 3-lobed, up to 2 ¾" broad and long.
Flowers: Radial, purplish to white, up to 2" broad; sepals 5, petals 5, partially covered by several rings of threadlike appendages.
Fruit: Roundish, greenish berries up to 2" broad, splitting open when dried. The berries are enclosed in the enlarged calyxes, which have highly fringed margins.
Bloom Period: Spring, summer, fall.
Distribution: Cameron, Hidalgo, Willacy, and Starr counties.
Comments: This is our showiest passion flower, easily recognized by the size and color of the flowers. It is a host plant for Heliconia, Gulf Fritillary, and Mexican Silverspot butterflies. An introduced cultivar, *Passiflora foetida* var. *hirsuta,* is more vigorous and has red fruit. The specific epithet *foetida,* refers to an unpleasant smell from crushed leaves. The Spanish common name translates as "crown of Christ."

Corona De Cristo *Passiflora foetida* var. *gossypifolia*

Passifloraceae

Corky Stemmed Passion Flower

Family: Passifloraceae
Scientific Name: *Passiflora suberosa* [*P. pallida*]
Habit: Vines with tendrils, usually but not always hairless; older stems corky.
Leaves: Alternate; blades up to 3 ½" long and broad, usually deeply 3-lobed; petioles with a pair of stalked glands.
Flowers: Radial, greenish yellow, about 1 ⅛" broad; sepals partially covered by a ring of purple and yellow threadlike appendages; petals none.
Fruit: Roundish berries, purplish to black when ripe, up to ⅝" broad, splitting open when dried.
Bloom Period: All seasons.
Distribution: Cameron and Hidalgo counties.
Comments: On younger plants the glands with stalks on the petioles, combined with the lower leaf surfaces without raised veins, serve to identify this species. on older plants the corky stems are an identifying characteristic. The species is a host plant for Heliconia, Gulf Fritillary, and Mexican Silverspot butterflies. This vine is not found elsewhere in Texas. It is more abundant in Mexico.

Corky Stemmed Passion Flower *Passiflora suberosa*

Corky Stemmed Passion Flower
P. suberosa leaf with paired glands on the petiole

Slender Lobe Passion Flower, Longhorn Passion Flower

Family: Passifloraceae
Scientific Name: *Passiflora tenuiloba*
Habit: Vines with tendrils, mostly hairless.
Leaves: Alternate; blades up to 1 ⅛" long and 6 ¾" broad, 3-lobed, the lobes variable, sometimes lobed again, seeming to have 5 lobes.
Flowers: Radial, up to ¾" broad; sepals greenish, partially covered by a ring of threadlike greenish red appendages; petals absent.
Fruit: Roundish berries up to ⅜" broad, black when ripe, splitting open when dried.
Bloom Period: Spring, summer, fall.
Distribution: Hidalgo and Starr counties.
Comments: This species is more abundant in West Texas. It is a host plant for Heliconia, Gulf Fritillary, and Mexican Silverspot butterflies. All the passion flowers have a similar flower pattern (though some lack petals). There is much symbolism for the passion of Christ taken from the flowers, leaves, and tendrils.

Slender Lobe Passion Flower, Longhorn Passion Flower
Passiflora tenuiloba

Devil's Claw

Family: Pedaliaceae (Martyniaceae)
Scientific Name: *Proboscidea louisianica* subsp. *fragrans* [*P. fragrans*]
Habit: Large stout annuals covered with sticky glands, forming mats; young stems erect, later falling over.
Leaves: Mostly alternate, some opposite; blades broadly ovate, up to 5 ½" long and about as broad.
Flowers: Bilateral, in terminal inflorescences held above the foliage; petals united, purplish or almost white, up to 2" long.
Fruit: A large woody capsule, the main body about 4" long, with a pair of pointed, springy, incurved extensions.
Bloom Period: Summer, fall.
Distribution: Hidalgo and Starr counties.
Comments: The common name is appropriate: The fruit is built to snare a large animal's foot. The animal then carries it to various places, releasing seeds as it goes, until the fruit falls away. The young fruit are said to be edible. The dried fruit are used to make whimsical craft items. Fibers of the mature fruit are used to make baskets.

Devil's Claw *Proboscidea louisianica* subsp. *fragrans*

Devil's Claw *P. louisianica* subsp. *fragrans* fruit

Small Flowered Devil's Claw

Family: Pedaliaceae (Martyniaceae)
Scientific Name: *Proboscidea parviflora*
Habit: Annuals forming mats; vegetative parts glandular, producing a watery substance.
Leaves: Alternate or sometimes opposite; blades roundish to somewhat kidney shaped, sometimes lobed, up to 10" long and broad.
Flowers: Bilateral, crowded, in inflorescences that do not extend above the foliage; corollas pinkish to purplish, ¾" to 1 ½" long.
Fruit: A large woody capsule, the main body up to 4" long, with a pair of pointed, springy, incurved extensions.
Bloom Period: Spring, summer, and fall.
Distribution: Starr County.
Comments: This species is easily distinguished from the more common devil's claw, *Proboscidea louisianica* subsp. *fragrans*. The glands of *P. parviflora* produce a watery rather than a sticky substance, and the flowers do not extend above the foliage. In Texas this species has previously been reported to occur only in West Texas.

Small Flowered Devil's Claw *Proboscidea parviflora*

Small Flowered Devil's Claw *P. parviflora* immature fruit

Phytolaccaceae

Garlic Weed, Garlic Guinea Weed

Family: Phytolaccaceae (Petiveriaceae)
Scientific Name: *Petiveria alliacea*
Habit: Erect perennial up to 40" tall, smelling of garlic when bruised.
Leaves: Alternate; blades elliptic, up to 6" long.
Flowers: Sessile on terminal inflorescences; sepals whitish, about 3/16" long; petals absent.
Fruit: Dry, one-seeded, about 5/16" tall.
Bloom Period: Summer.
Distribution: Cameron and Hidalgo counties.
Comments: The garlic scent and the small flowers are important characteristics of this species. In Texas this species is limited to our area, but it ranges widely in the American tropics.

Garlic Weed, Garlic Guinea Weed *Petiveria alliacea*

Pigeon Berry, Rouge Plant

Family: Phytolaccaceae
Scientific Name: *Rivina humilis*
Habit: Erect to sprawling perennials, usually about 12–24" tall.
Leaves: Alternate; blades ovate, up to 6" long.
Flowers: Inflorescences terminal and axillary; sepals pink to white, up to 1/8" long; petals absent.
Fruit: A red, orange, or yellow globose berry about 3/16" broad.
Bloom Period: All seasons.
Distribution: Cameron, Hidalgo, Willacy, and Starr counties.
Comments: This plant is variable. Some forms have white flowers, while others have pink or almost red flowers; some forms have red berries, others have orange ones. This is one of our few shade-tolerant natives.

Pigeon Berry, Rouge Plant *Rivina humilis*

Shiny Bush

Family: Piperaceae
Scientific Name: *Peperomia pellucida*
Habit: Herbaceous annuals; stems falling over, the ends growing upright.
Leaves: Alternate; blades mostly broadly ovate, up to 1 5/8" long, with 5–7 veins originating from the base.
Flowers: Tiny, embedded in a rod-shaped receptacle about 1/16" broad.
Fruit: Very small, embedded in the receptacle.
Bloom Period: All seasons.
Distribution: Hidalgo County.
Comments: This is an introduced plant that arrived from the American tropics through the nursery trade. In Texas it has previously been reported only for Harris County. We found the plants growing vigorously in two nurseries, both in pots and in soil around the pots. We expect they will do equally well in the landscape where they have some shade and a good supply of water. The species is widespread, growing in Florida, Georgia, Louisiana, Texas, Mexico, the West Indies, Central America, and northern South America. Extracts from this species have been shown to inhibit growth of *Staphylococcus aureus, Bacillus subtilis, Pseudomonas aeruginosa,* and *Escherichia coli.* Possibly it could be used as a broad-spectrum antibiotic (*Flora of North America,* vol. 3).

Shiny Bush *Peperomia pellucida*

Tallow Weed, Hooker's Plantain

Family: Plantaginaceae
Scientific Name: *Plantago hookeriana*
Habit: Annuals with leaves in basal rosettes.
Leaves: Blades oblanceolate to linear, sometimes hairy, up to 8" long; petioles absent or tiny.
Flowers: Very tiny, mostly hidden by bracts, in rod-shaped spikes up to 4 1/2" or taller.
Fruit: Capsules globose, about 3/16" broad, splitting around the middle, with 2 brown seeds.
Bloom Period: Spring, summer.
Distribution: Cameron, Hidalgo, Willacy, and Starr counties.
Comments: We have three species of *Plantago*: *P. hookeriana, P. rhodosperma,* and *P. virginica* (not pictured). They are all well-adapted to hot, dry weather. This species is distinguished from the others by the hairs on the upper part of the inflorescence, pointing upward or pressed flat. It grows in sandy soils.

Tallow Weed, Hooker's Plantain *Plantago hookeriana*

Plumbaginaceae

Red Seeded Plantain
Family: Plantaginaceae
Scientific Name: *Plantago rhodosperma*
Habit: Annuals with leaves in basal rosettes.
Leaves: Blades oblanceolate to obovate, up to 12" or longer; petioles absent or tiny.
Flowers: Very tiny, mostly hidden by bracts, in rod-shaped spikes up to 8" or longer.
Fruit: Capsules globose, about 3/16" broad, splitting around the middle, with 2 red to reddish black seeds.
Bloom Period: Spring.
Distribution: Cameron, Hidalgo, Willacy, and Starr counties.
Comments: The specific epithet *rhodosperma* is made up of two Greek words for "red" and "seed." The red to reddish black color of the seeds identifies this species.

Red Seeded Plantain *Plantago rhodosperma*

Sea Lavender
Family: Plumbaginaceae
Scientific Name: *Limonium carolinianum* [*L. nashii*]
Habit: Perennial herbs.
Leaves: Basal, leathery; blades spatulate, up to 8 3/8" long.
Flowers: In highly branched inflorescences; petals lavender, up to 5/16" long, falling away and leaving the whitish sepals.
Fruit: Dry, one-seeded.
Bloom Period: Summer, fall.
Distribution: Cameron and Willacy counties.
Comments: This species occurs on coastal sands and inland in saline areas. Although the petals fall away, the whitish calyxes remain for quite some time. Because of this long-lasting quality, the inflorescences are often cut and used in floral arrangements.

Sea Lavender *Limonium carolinianum*

Hierba De Alacrán, White Plumbago

Family: Plumbaginaceae
Scientific Name: *Plumbago scandens*
Habit: Sprawling woody plants; stems up to 40" or longer.
Leaves: Alternate; blades ovate, up to 5" or longer.
Flowers: Calyxes with many stalked glands; corollas white, occasionally pale bluish, about 1" long.
Fruit: Capsules about ¼" long.
Bloom Period: Spring, summer, fall.
Distribution: Cameron and Hidalgo counties.
Comments: Although native, this species is not often found growing naturally but is frequently seen growing as an ornamental. It is easy to recognize by its sprawling habit, the glandular calyxes, and the white corollas. Because of the glands, the flower and seed areas are very tacky. This is an excellent butterfly nectar plant. The Spanish common name translates as "scorpion plant."

Hierba De Alacrán, White Plumbago *Plumbago scandens*

Cut Leaf Gilia

Family: Polemoniaceae
Scientific Name: *Gilia incisa*
Habit: Perennials up to 22" tall.
Leaves: Alternate; blades more or less ovate, up to 2 ⅛" long; margins toothed, usually lobed.
Flowers: Radial; corollas bluish to almost white, funnel shaped, up to ⅜" broad.
Fruit: Capsules about ⅛" tall.
Bloom Period: Spring, summer.
Distribution: Cameron, Hidalgo, Willacy, and Starr counties.
Comments: This plant was named for Filippo Luigi Gilii, an astronomer in the late eighteenth and early nineteenth centuries.

Cut Leaf Gilia *Gilia incisa*

Rio Grande Phlox

Family: Polemoniaceae
Scientific Name: *Phlox drummondii* subsp. *drummondii* [*P. drummondii* var. *peregrina*]
Habit: Hairy annuals with weak stems up to 12" tall.
Leaves: Opposite at the base, alternate higher on the stems; blades narrowly ovate to lanceolate, up to 2 ¾" long.
Flowers: Radial; corollas purple to pink with light centers, up to 1" broad.
Fruit: Capsules about 3⁄16" tall.
Bloom Period: Spring, summer.

Rio Grande Phlox *Phlox drummondii* subsp. *drummondii*

Polygalaceae

Distribution: Hidalgo, Willacy, and Starr counties.
Comments: This is one of our showy wildflowers, preferring sandy soils. The taxonomy has been somewhat confused by the sowing of seeds along Texas highways. This practice sometimes introduces variations that would not normally exist in a given area. *P. drummondii* was named in honor of Thomas Drummond, who made important contributions to the knowledge of Texas flora, especially that of Central Texas.

White Milkwort
Family: Polygalaceae
Scientific Name: *Polygala alba*
Habit: Erect perennials, up to 12" tall.
Leaves: Alternate; blades linear, about ¾" long.
Flowers: Bilateral; petals 3, white, less than ⅛" long, united basally.
Fruit: Capsules about ⅛" tall.
Bloom Period: Spring, summer, fall.
Distribution: Cameron, Hidalgo, Willacy, and Starr counties.
Comments: This species prefers sandy soils. The genus name comes from the Greek *polys* (much) and *galon* (milk). Some European species were believed to enhance milk production in cattle.

White Milkwort *Polygala alba*

Glandular Milkwort *Polygala glandulosa*

Glandular Milkwort
Family: Polygalaceae
Scientific Name: *Polygala glandulosa*
Habit: Small woody perennials; stems much branched, prostrate or leaning.
Leaves: Alternate; blades glandular, obovate to spatulate, up to ⅜" long.
Flowers: Bilateral, axillary; petals 3, pink to purple or white, united basally, about ⅛" long.
Fruit: Capsules about ⅛" or less tall.
Bloom Period: Spring, summer, fall.
Distribution: Hidalgo and Starr counties.
Comments: The more common flower color is purple, but lighter-colored forms sometimes occur. This attractive little plant is most abundant in our area of Texas, although it is occasionally found as far north as San Patricio County. It commonly grows on gravelly hillsides under thorny brush.

Glandular Milkwort *P. glandulosa* white-flowering form

Queen's Wreath, Corona De Reina

Family: Polygonaceae
Scientific Name: *Antigonon leptopus*
Habit: Perennial vines with tendrils, forming tubers at the roots.
Leaves: Alternate; blades heart shaped, up to 5 ⅛" long.
Flowers: Pendulous, pink, sometimes white, in branched, showy inflorescences.
Fruit: Calyxes persisting, about ⅜" long, enclosing the achenes.
Bloom Period: Spring, summer, and fall.
Distribution: Cameron, Hidalgo, and Starr counties.
Comments: Native to Mexico, these ornamental vines have been widely cultivated in warm areas. They often escape and are seen growing in vacant lots and sometimes in native habitats. They are attractive to bees and butterflies. The Spanish common name translates as "queen's crown."

Queen's Wreath, Corona De Reina *Antigonon leptopus*

Gregg's Buckwheat

Family: Polygonaceae
Scientific Name: *Eriogonum greggii*
Habit: Erect perennials up to 16" tall, with glands on all parts.
Leaves: Alternate; basal blades broadly spatulate, up to 4" long, with hairs along the margins; upper leaves smaller.
Flowers: Very small, less than ⅛" broad, subtended by united bracts up to ³⁄₁₆" broad; sepals 6, reddish at the edges; petals none.
Fruit: Achenes about ⅛" tall.
Bloom Period: Spring.
Distribution: Hidalgo and Starr counties.
Comments: In Texas this species is considered a rare plant, known only from Starr and western Hidalgo counties, growing on bluffs overlooking the Rio Grande.

Gregg's Buckwheat *Eriogonum greggii*

Gregg's Buckwheat *E. greggii* flowers

Polygonaceae

Wild Buckwheat

Family: Polygonaceae
Scientific Name: *Eriogonum riograndensis*
  [*E. multiflorum*]
Habit: Annual hairy herbs up to 6 ½' tall.
Leaves: Alternate; blades lanceolate to ovate, up to 3 ⅛" long; lower surfaces white or tan in color.
Flowers: Small, white, subtended by united bracts, in inflorescences arranging the flowers in flattened or saucerlike groups; sepals about ⅛" long; petals none.
Fruit: Achenes less than ⅛" tall.
Bloom Period: Summer, fall.
Distribution: Hidalgo and Willacy counties.
Comments: This species seems to prefer sandy soils.

Wild Buckwheat *Eriogonum riograndensis*

Water Smartweed

Family: Polygonaceae
Scientific Name: *Persicaria amphibia*
  [*Polygonum amphibium*]
Habit: Stems erect to falling over, up to 4' tall, growing in or near water. Stems and leaves tacky.
Leaves: Alternate; blades ovate to elliptic, variable, up to 9" long. Stipules grown together to form a noticeable fringed ring at each node.
Flowers: Perfect or unisexual, sepals white to reddish, up to ¼" long; petals none.
Fruit: Achenes about ⅛" tall.
Bloom Period: Summer, fall.
Distribution: Cameron and Starr counties.
Comments: Noticeable characteristics of this species are its tackiness and its stipules joined into a collar around the node. Various reference books may refer to this species as either *Persicaria* or *Polygonum*.

Water Smartweed *Persicaria amphibia*

Lady's Thumb

Family: Polygonaceae
Scientific Name: *Persicaria maculosa* [*Polygonum persicaria*]
Habit: Erect or leaning annuals up to 40" tall, usually about 24"; stems green, often with red marks.
Leaves: Alternate, up to 6" long; stipules forming sheaths with bristles.
Flowers: Petals absent; sepals white to purple, up to 1/8" or slightly longer.
Fruit: Achenes flattened or 3-sided, about 1/8" tall.
Bloom Period: Spring, summer, fall.
Distribution: Cameron and Hidalgo counties.
Comments: This is a plant of wet places or recently wet areas. Where it is found, it is usually abundant.

Lady's Thumb *Persicaria maculosa*

Pink Smartweed

Family: Polygonaceae
Scientific Name: *Persicaria pensylvanica* [*Polygonum pensylvanicum*]
Habit: Erect annuals up to 80" tall, usually less; stems often red.
Leaves: Alternate; stipules large, forming sheaths, fringed; blades lanceolate, up to 7" long.
Flowers: Small, pink to white, in groups subtended by bracts; peduncles glandular; sepals about 3/16" long; petals none.
Fruit: Achenes about 1/8" tall.
Bloom Period: All seasons.
Distribution: Cameron, Hidalgo, and Willacy counties.
Comments: This is a species preferring wet places. It is an important waterfowl food plant.

Pink Smartweed *Persicaria pensylvanica*

Polygonaceae

Golden Fruited Dock

Family: Polygonaceae
Scientific Name: *Rumex chrysocarpus*
Habit: Perennial herbs up to 28" tall; stems often reddish.
Leaves: Alternate; blades lanceolate to oblong, up to 4 ¾" long.
Flowers: Small, greenish, in somewhat condensed inflorescences; sepals 6, the inner 3 growing in size to surround the fruit.
Fruit: Achenes about ⅛" tall.
Bloom Period: Spring, summer, fall.
Distribution: Cameron, Hidalgo, Willacy, and Starr counties.
Comments: This species is commonly seen in seasonally wet places. It is abundant, particularly in counties along the Texas coast. The specific epithet *chrysocarpus* comes from two Greek words for "golden" and "fruit," referring to the color of the dried sepals surrounding the fruit.

Golden Fruited Dock *Rumex chrysocarpus*

Fiddle Dock

Family: Polygonaceae
Scientific Name: *Rumex pulcher*
Habit: Perennial herbs about 2' or taller.
Leaves: Alternate; blades narrowly ovate to elliptic, heart shaped basally, up to 6" or longer, becoming smaller up the stem.
Flowers: Small, greenish, in somewhat condensed inflorescences; sepals 6, the inner 3 growing in size to surround the fruit.
Fruit: Achenes about ⅛" tall.
Bloom Period: Spring, summer.
Distribution: Cameron, Hidalgo, and Willacy counties.
Comments: This is an introduced species, native of Europe. It is seen in seasonally wet places.

Fiddle Dock *Rumex pulcher*

Fiddle Dock *R. pulcher* inflorescence and fruit

Yellow Flameflower

Family: Portulacaceae
Scientific Name: *Phemeranthus aurantiacus* [*Talinum aurantiacum, T. angustissimum*]
Habit: Herbs up to 16" tall (usually less), with tuberous roots.
Leaves: Alternate, linear to cylindrical, up to 2 ⅜" long.
Flowers: Radial, solitary from the leaf axils; petals yellow, up to ½" long.
Fruit: Capsules roundish, about ³⁄₁₆" broad, splitting down the length.
Bloom Period: Spring, summer, fall.
Distribution: Cameron, Hidalgo, Willacy, and Starr counties.
Comments: In our area, this species is distinguished from pink baby's breath, *Talinum paniculatum,* by its linear leaves and yellow flowers. A third species, *Phemeranthus parviflorus,* occurs in our area. It has linear leaves and pink to red flowers. The name *Talinum aurantiacum* originally included plants with orange-colored flowers. *T. angustissimum,* which has yellow flowers, was combined with *T. aurantiacum* to give a widespread species with orange or yellow flowers. Subsequently, the species was placed into the genus *Phemeranthus.* Our populations of *P. aurantiacus* all have yellow flowers.

Yellow Flameflower *Phemeranthus aurantiacus*

Common Purslane

Family: Portulacaceae
Scientific Name: *Portulaca oleracea* [*P. retusa*]
Habit: Prostrate, hairless succulent herbs.
Leaves: Alternate; blades obovate to spatulate, up to 1 ⅛" long.
Flowers: Radial, solitary or in clusters at the stem tips; petals yellow, about ³⁄₁₆" long.
Fruit: Capsules splitting around the equator.
Bloom Period: Spring, summer, fall.
Distribution: Cameron, Hidalgo, Willacy, and Starr counties.
Comments: This species is said to be edible, and is sometimes used in salads. Although leaves of this species have reportedly been used in salads, they can be dangerous if eaten in quantity.

Common Purslane *Portulaca oleracea*

Portulacaceae

Chisme

Family: Portulacaceae
Scientific Name: *Portulaca pilosa* [*P. mundula*]
Habit: Prostrate herbs.
Leaves: Alternate, the axils woolly; blades linear, up to ⅝" long.
Flowers: Radial, in clusters with long hairs; petals purple, up to ¼" long.
Fruit: Capsules splitting around the equator.
Bloom Period: Spring, summer, fall.
Distribution: Cameron, Hidalgo, Willacy, and Starr counties.
Comments: This species is beautiful in the garden. It is related to the "moss rose" and the purslanes sold in nurseries.

Chisme *Portulaca pilosa*

Wing Pod Portulaca

Family: Portulacaceae
Scientific Name: *Portulaca umbraticola* subsp. *lanceolata*
Habit: Prostrate somewhat succulent annuals, the stems angled.
Leaves: Variable in position and shape; blades up to 1 ⅛" long.
Flowers: Radial, clustered; petals yellow to orange, with red tips, up to ⅜" long.
Fruit: Capsules splitting around the equator, with a circular collar around the lower half.
Bloom Period: Spring, summer, fall.
Distribution: Cameron, Hidalgo, and Willacy counties.
Comments: *P. umbraticola* subsp. *coronata* has been reported for the LRGV. Others say it is absent. Its petals are yellow, not tipped with red as in subsp. *lanceolata*.

Wing Pod Portulaca *Portulaca umbraticola* subsp. *lanceolata*

Pink Baby's Breath

Family: Portulacaceae
Scientific Name: *Talinum paniculatum*
Habit: Herbs up to 30" tall, sometimes woody basally, with tuberous roots. The above-ground parts often disappear during cold seasons.
Leaves: Alternate, blades ovate, up to 4" long but usually less.
Flowers: Radial, in terminal inflorescences; petals red or pink, about 3/16" long.
Fruit: Capsules roundish, about 3/16" broad, splitting down the length.
Bloom Period: Summer, fall.
Distribution: Cameron, Hidalgo, Willacy, and Starr counties.
Comments: This species is distinguished from yellow flameflower, *Phemeranthus aurantiacus,* by its ovate leaves and pink to red flowers. A third species in our area, *P. parviflorum,* also has pink to red flowers but has linear leaves.

Pink Baby's Breath *Talinum paniculatum* leaves

Pink Baby's Breath *T. paniculatum* inflorescence

Scarlet Pimpernel

Family: Primulaceae
Scientific Name: *Anagallis arvensis*
Habit: Prostrate annuals.
Leaves: Opposite; blades ovate, up to 3/4" long.
Flowers: Radial, solitary on slender pedicels; petals 5, blue or salmon colored, about 3/16" long.
Fruit: Capsules roundish, up to 1/8" broad, splitting along the equator.
Bloom Period: Winter, spring.
Distribution: Cameron and Hidalgo counties.
Comments: The more common flower color in our area is blue; salmon is less common. Scarlet color is not reported. This is an introduced species, native of Europe.

Scarlet Pimpernel *Anagallis arvensis*

Scarlet Pimpernel *A. arvensis* blue-flowering form

Primulaceae 361

Chaffweed

Family: Primulaceae
Scientific Name: *Anagallis minima* [*Centunculus minimus*]
Habit: Small annual herbs forming clumps or mats; stems erect or leaning, up to 4 ¾" long.
Leaves: Alternate; blades variable, mostly broader toward the apices, up to ⅜" long.
Flowers: Radial, from the leaf axils; petals usually 4, white or pinkish, about 1/16" long.
Fruit: Capsules roundish, less than ⅛" broad, splitting along the equator.
Bloom Period: Winter, spring.
Distribution: Cameron and Willacy counties.
Comments: This small plant is usually found in low, damp or wet places. In unusually rainy seasons it is occasionally found in higher localities, which normally would quickly become dry. It is an introduced species, native of Europe.

Chaffweed *Anagallis minima*

Chaffweed *A. minima* flower

Bractless Brookweed

Family: Primulaceae
Scientific Name: *Samolus ebracteatus* subsp. *ebracteatus*
Habit: Perennials up to 22" tall, the stems often reddish.
Leaves: In basal rosettes and alternate on the stems; blades spatulate, up to 3 ½" long.
Flowers: Radial, in terminal inflorescences, the pedicels without bracts; petals 5, white, up to 5/16" long.
Fruit: Capsules roundish, about 3/16" broad, splitting open at the tops.
Bloom Period: Spring, summer.
Distribution: Cameron, Hidalgo, and Starr counties.
Comments: The prefixes *a-* and *e-* usually mean "without," or a similar negative term. We have three similar-appearing named subspecies. Subsp. *ebracteatus* is reported for Cameron County, growing in sandy areas. Subsp. *cuneatus* is reported for Cameron and Hidalgo counties. Subsp. *alysoides* is reported for Cameron and Willacy counties.

Bractless Brookweed *Samolus ebracteatus* subsp. *ebracteatus*

Thin Leaf Brookweed

Family: Primulaceae
Scientific Name: *Samolus valerandi* subsp. *parviflorus* [*S. parviflorus*]
Habit: Light green, hairless perennials up to 24" tall.
Leaves: In basal rosettes and alternate on the stems; blades variable shapes, broader toward the apices, up to 6" long.
Flowers: Radial, white, about 1/8" broad, in terminal inflorescences; petals grown together basally.
Fruit: Capsules roundish, about 1/8" broad, opening at the tops.
Bloom Period: Spring, summer, fall.
Distribution: Cameron, Hidalgo, and Starr counties.
Comments: This species is usually found in wet or damp places.

Thin Leaf Brookweed *Samolus valerandi* subsp. *parviflorus*

Old Man's Beard, Barbas De Chivato

Family: Ranunculaceae
Scientific Name: *Clematis drummondii*
Habit: Robust, hairy vines.
Leaves: Opposite, compound; leaflets up to 1" long.
Flowers: Unisexual, on separate plants. Corollas radial; sepals usually 4, whitish, about 1/2" long; petals absent, or looking like the stamens.
Fruit: Achenes with white plumes up to 4" long.
Bloom Period: Spring, summer, fall.
Distribution: Cameron, Hidalgo, Willacy, and Starr counties.
Comments: The separate male and female flowers have sometimes been mistaken as coming from two different species. It has been reported that sometimes the plants are monoecious (both sexes on the same plant); we have not observed this, but when several vines grow entwined together on a support, it can be difficult to determine the origin of a given group of flowers. These vines are especially noticeable growing on fences in the summer when the fruit are mature. They are almost covered with the white or silvery plumes. The plumes serve as "parachutes" to carry the seed away from the parent plant. *C. drummondii* was named in honor of Thomas Drummond, an immigrant from Scotland, who made important contributions to the knowledge of the flora of Texas. The Spanish name translates as "goat's beard."

Old Man's Beard, Barbas De Chivato *Clematis drummondii* in fruit

Old Man's Beard, Barbas De Chivato *C. drummondii* male flowers

Old Man's Beard, Barbas De *C. drummondii* female flowers

Resedaceae

Celery Leaf Buttercup

Family: Ranunculaceae
Scientific Name: *Ranunculus sceleratus*
Habit: Usually hairless annuals, sometimes living a few years; stems erect, up to 39" tall.
Leaves: Stem leaves alternate; blades ovate or kidney shaped in outline, up to 1 ⅜" long, deeply cut along the margins.
Flowers: Radial; petals 5 or more, bright yellow, up to ⅜" long
Fruit: Achenes about ⅛" tall.
Bloom Period: Spring, summer.
Distribution: Cameron County.
Comments: This beautiful plant has not previously been reported for the LRGV, although it has been reported for Zapata County. A small population was recently seen in a drainage ditch in Harlingen, Cameron County. A few plants were also observed in Kenedy County, adjacent to Willacy County, near a wetland.

Celery Leaf Buttercup *Ranunculus sceleratus*

Desert Spikes

Family: Resedaceae
Scientific Name: *Oligomeris linifolia*
Habit: Somewhat succulent, erect annuals up to 14" tall.
Leaves: Alternate or in groups; blades linear, up to 10" long.
Flowers: Tiny, in usually tight clusters on the spike.
Fruit: Capsules roundish, about ⅛" tall.
Bloom Period: Spring, summer.
Distribution: Cameron and Starr counties.
Comments: This species is found in the LRGV and also in West Texas, with no plants reported between the two general localities. This is true of several of our species. Cattle have been poisoned by eating these plants.

Desert Spikes *Oligomeris linifolia*

Gregg's Colubrina

Family: Rhamnaceae

Scientific Name: *Colubrina greggii*

Habit: Shrubs or small trees, without thorns, up to 30' tall. In our area, they are shrubs.

Leaves: Alternate; blades ovate, with prominent veins, up to 9 1/8" long, the margins toothed.

Flowers: Small, greenish yellow, usually in large groups.

Fruit: Capsules hard, about 5/16" broad, remaining on the plant indefinitely.

Bloom Period: Summer, fall.

Distribution: Cameron County.

Comments: In Texas this plant is known only in Cameron County. A few individuals exist in the remnant groves of Texas sabal palm, *Sabal mexicana*. Gregg's colubrina also grows in Mexico.

Gregg's Colubrina *Colubrina greggii*

Hog Plum, Texas Colubrina

Family: Rhamnaceae

Scientific Name: *Colubrina texensis*

Habit: Hairy shrubs without thorns, up to 6 1/2' tall.

Leaves: Alternate; blades ovate, up to 1 1/8" long, the margins minutely toothed.

Flowers: Small, greenish yellow, few to one from the leaf axils.

Fruit: Capsules hard, about 5/16" broad, persisting on the plants.

Bloom Period: Spring, fall.

Distribution: Cameron, Hidalgo, and Starr counties.

Comments: The flowers look the same in all species in the family. The species are recognized by presence or absence of thorns and by leaf characteristics. This thornless species has alternate leaves, and the veins on the lower leaf surfaces are not prominent. It is more abundant on sandy soils.

Hog Plum, Texas Colubrina *Colubrina texensis*

Rhamnaceae

Brasil

Family: Rhamnaceae
Scientific Name: *Condalia hookeri*
Habit: Shrubs or small trees, with thorns, up to 30' tall, usually shorter.
Leaves: Alternate or in tight groups; blades thin, obovate to spatulate, up to 1" long.
Flowers: Small, greenish yellow.
Fruit: Drupes, black when mature, up to $5/16$" broad.
Bloom Period: Spring, summer.
Distribution: Cameron, Hidalgo, Willacy, and Starr counties.
Comments: *Condalia* is distinguished from other members of the family by its thorns and its leaves, which are broadest above the middle. The leaves of *C. hookeri* do not have prominent veins; those of squaw bush, *C. spathulata*, have somewhat prominent veins. The ripe drupes of brasil are an important food source for birds and have been used to produce a purple dye. The heartwood is red, dense, and heavy, sinking in water.

Brasil *Condalia hookeri*

Squaw Bush

Family: Rhamnaceae
Scientific Name: *Condalia spathulata*
Habit: Shrubs with thorns, usually about 3' tall or less.
Leaves: Mostly in tight groups; blades spatulate, ½" long or less; veins somewhat prominent.
Flowers: Small, greenish yellow.
Fruit: Drupes, black when mature, ⅛" or slightly broader.
Bloom Period: Summer, fall.
Distribution: Starr County.
Comments: Squaw bush is smaller than brasil, *C. hookeri*, and its leaves have somewhat prominent veins.

Squaw Bush *Condalia spathulata*

Coyotillo

Family: Rhamnaceae
Scientific Name: *Karwinskia humboldtiana*
Habit: Thornless shrubs up to 6 ½' tall.
Leaves: Opposite; blades ovate, up to 2 ¾" long, usually smaller, with prominent veins.
Flowers: Small, greenish yellow.
Fruit: Drupes black when mature, about 3/16" broad.
Bloom Period: Spring, summer.
Distribution: Cameron, Hidalgo, Willacy, and Starr counties.
Comments: This thornless member of the Rhamnaceae is easy to recognize by the prominent veins on the leaves. The seeds and leaves are poisonous. Many years ago, someone caged a number of monkeys in native brush of the western LRGV. The monkeys ate the fruit and seeds of coyotillo, were paralyzed, and died. The plants are hosts for the Two-barred Flasher butterfly.

Coyotillo *Karwinskia humboldtiana*

Lotebush

Family: Rhamnaceae
Scientific Name: *Ziziphus obtusifolia*
Habit: Thorny shrubs with whitish stems. The stems usually have longitudinal markings.
Leaves: Alternate; blades ovate, up to 1 ¼" long; margins often toothed.
Flowers: Small, greenish yellow.
Fruit: Drupes black when mature, about 3/8" broad.
Bloom Period: Spring, summer, fall.
Distribution: Cameron, Hidalgo, Willacy, and Starr counties.
Comments: The prominent thorns and whitish stems with longitudinal striations help to identify this species. It is common in the LRGV. The western populations have narrower leaves. In dry seasons, only the greenish stems are seen. After copious rains, leaves (sometimes unusually large ones) appear. The fruit are an important wildlife food source.

Lotebush *Ziziphus obtusifolia*

Rhizophoraceae

Red Mangrove

Family: Rhizophoraceae
Scientific Name: *Rhizophora mangle*
Habit: Trees or shrubs up to 23' tall, usually less; young branches reddish. Trunk and branches produce prop roots.
Leaves: Opposite; blades obovate to elliptic, leathery, up to 5" long.
Flowers: Radial, in pairs from the axils of the leaves; petals 4, white to pale yellow, hairy, up to 5/16" long.
Fruit: One-seeded, cone shaped, up to 1' long, germinating while still on the tree, then reaching up to 10" in length.
Bloom Period: This is a tropical species, and it probably blooms much of the year.
Distribution: Cameron County.
Comments: We have two other species of mangroves that are not related (black mangrove, *Avicennia germinans*, and button mangrove, *Conocarpus erectus*). They grow in similar habitats. The red mangrove is widespread in the tropics, but it is not common in our area. It is reported only from near the mouth of the Rio Grande. The germinated seeds float, and water currents carry them to various locales. It is not uncommon to see them washed up on our beaches. The prop roots make this an easy species to recognize.

Red Mangrove *Rhizophora mangle* with prop roots

Red Mangrove *R. mangle* flowers

Red Mangrove *R. mangle* fruit

Peach Bush, Duraznillo

Family: Rosaceae

Scientific Name: *Prunus texana*

Habit: Small shrubs up to 4' or taller, with grayish bark; young branches crooked, very hairy.

Leaves: Alternate, glandular basally; blades ovate to elliptic, up to 1 ¼" long.

Flowers: Radial, solitary or in pairs, up to ½" broad; petals white

Fruit: Drupes orange to reddish when ripe, densely hairy, about ¾" long.

Bloom Period: Spring.

Distribution: Hidalgo and Starr counties.

Comments: This species is endemic to Texas. In our area it grows in sandy soils. The fruit are small for eating but have been used for making jelly and wine. When the glandular leaves are touched, a tacky substance remains on the hands.

Peach Bush, Duraznillo *Prunus texana*

Dewberry, Zarzamora

Family: Rosaceae

Scientific Name: *Rubus riograndis* [*R. trivialis*]

Habit: Woody perennials with very long clambering canes; canes and leaves with many curved prickles.

Leaves: Alternate, compound with 3 leaflets with toothed margins.

Peach Bush, Duraznillo *P. texana* immature fruit

Flowers: Radial, resembling old-fashioned roses; petals 5, white, about ⅜" long.

Fruit: Roundish, black when mature.

Bloom Period: Spring.

Distribution: Cameron and Hidalgo counties.

Comments: The fruit, similar to blackberries, are juicy, sweet and edible. The Spanish common name translates as "blackberry."

Common Buttonbush

Family: Rubiaceae

Scientific Name: *Cephalanthus occidentalis* var. *californicus*

Habit: Shrubs up to 16' tall, sometimes becoming taller trees.

Dewberry, Zarzamora *Rubus riograndis*

Leaves: Opposite; blades mostly ovate, up to about 5 ⅛" long.

Flowers: In dense, globose heads about ⅝" broad; corollas white.

Fruit: Capsules about 5/16" tall.

Bloom Period: Summer.

Rubiaceae

Distribution: Hidalgo and Starr counties.
Comments: In common buttonbush, the outside of the calyx is not densely hairy. This characteristic can be used to distinguish it from willow leaf buttonbush, *C. salicifolius,* in which the outside of the calyx is densely hairy. Common buttonbush is often seen growing in wet places or near water. The flowers are a good butterfly nectar source, and the shrubs provide nest areas for Green Herons.

Common Buttonbush *Cephalanthus occidentalis* var. *californicus*

Willow Leaf Buttonbush, Mexican Buttonbush
Family: Rubiaceae
Scientific Name: *Cephalanthus salicifolius*
Habit: Shrubs up to 16' tall, sometimes growing into taller trees.
Leaves: Opposite; blades lanceolate or narrowly oblong, up to 4 ¾" long.
Flowers: In dense globose heads about ⅝" broad; corollas white.
Fruit: Capsules about 3/16" tall.
Bloom Period: Spring, summer.
Distribution: Cameron and Hidalgo counties.
Comments: This rarely seen species usually grows in wet soils near the Rio Grande. It is a good butterfly nectar source. Willow leaf buttonbush has calyxes that are densely hairy on the outer surfaces; common buttonbush, *C. occidentalis,* does not.

Willow Leaf Buttonbush, Mexican Buttonbush *Cephalanthus salicifolius*

David's Milkberry
Family: Rubiaceae
Scientific Name: *Chiococca alba*
Habit: Mostly clambering shrubs, sometimes erect; stems about 6' long.
Leaves: Opposite; blades ovate, up to 2 ⅝" long.
Flowers: Inflorescences aromatic, short, arching, with flowers hanging downward; corollas bell-shaped, white to very pale yellow, about ⅜" long.
Fruit: Fleshy, white, roundish, about 3/16" broad.
Bloom Period: Spring, summer, fall.
Distribution: Cameron and Hidalgo counties.
Comments: This is a shrub of the tropics, known to grow as far north as Cameron and Hidalgo counties. It also occurs in Florida. It is easily recognized by the white fruit, which inspired its common name.

David's Milkberry *Chiococca alba*

David's Milkberry *C. alba* fruit

Rough Buttonweed

Family: Rubiaceae
Scientific Name: *Diodia teres*
Habit: Herbs with erect or leaning stems.
Leaves: Opposite; blades lanceolate, up to 1 ½" long. Stipular hairs prominent.
Flowers: Usually one or few in the axils of the leaves; corollas white to purplish, usually 4-lobed, up to ¼" long
Fruit: Hard capsules about 3/16" tall, topped with 4 calyx teeth.
Bloom Period: Spring, summer, fall.
Distribution: Cameron, Hidalgo, Willacy, and Starr counties.
Comments: Rough buttonweed grows in sandy soils. In this species, the capsules have no ribs and are topped with 4 calyx teeth. These characteristics separate it from Virginia buttonweed, *D. virginiana,* which has ribbed capsules topped with 2 calyx teeth.

Rough Buttonweed *Diodia teres*

Virginia Buttonweed

Family: Rubiaceae
Scientific Name: *Diodia virginiana*
Habit: Perennials, prostrate or almost so.
Leaves: Opposite; blades elliptic to lanceolate, up to 3 ½" long; margins minutely toothed.
Flowers: Usually one or few in the axils of the leaves; corollas white, usually 4-lobed, up to 3/8" long.
Fruit: Leathery, ribbed capsules about 5/16" tall, topped with 2 calyx teeth.
Bloom Period: Spring, summer, fall.
Distribution: Cameron County.
Comments: This species has leathery ribbed capsules topped with 2 calyx teeth. This characteristic separates it from rough buttonweed, *D. teres,* which has capsules with no ribs, topped with 4 calyx teeth.

Virginia Buttonweed *Diodia virginiana*

Rubiaceae

Catchweed, Bedstraw

Family: Rubiaceae
Scientific Name: *Galium aparine*
Habit: Herbs; stems sprawling, with stiff, curved bristles.
Leaves: Whorled, in groups of 6–8; blades spatulate, up to 1 ⅛" long.
Flowers: Radial, white, less than ⅛" broad.
Fruit: About 3⁄16" broad, bristly.
Bloom Period: Spring.
Distribution: Cameron and Hidalgo counties.
Comments: The stem bristles, while not strong enough to damage the skin, are efficient in clinging to clothing.

Catchweed, Bedstraw *Galium aparine*

Southwest Bedstraw

Family: Rubiaceae
Scientific Name: *Galium virgatum*
Habit: Slender erect annuals up to 16" tall (usually much shorter); stems smooth.
Leaves: Whorled, usually in groups of 4; blades oblong, bristly, up to ⅜" long.
Flowers: Radial, white, less than ⅛" broad.
Fruit: About 3⁄16" broad, with hooked bristles.
Bloom Period: Spring, summer.
Distribution: Cameron, Hidalgo, and Starr counties.
Comments: We have observed this species growing on hillsides and ravines. The smooth stems and smaller size serve to distinguish it from catchweed, *G. aparine*.

Southwest Bedstraw *Galium virgatum*

Prairie Bluets

Family: Rubiaceae
Scientific Name: *Hedyotis nigricans*
Habit: Perennials with usually leaning stems, often falling over.
Leaves: Opposite; blades linear to lanceolate, up to ⅞" long.
Flowers: Inflorescences with up to 15 flowers and leaflike bracts; corollas white to pale purplish, 4-lobed, up to 5⁄16" long.
Fruit: Capsules about ⅛" broad, dividing into 2 parts.
Bloom Period: Spring, summer.
Distribution: Cameron and Willacy counties.
Comments: Prairie bluets flowers resemble those of polly prim, *Polypremum procumbens,* which grows lower to the ground and has a superior ovary. Prairie bluets have an inferior ovary.

Prairie Bluets *Hedyotis nigricans*

White Girdlepod

Family: Rubiaceae
Scientific Name: *Mitracarpus breviflorus*
Habit: Annuals up to 16" or sometimes taller.
Leaves: Opposite; blades lanceolate, up to 1 3/16" long.
Flowers: In clusters in the leaf axils; corollas 4-lobed, white, about 1/8" long.
Fruit: Capsules less than 1/8" broad, splitting around the middle.
Bloom Period: Summer, fall.
Distribution: Cameron County
Comments: This species has not been reported elsewhere in Texas. It is more abundant in adjacent Mexico.

White Girdlepod *Mitracarpus breviflorus*

Crucillo

Family: Rubiaceae
Scientific Name: *Randia rhagocarpa*
Habit: Shrubs up to 6' or taller, with paired stout spines up to 1/2" long.
Leaves: Opposite; blades mostly obovate, often notched apically, up to 1 5/8" long.
Flowers: Single or few in the leaf axils; corollas white, usually 4-lobed, sometimes 5-lobed, up to 5/16" long.
Fruit: Leathery, green-striped, ripening to blue-black, round, up to 3/8" broad.
Bloom Period: Mainly spring, summer, following rains.
Distribution: Cameron, Hidalgo, and Willacy counties.
Comments: In spite of the spines (which are much less of a problem than those of many other natives), this is a beautiful plant in the garden. When it is in full bloom, the branches are almost covered with the delicate white flowers. Like many of our natives, with a little care it becomes a more attractive plant and often flowers more than once a year. Definition of the Spanish common name is not certain. It translates as "stick-pin," which might refer to the many spines, but it may be a derivative of *cruz,* "cross," referring to the paired spines making small crosses.

Crucillo *Randia rhagocarpa*

Rubiaceae

Tropical Mexican Clover

Family: Rubiaceae
Scientific Name: *Richardia brasiliensis*
Habit: Stems prostrate or sometimes leaning; stems and leaves hairy.
Leaves: Opposite, hairy; blades ovate, up to 1 ⅝" long.
Flowers: In clusters at the stem tips; corollas white, 5- to 7-lobed, up to ¼" long.
Fruit: Capsules obovoid, about ⅛" tall, splitting top to bottom.
Bloom Period: Spring, summer, fall.
Distribution: Hidalgo, Willacy, and Starr counties.
Comments: This species prefers sandy soils. We have three species of *Richardia*. The leaves of *R. scabra* (not pictured) are hairy only on the margins and veins; prairie Mexican clover, *R. tricocca*, usually has 3- to 4-lobed corollas. These characteristics distinguish them from *R. brasiliensis,,* which has hairy leaves and 5- to 7-lobed corollas.

Tropical Mexican Clover *Richardia brasiliensis*

Prairie Mexican Clover

Family: Rubiaceae
Scientific Name: *Richardia tricocca*
[*Diodia tricocca*]
Habit: Prostrate perennials.
Leaves: Opposite; blades lanceolate, up to 1" long.
Flowers: Mostly single or few from the leaf axils; corollas white, 3- to 4-lobed, up to 3/16" long.
Fruit: Capsules roundish, less than ⅛" tall.
Bloom Period: Spring, summer, fall.
Distribution: Hidalgo and Willacy counties.
Comments: This species prefers sandy soils. The 3- to 4-lobed corolla separates this species from tropical Mexican clover, *R. brasiliensis*. Its calyx teeth (at least one) remain on the ripe fruit. On *R. scabra,* they fall away.

Prairie Mexican Clover *Richardia tricocca*

Florida Buttonweed

Family: Rubiaceae
Scientific Name: *Spermacoce floridana*
Habit: Erect, leaning, or prostrate plants.
Leaves: Opposite; blades ovate to elliptic, up to 2" long.
Flowers: In small clusters, in the leaf axils; corollas white or greenish, 4-lobed, about 1/16" long.
Fruit: Capsules about 1/16" long, splitting from top to bottom into 2 parts.
Bloom Period: All seasons, but mostly spring and summer.
Distribution: Cameron County
Comments: This species is often found in low, wet areas. The smaller flowers and fruit of this species distinguish it from smooth buttonweed, *S. glabra*. Otherwise, the two species are similar.

Florida Buttonweed *Spermacoce floridana*

Smooth Buttonweed

Family: Rubiaceae
Scientific Name: *Spermacoce glabra*
Habit: Perennials with prostrate or leaning stems.
Leaves: Opposite; blades lanceolate, up to 1 5/8" long.
Flowers: In clusters, crowded in the leaf axils; corollas white or greenish, 4-lobed, about 1/8" long.
Fruit: Capsules about 1/8" long, splitting from top to bottom into 2 parts.
Bloom Period: Spring, summer, fall.
Distribution: Cameron and Hidalgo counties.
Comments: This species is somewhat similar to *Richardia scabra* (not pictured). It produces flowers in the leaf axils, whereas *R. scabra* produces them in terminal clusters. The larger flowers and fruit distinguish it from Florida buttonweed, *Spermacoce floridana*.

Smooth Buttonweed *Spermacoce glabra*

Rutaceae

Torchwood, Sierra Madre Torchwood

Family: Rutaceae
Scientific Name: *Amyris madrensis*
Habit: Thornless shrubs or small trees with very upright growth, up to 10' tall or more; branches silvery colored.
Leaves: Opposite and compound with 5 or more leaflets; leaflets up to 1 ⅛" long.
Flowers: Inflorescences aromatic, produced from the branch tips; corollas white, about ⅛" long, with 4 to 5 petals.
Fruit: Roundish, juicy, blue-black when ripe, up to ½" broad.
Bloom Period: Spring, summer, fall.
Distribution: Cameron, Hidalgo, and Willacy counties.
Comments: This is a host plant for the Giant Swallowtail butterfly. It is an attractive ornamental for small areas. The family Rutaceae is the citrus family. Like many members of the family, the leaves of this species have a citrus scent when crushed. Torchwoods are so called because the green branches can be ignited.

Torchwood, Sierra Madre Torchwood *Amyris madrensis*

Texas Torchwood

Family: Rutaceae
Scientific Name: *Amyris texana*
Habit: Slow-growing, thornless shrubs 3–6' tall.
Leaves: Alternate and compound, with 3 leaflets; leaflets with wavy edges, up to 1 ¼" long.
Flowers: Inflorescences aromatic, produced from the branch tips; corollas white, about ⅛" long, with 4 to 5 petals.
Fruit: Roundish, juicy, blue-black when ripe, about 3⁄16" broad.
Bloom Period: Spring, summer, fall.
Distribution: Cameron, Hidalgo, Willacy, and Starr counties.
Comments: This is a member of the citrus family. Its leaves emit a citrus odor when crushed, as is true with many members of the family. The Texas state fruit is in this family. It is widely sold as the Texas "Ruby Red" grapefruit. Texas torchwood blooms attract evening pollinators, such as Hawk moths.

Texas Torchwood *Amyris texana*

Limoncillo, Runyon's Esenbeckia

Family: Rutaceae
Scientific Name: *Esenbeckia berlandieri* [*E. runyonii*]
Habit: Trees up to 20' or taller, with peeling bark.
Leaves: Alternate, compound with 3 leaflets; leaflets up to 6 ¾" long.
Flowers: Radial, in terminal clusters; petals 4 or 5, whitish, about ⅛" long.
Fruit: Roundish capsules up to 1 ⅜" broad, with 4 or 5 one-seeded compartments.
Bloom Period: Spring, fall.
Distribution: Cameron County.
Comments: This species is more common in Mexico. Only about a dozen trees are known to be growing naturally in our area, in Cameron County. Seeds from Cameron County trees and Mexican populations have been planted in various parts of the Lower Rio Grande Valley. This tree was first reported in our area by Robert Runyon, former Brownsville mayor, photographer, and self-taught botanist. The Spanish common name translates as "little lemon." Lemons are in the same plant family as *Esenbeckia*.

Limoncillo, Runyon's Esenbeckia *Esenbeckia berlandieri*

Barreta

Family: Rutaceae
Scientific Name: *Helietta parvifolia*
Habit: Thornless shrubs up to 13' tall.
Leaves: Aromatic when bruised, opposite, compound with 3 leaflets; leaflets usually spatulate, up to 1 ⅝" long.
Flowers: Radial, in terminal inflorescences; petals 3 or 4, whitish, about ⅛" long.
Fruit: Dry, up to ⅝" long, separating into 3 or 4 winged, one-seeded parts.
Bloom Period: Spring, summer.
Distribution: Cameron, Hidalgo, and Starr counties.
Comments: These attractive shrubs are particularly noticeable growing on gravelly hills in Starr County. The reports of Cameron County populations are questionable. Barretas get much larger in Mexico and are preferred for fence posts.

Barreta *Helietta parvifolia*

Barreta *H. parvifolia* dry fruit

Rutaceae

Dutchman's Breeches

Family: Rutaceae
Scientific Name: *Thamnosma texanum*
Habit: Low perennials about 10" tall or less.
Leaves: Alternate, mostly linear, up to ⅝" long, smelling like citrus when bruised.
Flowers: In unbranched inflorescences; petals 4, whitish with yellow or purple, about ⅛" long.
Fruit: Dry, partially divided into 2 "legs" and swollen, about ¼" long.
Bloom Period: Spring, summer.
Distribution: Cameron, Hidalgo, and Starr counties.
Comments: This species is a host plant for the Black Swallowtail butterfly. The genus name comes from two Greek words meaning "shrub" and "smell." The specific epithet is usually spelled *texana,* but according to the *Flora of North Central Texas* (Diggs et al. 1999), the correct spelling is *texanum.*

Dutchman's Breeches *Thamnosma texanum*

Colima

Family: Rutaceae
Scientific Name: *Zanthoxylum fagara*
Habit: Shrubs up to 29' tall, with curved, sharp prickles.
Leaves: Alternate, compound; petioles and rachises winged; leaflets 9 or fewer, obovate to ovate, up to 1" long; when bruised, giving off a citrus odor.
Flowers: Unisexual; petals 4, greenish, less than ⅛" long.
Fruit: Dry, roundish, reddish color, turning black when dry, splitting open, one-seeded, about ⅛" broad.
Bloom Period: Spring.
Distribution: Cameron, Hidalgo, and Willacy counties.
Comments: The curved prickles and the citrus odor of the bruised leaves, combined with the winged petioles and rachises, help to identify this species. Colima, with its defensive prickles, provides shelter for birds to nest. It is a widespread member of our flora and a host plant for the Giant Swallowtail butterfly.

Colima *Zanthoxylum fagara* in fruit

Colima *Z. fagara* plant in flower

Hairy Tickle Tongue

Family: Rutaceae

Scientific Name: *Zanthoxylum hirsutum*

Habit: Aromatic shrubs or small trees up to 13' tall, with many curved prickles.

Leaves: Alternate, compound with up to 7 leaflets; leaflets varying from elliptic to oval, up to 1 ⅜" long.

Flowers: Tiny, unisexual, in terminal inflorescences; petals greenish.

Fruit: Dry, reddish, one-seeded, about ⅛" broad, splitting open.

Bloom Period: Spring.

Distribution: Hidalgo and Willacy counties.

Comments: Hairy tickle tongue is not as commonly seen as colima, *Z. fagara*. It seems to prefer sandy soils. It has more prickles and bigger leaves than colima. The leaves, when chewed, produce numbness in the mouth and have been used for toothache. The illustration shows the plant with buds but no open flowers.

Hairy Tickle Tongue *Zanthoxylum hirsutum*

Hairy Tickle Tongue *Z. hirsutum* fruit

Narrow Leaved Sandbar Willow

Family: Salicaceae

Scientific Name: *Salix exigua* subsp. *interior* [*S. interior*]

Habit: Deciduous thornless trees up to 13' or taller.

Leaves: Alternate; blades linear, up to 4 ¾" long; margins almost entire to minutely toothed, the teeth mostly ⅛" or more apart.

Flowers: Tiny, unisexual, male and female on separate plants.

Fruit: Capsules gray-hairy, releasing seeds with silky hairs.

Bloom Period: Spring, summer, fall.

Distribution: Cameron and Hidalgo counties.

Comments: Cone-shaped galls (abnormal growths caused by insects) are often present and mistaken for fruit. The willows grow around resacas and other bodies of water. Our two species are very similar. Black willow, *S. exigua*, has hairy capsules, and its leaf margins have teeth ⅛" or more apart; *S. nigra* has hairless capsules, and its leaf margins have teeth less than ⅛" apart. Twigs cut and placed into water produce a solution used to promote rooting of stem cuttings of ornamentals.

Narrow Leaved Sandbar Willow *Salix exigua* subsp. *interior*

Narrow Leaved Sandbar Willow *S. exigua* subsp. *interior* stem with galls

Sapindaceae

Black Willow, Sauz

Family: Salicaceae
Scientific Name: *Salix nigra*
Habit: Deciduous thornless trees up to 33' or taller.
Leaves: Alternate; blades lanceolate, the margins entire or (seen with magnifying lens) finely toothed, the teeth at least 1/16" apart.
Flowers: Tiny, unisexual, male and female on separate plants.
Fruit: Capsules hairless, releasing seeds with silky hairs.
Bloom Period: Spring, summer.
Distribution: Cameron, Hidalgo, Willacy, and Starr counties.
Comments: This species is usually found growing around bodies of fresh water. Its capsules are hairless, and its leaf margins have teeth less than 1/8" apart. The similar narrow leaved sandbar willow, *S. exigua,* has hairy capsules, and its leaf margins have teeth 1/8" or more apart. Twigs cut and placed into water produce a solution used to promote rooting of stem cuttings of ornamentals. The bark of this species is the original source of salicylic acid (aspirin).

Black Willow, Sauz *Salix nigra*

Chihuahuan Balloon Vine

Family: Sapindaceae
Scientific Name: *Cardiospermum dissectum*
Habit: Hairless vines with tendrils.
Leaves: Alternate, compound, the leaflets mostly linear, mostly less than 1/16" broad.
Flowers: Inflorescence branches originating from the same point; flowers bilateral; petals 4, white, about 3/16" long.
Fruit: An inflated pod with 3 seeds, up to 1 1/8" broad.
Bloom Period: Spring, summer.
Distribution: Hidalgo and Starr counties.
Comments: This species is easily recognized by its hairless condition and its highly dissected leaves. It is considered to be uncommon to rare. The genus name comes from Greek words for "heart" and "seed," referring to a heart-shaped spot on the seed.

Chihuahuan Balloon Vine *Cardiospermum dissectum*

Common Balloon Vine

Family: Sapindaceae
Scientific Name: *Cardiospermum halicacabum*
Habit: Hairless vines with tendrils.
Leaves: Alternate, compound, with 3 ovate, lobed leaflets up to 1 ¾" long. Sometimes there are more than 3 leaflets, but if so, then the margins are toothed.
Flowers: Inflorescence branches originating from the same point; flowers bilateral; petals 4, white, about 3/16" long.
Fruit: An inflated pod with 3 seeds, up to 1 ½" broad but usually less.
Bloom Period: All seasons.
Distribution: Cameron, Hidalgo, and Willacy counties.
Comments: This is a host plant for the Silver-banded Hairstreak butterfly. The broader leaflets distinguish this species from Chihuahuan balloon vine, *C. dissectum,* and its lack of hairs distinguishes it from *C. corindum* (not pictured). The similar vine Mexican urvillea, *Urvillea ulmacea,* has football-shaped fruit that break up into 3 winged seeds, with the seeds located at the edge, about the midpoint. Another similar vine, serjania, *Serjania brachycarpa,* also has fruit that break up into 3 winged seeds, which turn reddish and can be mistaken at a distance for flowers. Its seeds are located at the end of the wing.

Common Balloon Vine *Cardiospermum halicacabum* with fruit

Common Balloon Vine *C. halicacabum* flowers

Soapberry, Jaboncillo

Family: Sapindaceae
Scientific Name: *Sapindus saponaria* var. *drummondii*
Habit: Trees up to 32' or taller.
Leaves: Alternate, compound; leaflets mostly ovate, up to 2 ¾" long, turning yellow in the fall.
Flowers: Inflorescences at the tips of the branches; flowers bilateral; corollas white, about 3/16" long.
Fruit: Round, about ½" broad, in clusters, orange and hardening when ripe, fleshy with one seed.
Bloom Period: Winter, spring, summer.
Distribution: Cameron, Hidalgo, and Starr counties.
Comments: This attractive tree is often cultivated. The fruit has been used to make a kind

Soapberry, Jaboncillo *Sapindus saponaria* var. *drummondii*

Soapberry, Jaboncillo *S. saponaria* var. *drummondii* fruiting branch

of soap. Indians have used the fruit in water to kill fish. Soapberry is one of our few native trees that produce colored leaves in the fall. The genus name and the specific epithet both contain the Latin word *sapo,* for "soap."

Serjania

Family: Sapindaceae
Scientific Name: *Serjania brachycarpa*
Habit: Vines with tendrils.
Leaves: Alternate, hairy, usually 2 times compound; leaflets ovate, about 1 ⅜" long.
Flowers: Inflorescences up to 2" long, the branches not originating from the same point; flowers bilateral; petals 5, whitish or yellowish, less than ⅛" long.
Fruit: Dry, up to ½" long, breaking up into 3 reddish winged seeds.
Bloom Period: Summer, fall.
Distribution: Cameron, Hidalgo, and Starr counties.
Comments: When seen at a distance, the reddish winged seeds are sometimes mistaken for flowers. The seeds located at the ends of the wings are a good indicator for this species. The genus was named in 1703 in honor of a French monk, Philippe Sergeant, who was known for his ability in botany and medicine.

Serjania *Serjania brachycarpa*

Mexican Urvillea, Elm Leaved Urvillea

Family: Sapindaceae
Scientific Name: *Urvillea ulmacea*
Habit: Vines with tendrils.
Leaves: Alternate, hairy, compound with 3 ovate leaflets up to 3 ⅛" long, the margins toothed.
Flowers: Bilateral; petals 4, whitish, about 1/16" long.
Fruit: Dry, hollow, football shaped, up to ½" long, breaking up into 3 winged seeds.
Bloom Period: Spring, summer, and fall.
Distribution: Cameron, Hidalgo, and Willacy counties.
Comments: The 3 leaflets with toothed margins separate this species from other vines in the family. The seed is located at the edge of the wing, about halfway down it, another good indicator. The fruit are the best characters to identify the species.

Mexican Urvillea, Elm Leaved Urvillea *Urvillea ulmacea*

Coma

Family: Sapotaceae
Scientific Name: *Sideroxylon celastrinum*
[*Bumelia celastrina*]
Habit: Shrubs or trees up to 30' tall (usually less), with milky sap, forming thickets from root sprouts. Side branches usually long slender thorns.
Leaves: Alternate; blades spatulate to ovate, tapering onto the petioles, up to 1 5/8" long.
Flowers: Radial, fragrant, in clusters from the leaf axils; petals 5, white, up to 3/16" long.
Fruit: Drupes, black when mature, exuding drops of white latex, 7/16" long and 1/4" or less broad.
Bloom Period: Spring, summer, fall.
Distribution: Cameron, Hidalgo, and Starr counties.
Comments: In less prosperous times the fruit have been used by children as chewing gum. The sweet fragrance of plants in bloom can be detected from long distances.

Coma *Sideroxylon celastrinum*

Coma *S. celastrinum* fruiting branch

Prairie Agalinis

Family: Scrophulariaceae
Scientific Name: *Agalinis heterophylla*
[*Gerardia heterophylla*]
Habit: Erect annuals up to 24" tall.
Leaves: Opposite on lower parts, sometimes alternate higher; blades more or less linear, up to 2 1/2" long.
Flowers: Bilateral, pink or almost white, up to 1 1/4" long.
Fruit: Capsules up to 5/16" tall.
Bloom Period: Spring, summer.
Distribution: Cameron and Willacy counties.
Comments: This species is a host plant for the Common Buckeye butterfly. We have three *Agalinis* species, all from the coastal sands. *A. heterophylla* has calyx lobes about half the length of the calyx; the other species have very short calyx lobes. The leaves of seaside agalinis, *A. maritima*, are succulent and crowded; those of erect leaf agalinis, *A. strictifolia*, are not.

Prairie Agalinis *Agalinis heterophylla*

Scrophulariaceae

Seaside Agalinis

Family: Scrophulariaceae
Scientific Name: *Agalinis maritima* var. *grandiflora* [*Gerardia maritima*]
Habit: Annuals up to 18" tall.
Leaves: Opposite on lower parts, sometimes alternate higher; blades succulent, linear, up to 1½" long.
Flowers: Bilateral, pink, up to ¾" long.
Fruit: Capsules about ¼" tall.
Bloom Period: Spring, summer.
Distribution: Cameron and Willacy counties.
Comments: This species is a host plant for the Common Buckeye butterfly. Of our three *Agalinis* species, this one is easy to distinguish by its succulent leaves and short calyx lobes. The flowers are usually somewhat smaller than those of the other two species. All three species are showy in bloom and bloom intermittently whenever conditions are right. A few can sometimes be seen blooming in fall and winter. The variety name *grandiflora* can be misleading. The flowers are bigger than those of other varieties of the same species but still smaller than those of our other two species.

Seaside Agalinis *Agalinis maritima var. grandiflora*

Erect Leaf Agalinis

Family: Scrophulariaceae
Scientific Name: *Agalinis strictifolia* [*Gerardia strictifolia*]
Habit: Erect annuals up to 32" tall.
Leaves: Opposite on lower parts, sometimes alternate higher; blades more or less linear, up to ¾" long.
Flowers: Bilateral, pink, up to 1" long.
Fruit: Capsules about ¼" tall.
Bloom Period: Spring, summer, fall.
Distribution: Cameron and Willacy counties.
Comments: This species is a host plant for the Common Buckeye butterfly. We have three species of *Agalinis,* all growing in coastal sands. Prairie agalinis, *A. heterophylla,* has calyx lobes about half the length of the calyx. The calyx lobes of *A. strictifolia* and seaside agalinis, *A. maritima,* are about 1/16" long. The leaves of *A. maritima* are succulent and crowded while those of *A. strictifolia* are not.

Erect Leaf Agalinis *Agalinis strictifolia*

Coastal Water Hyssop

Family: Scrophulariaceae
Scientific Name: *Bacopa monnieri*
Habit: Creeping perennials, the stem ends sometimes leaning upward.
Leaves: Opposite; blades spatulate, up to ⅝" long.
Flowers: From the leaf axils; corollas only slightly bilateral, white to bluish, up to ⅜" long.
Fruit: Capsules 5⁄16" tall.
Bloom Period: Spring, summer, fall.
Distribution: Cameron, Hidalgo, Willacy, and Starr counties.
Comments: This is a common wetland plant, often seen in or near various low places where water stands for a time. It is a food plant for the larvae of the White Peacock butterfly.

Coastal Water Hyssop *Bacopa monnieri*

Disc Water Hyssop

Family: Scrophulariaceae
Scientific Name: *Bacopa rotundifolia*
Habit: Creeping perennials of low, wet localities, the leaves floating when the plants are submerged.
Leaves: Opposite, without petioles; blades clasping the stems, rounded to broadly ovate, up to 1" long.
Flowers: From the leaf axils; corollas seemingly radial but very slightly bilateral; white with yellow centers, up to ⅜" long.
Fruit: Capsules rounded, about ¾" tall.
Bloom Period: Summer, fall.
Distribution: Cameron, Hidalgo, and Willacy counties.
Comments: This species is seldom encountered in our area. It has been seen in Cameron and Hidalgo counties, and in southern Willacy County in a small wetland that has since become a plowed field.

Disc Water Hyssop *Bacopa rotundifolia*

Scrophulariaceae

American Bluehearts

Family: Scrophulariaceae
Scientific Name: *Buchnera americana*
 [*B. floridana*]
Habit: Perennials herbs up to 24" tall.
Leaves: Usually opposite, without petioles; blades spatulate, up to 3 ¾" long and ⅝" broad.
Flowers: Inflorescences apical; corollas slightly bilateral, violet or rarely white, 5-lobed, with narrow tubes, up to ½" long.
Fruit: Capsules about 3/16" tall.
Bloom Period: Summer, fall.
Distribution: Cameron County.
Comments: This species is not commonly seen in the Lower Rio Grande Valley. It is sometimes seen in the sandy soils on South Padre Island. Stems that are broken leave a black dye on the hands. The plants are partially parasitic on roots of nearby plants.

American Bluehearts *Buchnera americana*

Mexican Capraria

Family: Scrophulariaceae
Scientific Name: *Capraria mexicana*
Habit: Shrubs up to 6' or taller.
Leaves: Alternate; blades lanceolate, toothed toward the tip, up to 5" long; petioles tiny or absent.
Flowers: Borne from the leaf axils; petals white, partially grown together, the lobes about ⅜" long.
Fruit: Capsules cylindrical, up to ⅜" long.
Bloom Period: Winter, spring.
Distribution: Cameron County.
Comments: A small population of this species near Falcon Reservoir in Zapata County was reported by Joe Ideker (Ideker 1996). It was not known in the United States until that time. We have seen one plant in Cameron County at Laguna Atascosa National Wildlife Refuge, in one of the gardens. According to Ellie Thompson, volunteer manager of the butterfly garden, it was not planted there but grew voluntarily from seed. Recently we found a fairly large population in Brownsville, Cameron County (Richardson & King, 2006). Later, Mike Heep, Diane Ballesteros, and Christina Mild found a colony near Ramsey Park in Harlingen, Cameron County. The species is more abundant in Mexico. It has probably been overlooked here because of its resemblance to depression weed, *Baccharis neglecta*.

Mexican Capraria *Capraria mexicana*

Texas Paintbrush

Family: Scrophulariaceae
Scientific Name: *Castilleja indivisa*
Habit: Hairy annuals up to 16" tall.
Leaves: Alternate; blades lanceolate or linear, up to 4" long.
Flowers: Inflorescences with colorful bracts; corollas bilateral, white to greenish, up to 1 ⅛" long; bracts bright orange-red, or occasionally white.
Fruit: Capsules ovoid, small.
Bloom Period: Spring.
Distribution: Cameron, Hidalgo, and Willacy counties.
Comments: This is one of our showier wildflowers. Many of the populations in Cameron County are from sown seed, and they will persist only a few years. Although the leaves are green and photosynthetic, the roots of Texas paintbrush extract some nutrients from roots of nearby plants. It is therefore called semiparasitic. The flowers are fairly inconspicuous. It is the showy bracts that give the plant its color.

Texas Paintbrush *Castilleja indivisa*

Texas Paintbrush *C. indivisa* inflorescence with the red bracts and pinkish flowers

Cenizo, Texas Purple Sage

Family: Scrophulariaceae
Scientific Name: *Leucophyllum frutescens*
Habit: Shrubs up to 6 ½' tall, sometimes taller.
Leaves: Alternate, usually densely hairy and silvery colored; blades obovate, up to ¾" long.
Flowers: Borne singly from the leaf axils; corollas bilateral, pinkish to purple, occasionally white, up to ¾" long.
Fruit: Capsules about ⁵⁄₁₆" tall.
Bloom Period: Spring, summer, fall, usually after rain.
Distribution: Cameron, Hidalgo, Willacy, and Starr counties.
Comments: The Spanish common name, meaning "ash," refers to the ashy colored leaves. The scientific name comes from two Greek words meaning "white" and "leaf." Cenizo is one of our most beautiful shrubs. It has no thorns, and it blooms several times a year, with the rains. In areas where the shrubs are abundant, they produce a stunning effect. Cenizo is often cultivated, and there are various forms, including some with green leaves. Attempts to collect seed often fail because the ripe capsule snaps open and throws them some distance away. This species is a host plant for the Theona Checkerspot butterfly.

Cenizo, Texas Purple Sage *Leucophyllum frutescens*

Scrophulariaceae

Clasping False Pimpernel

Family: Scrophulariaceae
Scientific Name: *Lindernia dubia* var. *anagallidea*
Habit: Erect or leaning annuals.
Leaves: Opposite; blades ovate, up to ⅝" long.
Flowers: from the leaf axils; corollas bilateral, 5-lobed, whitish or lavender, up to ⅜" long.
Fruit: Capsules about 3/16" tall.
Bloom Period: Spring, summer, fall.
Distribution: Cameron, Hidalgo, and Willacy counties.
Comments: This species prefers damp places. It is easy to miss because the plants are low, and the flowers are small and not brightly colored.

Clasping False Pimpernel *Lindernia dubia* var. *anagallidea*

Snapdragon Vine

Family: Scrophulariaceae
Scientific Name: *Maurandya antirrhiniflora*
Habit: Twining perennial vines.
Leaves: Alternate, sometimes nearly opposite; blades arrowhead shaped, with 2 basal lobes turning outward, up to 1" long.
Flowers: From the leaf axils; corollas bilateral, 5-lobed, purplish, sometimes almost white, up to 1" long.
Fruit: Capsules about ¼" tall.
Bloom Period: Spring, summer.
Distribution: Cameron, Willacy, and Starr counties.
Comments: This is an attractive little vine and would be beautiful growing on a fence or some other support in the garden. It is a host plant for the Common Buckeye butterfly. In Cameron County, populations of this species grow in coastal sands and along ditches and waterways. They are widely separated from the western populations, which grow in caliche or sandy soil.

Snapdragon Vine *Maurandya antirrhiniflora*

Prostrate Water Hyssop, Yellow Flowered Mecardonia

Family: Scrophulariaceae
Scientific Name: *Mecardonia procumbens* [*M. vandelloides*]
Habit: Erect or leaning perennials.
Leaves: Opposite; blades ovate to obovate, up to ⅞" long.
Flowers: from the leaf axils; corollas bilateral, 5-lobed, yellow, up to ½" long.
Fruit: Capsules about ¼" tall or less.
Bloom Period: All seasons.
Distribution: Cameron, Hidalgo, Willacy, and Starr counties.
Comments: This species likes wet or damp places but can also survive in drier habitats.

Prostrate Water Hyssop, Yellow Flowered Mecardonia *Mecardonia procumbens*

Texas Toad Flax

Family: Scrophulariaceae
Scientific Name: *Nuttallianthus texanus* [*Linaria texana*]
Habit: Erect annuals up to 22" tall.
Leaves: Opposite at the base, alternate higher on the stem; blades linear, up to 1 ⅜" long.
Flowers: Inflorescences from the apex; corollas bilateral, 5-lobed, lavender-blue or pinkish, about ½" long, with a basal spur.
Fruit: Capsules up to 3/16" tall.
Bloom Period: Spring.
Distribution: Hidalgo and Willacy counties.
Comments: This species prefers sandy soils. The common name comes from the resemblance of the leaves to those of flax, *Linum*.

Texas Toad Flax *Nuttallianthus texanus*

Sweet Broomwort

Family: Scrophulariaceae
Scientific Name: *Scoparia dulcis*
Habit: Hairless branching annuals up to 40" tall.
Leaves: Opposite, sometimes whorled; blades mostly ovate or lanceolate, up to 1 ½" long, the margins toothed.
Flowers: Radial, from the leaf axils, with many white, beardlike filaments in the center; petals 4, white; sepals 4, stamens 4.
Fruit: Capsules roundish, many-seeded, about ⅛" broad.
Bloom Period: Spring, summer, and fall.
Distribution: Cameron County.
Comments: *Scoparia dulcis* is a weedy species

Sweet Broomwort *Scoparia dulcis*

known in southern East Texas. It was called to our attention by Mike Heep. This is the first report of its occurrence in the Lower Rio Grande Valley. It reproduces rapidly from seed, and a suitable habitat is rapidly populated by this species. It seems to prefer moist localities.

Woolly Stemodia

Family: Scrophulariaceae
Scientific Name: *Stemodia lanata* [*S. tomentosa*]
Habit: Prostrate perennials of sandy soils, forming mats.
Leaves: Opposite, silvery or white with dense hairs; blades obovate, up to ¾" long.
Flowers: Usually single from the axils of the leaves; corollas bilateral, 5-lobed, purplish, up to 5⁄16" long.
Fruit: Capsules about ⅛" tall.
Bloom Period: Spring, summer.
Distribution: Cameron, Hidalgo, and Willacy counties.
Comments: This is a beautiful plant in a natural setting. Beverly Wheelock in Olmito, Texas, tried it as a ground cover in the garden; it grew well, but became unattractive.

Woolly Stemodia *Stemodia lanata*

Schott's Twintip

Family: Scrophulariaceae
Scientific Name: *Stemodia schottii*
Habit: Erect or leaning perennials.
Leaves: Opposite; blades green, mostly obovate, up to ¾" long; margins toothed.
Flowers: Usually single from the axils of the leaves; corollas bilateral, 5-lobed, blue with yellow, up to ⅝" long.
Fruit: Capsules about 3⁄16" tall.
Bloom Period: All seasons.
Distribution: Cameron, Hidalgo, and Starr counties.
Comments: This species is uncommon here but probably more abundant in adjacent Mexico. According to the *Atlas of the Vascular Plants of Texas* (Turner et al. 2003), the only reported localities in Texas are our area and Zapata and Val Verde counties.

Schott's Twintip *Stemodia schottii*

Goat Bush, Amargosa

Family: Simaroubaceae

Scientific Name: *Castela erecta* subsp. *texana* [*C. texana*]

Habit: Shrubs up to 6 ½' tall with long whitish thorns.

Leaves: Alternate, shiny green on the upper surfaces, whitish on the lower surfaces, oblong, about ¾" long. The leaves are often appressed against the stems.

Flowers: Radial; petals 4, reddish, about ⅛" long.

Fruit: Bright red, fleshy, up to ⅜" broad.

Bloom Period: Spring.

Distribution: Cameron, Hidalgo, and Starr counties.

Comments: This is an attractive shrub with its whitish branches, shiny green leaves, and red flowers and fruit. The thorns could be a liability, but in the right place it would make an impressive specimen. The Spanish word *amargosa* translates as "bitter," referring to the bitter-tasting leaves, stems, and fruit.

Goat Bush, Amargosa *Castela erecta* subsp. *texana* in fruit

Goat Bush, Amargosa *C. erecta* subsp. *texana* flowering branch

Wild Petunia

Family: Solanaceae

Scientific Name: *Calibrachoa parviflora* [*Petunia parviflora*]

Habit: Prostrate annuals, the stems rooting at the nodes.

Leaves: Alternate; blades spatulate to linear, up to ⅝" long.

Flowers: Radial, 5-lobed, borne singly from the axils of the leaves; corollas purple, about 5/16" long.

Fruit: Capsules ovoid, about 3/16" long.

Bloom Period: Winter, spring.

Distribution: Cameron, Hidalgo, and Starr counties.

Comments: Wild petunia prefers moist places (low places, ditches, resaca banks) but is not limited to them, especially in seasons of abundant rain. The genus is named for the nineteenth-century Mexican botanist Antonio Cal y Bracho.

Wild Petunia *Calibrachoa parviflora*

Solanaceae

Chile Piquín, Chilipiquín

Family: Solanaceae
Scientific Name: *Capsicum annuum* var. *aviculare*
Habit: Shrubs about 3' tall, occasionally much taller.
Leaves: Alternate; blades ovate to elliptic, up to 2 ⅛" long.
Flowers: Radial, 5-lobed, from the leaf axils; corollas white, up to 5/16" broad.
Fruit: Berries ovoid to globose, about 5/16" long (some varieties much bigger), orange-red when mature. They are edible but fiery hot.
Bloom Period: All seasons.
Distribution: Cameron, Hidalgo, and Starr counties.
Comments: The name of the genus comes from a Greek word meaning "to bite," and the small peppers do have a bite. This is a prized and popular local plant. There are many variations in home cultivation. Mockingbirds and others enjoy the berries, often keeping a plant picked clean of ripe fruit. As if to thank the grower for the food, the birds return the seed from their perches, and soon many seedlings appear. People use the berries in various ways: mashing fresh berries with their food; cooking with them; pickling them in vinegar; and drying them and crumbling dried fragments onto their food. When the LRGV was more rural, a favorite game of older children was to pick a ripe berry, swallow it whole, pretend to chew it up, and tell the younger ones how good it was. Soon there was a lot of screaming and running for water.

Chile Piquín, Chilipiquín *Capsicum annuum* var. *aviculare*

Prostrate Ground Cherry

Family: Solanaceae
Scientific Name: *Chamaesaracha conioides*
Habit: Hairy perennials; stems erect or leaning over, up to 28" across.
Leaves: Alternate; blades more or less ovate, 1½–2" long; margins lobed.
Flowers: Radial, 5-lobed, from the leaf axils, yellowish, about ⅝" broad.
Fruit: Berries roundish, up to 5/16" broad, enclosed by the dry calyx.
Bloom Period: All seasons.
Distribution: Hidalgo and Starr counties.
Comments: This species is very similar to hairy false nightshade, *C. sordida*. The two species

Prostrate Ground Cherry *Chamaesaracha conioides*

were formerly lumped together and then split again into two distinct species. The leaves of *C. conioides* are more deeply lobed and have many simple hairs; the leaves of *C. sordida* are less lobed and have mostly small glandular hairs. Another species, green false nightshade, *C. coronopus*, has leaves with star-shaped hairs.

Green False Nightshade

Family: Solanaceae
Scientific Name: *Chamaesaracha coronopus*
Habit: Perennials with erect and arching stems with star-shaped hairs.
Leaves: Alternate, with many star-shaped hairs.
Flowers: Radial, 5-lobed, from the leaf axils, yellowish, about ⅝" broad.
Fruit: Berries roundish, up to ⁵⁄₁₆" broad, enclosed by the dry calyx.
Bloom Period: All seasons.
Distribution: Cameron, Hidalgo, and Willacy counties.
Comments: The star-shaped hairs distinguish this species from the other species of *Chamaesaracha*. It is the most abundant *Chamaesaracha* in our area.

Green False Nightshade *Chamaesaracha coronopus*

Hairy False Nightshade

Family: Solanaceae
Scientific Name: *Chamaesaracha sordida*
Habit: Hairy, somewhat sticky perennials, the stems leaning or erect.
Leaves: Alternate; blades ovate to obovate, up to 4" long, with simple hairs and sometimes glandular hairs, the margins entire to toothed or lobed.
Flowers: Radial, 5-lobed, from the leaf axils, white to yellowish, or with a purple caste, about ⅝" broad.
Fruit: Berries roundish, about ⁵⁄₁₆" broad.
Bloom Period: All seasons.
Distribution: Cameron, Hidalgo, and Starr counties.
Comments: The other species of *Chamaesaracha* are similar, but they have either star-shaped hairs or branched hairs. This species has unbranched hairs and often has hairs with glandular secretions at their tips.

Hairy False Nightshade *Chamaesaracha sordida*

Solanaceae

Shaggy False Nightshade

Family: Solanaceae
Scientific Name: *Chamaesaracha villosa*
Habit: Glandular perennials with branching hairs.
Leaves: Alternate; blades ovate to lanceolate, up to 1" long.
Flowers: Radial, 5-lobed, from the leaf axils, pale yellowish, sometimes with purplish centers, 3/8" or less broad.
Fruit: Berries roundish, about 3/16" broad, enclosed by the dry calyx.
Bloom Period: Spring, summer, fall.
Distribution: Starr County.
Comments: This species has not previously been reported for our area. It is more commonly found in West Texas. Its dense, branching hairs separate it from our other species of *Chamaesaracha*.

Shaggy False Nightshade *Chamaesaracha villosa*

Angel Trumpet, Jimson Weed, Moonflower

Family: Solanaceae
Scientific Name: *Datura wrightii*
Habit: Erect branching herbs from a perennial rootstock, up to 5' tall.
Leaves: Alternate; blades ovate about 6" long.
Flowers: Fragrant, opening at night or on cloudy days; corollas white, trumpet shaped, 6" or longer.
Fruit: Globose, prickly capsules about 1 5/8" broad.
Bloom Period: Spring, summer, fall.
Distribution: Cameron and Starr counties.
Comments: The leaves and seeds contain alkaloids that are narcotic, usually poisonous. This is true of many members of the family. Attempts to use *Datura* extracts as narcotics often result in poisoning and can cause death. The flowers, which open at night, are pollinated by Sphinx moths. *D. wrightii* was named in honor of Charles Wright, one of the best known American botanists in the nineteenth century.

Angel Trumpet, Jimson Weed, Moonflower *Datura wrightii*

Berlandier Wolfberry

Family: Solanaceae
Scientific Name: *Lycium berlandieri*
Habit: Thorny shrubs up to 8' tall.
Leaves: Alternate or in clusters; blades spatulate, up to 1" long.
Flowers: Radial, from the leaf axils; corollas 4- or 5-lobed, whitish, occasionally bluish, up to 5/16" broad.
Fruit: Berries roundish, about 3/16" broad, red when mature.
Bloom Period: Spring, summer, fall.
Distribution: Cameron, Hidalgo, Willacy, and Starr counties.
Comments: The berries are often eaten by small animals and were eaten by Native Americans. The smaller, usually white flowers and the smaller berries distinguish this species from Carolina wolfberry, *L. carolinianum*.
L. berlandieri was named in honor of Jean Louis Berlandier, who did important botanical work in Mexico and Texas in the 1820s.

Berlandier Wolfberry *Lycium berlandieri*

Berlandier Wolfberry *L. berlandieri* fruiting branch

Carolina Wolfberry

Family: Solanaceae
Scientific Name: *Lycium carolinianum* var. *quadrifidum*
Habit: Thorny shrubs up to 40" tall. Stems that have fallen over will root and produce new plants.
Leaves: Alternate or in clusters; blades spatulate, up to 1 5/8" long.
Flowers: Radial, from the leaf axils; corollas bluish, almost always 4-lobed, up to 1" broad.
Fruit: Berries roundish, about 3/8" broad, red when ripe.
Bloom Period: Spring, summer, fall.
Distribution: Cameron, Hidalgo, and Willacy counties.
Comments: The bright red berries are utilized by wildlife. This species is mostly found in the moist salty areas at or near the coast. Usually a single plant has a few scattered flowers, but occasionally it produces a large number and is quite attractive.

Carolina Wolfberry *Lycium carolinianum* var. *quadrifidum*

Carolina Wolfberry *L. carolinianum* var. *quadrifidum* fruiting branch

Solanaceae

Netted Globe Berry

Family: Solanaceae
Scientific Name: *Margaranthus solanaceus*
Habit: Erect annuals up to 24" tall.
Leaves: Alternate; blades ovate, up to 2 ⅜" long; margins sometimes toothed.
Flowers: Radial, from the leaf axils; corollas 5-lobed, greenish yellow, about 3/16" long. The flowers do not open broadly but remain almost closed.
Fruit: Berries roundish, about 3/16" broad, enclosed by the papery fruiting calyx.
Bloom Period: Spring, summer, fall, and winter.
Distribution: Cameron, Hidalgo, and Starr counties.
Comments: Four of our genera in this family have the fruit enclosed by the papery fruiting calyx (like that of tomatillos used in cooking). *Physalis* has yellowish corollas with dark purple centers; *Chamaesoracha* flowers open fully and are pale yellow, and the fruit are enclosed by an inflated calyx; *Margaranthus* has yellowish corollas that are greatly constricted, almost closed, at the mouth; and *Quincula* has broadly open, bluish to purple corollas. The illustration shows one flower and some fruit.

Netted Globe Berry *Margaranthus solanaceus*

Tree Tobacco

Family: Solanaceae
Scientific Name: *Nicotiana glauca*
Habit: Shrubs or small trees up to 10' tall.
Leaves: Alternate, leathery, ovate, up to 3 ¾" long.
Flowers: Radial; corollas yellowish, 5-lobed, up to 1 ⅝" long with a long tube.
Fruit: Capsules ovoid, about ½" long.
Bloom Period: Spring, summer, fall.
Distribution: Cameron, Hidalgo, Willacy, and Starr counties.
Comments: This is an introduced South American species that has become naturalized without being a pest. It occurs along waterways, caliche pits, and limestone cliffs. The yellow tubular flowers attract hummingbirds. All parts of the plant are toxic to livestock.

Tree Tobacco *Nicotiana glauca*

Fiddle Leaf Tobacco

Family: Solanaceae

Scientific Name: *Nicotiana repanda*

Habit: Annuals up to 39" tall, usually less, sticky with glandular hairs, malodorous when bruised.

Leaves: Mostly crowded basally; blades ovate to spatulate, up to 12" long.

Flowers: Radial; corollas white, 5-lobed, up to 3 ⅛" long with a long tube.

Fruit: Capsules about ⅜" long.

Bloom Period: Spring, summer, fall.

Distribution: Cameron, Hidalgo, Willacy, and Starr counties.

Comments: A similar species, *N. plumbaginifolia*, has corollas 1 ⅝" or shorter; *N. repanda* has corollas 2" or longer. The genus is named for Jean Nicot, who introduced tobacco into France.

Fiddle Leaf Tobacco *Nicotiana repanda*

Desert Tobacco

Family: Solanaceae

Scientific Name: *Nicotiana trigonophylla*

Habit: Hairy, glandular, sticky herbs up to 36" tall.

Leaves: Alternate; blades various shapes from obovate to elliptic, up to 8 ¾" long.

Flowers: Radial; corollas hairy, greenish or yellowish, up to ⅞" long, with a long tube.

Fruit: Capsules ovoid, up to 7/16" long.

Bloom Period: Spring, summer, fall.

Distribution: Cameron and Starr counties.

Comments: This species grows in rocky or gravelly disturbed areas. The greenish corollas distinguish it from the other *Nicotiana* species. All parts of the plants contain nicotine and other toxic materials. If eaten, death can result.

Desert Tobacco *Nicotiana trigonophylla*

Small Flower Ground Cherry

Family: Solanaceae

Scientific Name: *Physalis cinerascens* var. *cinerascens* [*P. viscosa* var. *cinerascens*]

Habit: Erect to sprawling perennial herbs with star-shaped hairs.

Leaves: Alternate; blades ovate to heart shaped, up to 2 ¾" long, with star-shaped hairs.

Flowers: Radial, hanging downward; corollas yellow with purple centers, up to ½" long.

Fruit: Berries up to ⅜" broad, enclosed loosely in the inflated calyx.

Bloom Period: All seasons.

Small Flower Ground Cherry *Physalis cinerascens* var. *cinerascens*

Distribution: Cameron, Hidalgo, Willacy, and Starr counties.
Comments: Yellowish flowers with purple centers and berries enclosed loosely in the inflated calyx characterize this species. Varieties are often difficult to distinguish, but the leaf shape easily distinguishes this variety from beach ground cherry, var. *spathulifolia*

Small Flower Ground Cherry *P. cinerascens* flower

Beach Ground Cherry
Family: Solanaceae
Scientific Name: *Physalis cinerascens* var. *spathulifolia*
Habit: Erect to sprawling perennial herbs with star-shaped hairs.
Leaves: Alternate; blades more or less spatulate, up to 3 ½" long, with star-shaped hairs.
Flowers: Radial, hanging downward; corollas yellow with purple centers, up to ½" long.
Fruit: Berries up to ⅜" broad, enclosed loosely in the inflated calyx.
Bloom Period: All seasons.
Distribution: Cameron and Willacy counties.
Comments: This is a coastal variety, often occurring in the dunes. The spatulate leaves identify the var. *spathulifolia*. Otherwise, it is like small flower ground cherry, var. *cinerascens*.

Beach Ground Cherry *Physalis cinerascens* var. *spathulifolia*

Fendler's Ivy Leaf Ground Cherry
Family: Solanaceae
Scientific Name: *Physalis hederifolia* var. *fendleri*
Habit: Herbs with weak stems, hairy with unbranched hairs.
Leaves: Alternate; blades ovate to rounded, up to 2 ¾" long.
Flowers: Radial, hanging downward; corollas yellow with purple centers, up to ⅝" long.
Fruit: Berries enclosed loosely in the inflated calyx.
Bloom Period: Spring, summer, fall.
Distribution: Cameron, Hidalgo, Willacy, and Starr counties.
Comments: The specific epithet *hederifolia* comes from Latin words meaning "ivy leaved." The leaves are variable in shape from ovate to heart shaped to rounded. A noticeable character of the genus is the inflated calyx enclosing the berries, although some other genera also have this characteristic.

Fendler's Ivy Leaf Ground Cherry *Physalis hederifolia* var. *fendleri*

Downy Ground Cherry

Family: Solanaceae
Scientific Name: *Physalis pubescens*
Habit: Hairy annuals; stems usually lying on the ground, occasionally partially erect.
Leaves: Alternate; blades ovate, 1 ½" to 3 ½" long, the margins usually toothed.
Flowers: Radial, hanging downward; corollas yellow with purple centers, about ⅜" long; anthers bluish.
Fruit: Berries up to ⅝" broad, enclosed loosely in the inflated calyx.
Bloom Period: Spring, summer, fall.
Distribution: Cameron, Hidalgo, and Willacy counties.
Comments: The 5-angled calyx combined with the bluish anthers are characteristics which separate this from the other *Physalis* species. The genus name comes from the Greek *physa,* for "bladder," referring to the inflated calyx surrounding the fruit.

Downy Ground Cherry *Physalis pubescens*

Texas Ground Cherry

Family: Solanaceae
Scientific Name: *Physalis virginiana* var. *texana*
Habit: Plants mostly sprawling. Stems from a deep rhizome.
Leaves: Alternate; blades lanceolate to ovate, sometimes slightly hairy.
Flowers: Radial, hanging downward; corollas yellow, ⅝–1" long; anthers yellow.
Fruit: Berries enclosed in the roundish inflated calyx.
Bloom Period: Spring, summer, fall.
Distribution: Cameron and Hidalgo counties.
Comments: The fruit inside the inflated calyx resembles a miniature form of the tomatillo used to make salsa. The hairless condition combined with the yellow anthers and the rounded rather than angular inflated calyx are important characteristics for identifying this plant.

Texas Ground Cherry *Physalis virginiana* var. *texana*

Solanaceae

Purple Ground Cherry

Family: Solanaceae
Scientific Name: *Quincula lobata* [*Physalis lobata*]
Habit: Perennials covered with small scales; stems mostly running along the ground.
Leaves: Alternate; blades covered with small scales, up to 4" long, the margins wavy or entire.
Flowers: Radial, blue to purple, up to ¾" broad.
Fruit: Berries about 5⁄16" broad, enclosed in the inflated calyx.
Bloom Period: All seasons.
Distribution: Cameron, Hidalgo, and Starr counties.
Comments: This is a beautiful little plant in a natural setting. It prefers full sun. Attempts to grow it in the garden have produced mixed results. Some growers have beautiful, showy plants; others are unsuccessful in growing them.

Purple Ground Cherry *Quincula lobata*

American Nightshade

Family: Solanaceae
Scientific Name: *Solanum americanum* [*S. nodiflorum*]
Habit: Perennials up to 20" tall, stems often falling over.
Leaves: Alternate; blades ovate, up to 4" long.
Flowers: Radial; corollas 5-lobed, white, often tinged with purple, up to 5⁄16" broad.
Fruit: Berries rounded, black when mature, up to ⅜" broad.
Bloom Period: All seasons.
Distribution: Cameron and Hidalgo counties.
Comments: As with many members of this family, the berries are poisonous. This species has calyx lobes partially grown together. It resembles *S. ptychanthum* (not pictured), in which the calyx lobes are not grown together.

American Nightshade *Solanum americanum*

Red Berry Nightshade

Family: Solanaceae

Scientific Name: *Solanum campechiense*

Habit: Spiny herbs up to 24" tall, often less, with star-shaped hairs.

Leaves: Alternate, spiny and with star-shaped hairs; blades oblong and lobed, up to 4" long.

Flowers: Radial; corollas 5-lobed, pale bluish, about ¾" broad.

Fruit: Berries rounded, green turning red, up to ⅝" broad.

Bloom Period: Summer, fall.

Distribution: Cameron and Hidalgo counties.

Comments: This spiny *Solanum,* with its bluish flowers, is easily distinguished from buffalo bur, *S. rostratum,* which is also spiny but has yellow flowers. Red berry nightshade fruit are reported to turn from green to red when mature, but we have not observed this color change to take place.

Red Berry Nightshade *Solanum campechiense*

Silver Leaf Nightshade

Family: Solanaceae

Scientific Name: *Solanum eleagnifolium*

Habit: Perennials spreading by rhizomes, up to 24" tall, occasionally with prickly stems and leaves.

Leaves: Alternate; blades ovate to oblong, with tiny hairs and sometimes prickles, up to 4 ⅛" long.

Flowers: Radial; corollas various purple shades or white, up to 1" broad.

Fruit: Berries roundish, black or brown when mature, up to ⅝" broad.

Bloom Period: Spring, summer, fall.

Distribution: Cameron, Hidalgo, Willacy, and Starr counties.

Comments: This species is widespread in Texas and common in our area. It spreads by rhizomes. It continues to bloom during dry periods when most other plants have stopped flowering. The plants are toxic to cattle, but birds consume the fruits.

Silver Leaf Nightshade *Solanum eleagnifolium*

Silver Leaf Nightshade *S. eleagnifolium* white-flowering form

Solanaceae

Potato Tree

Family: Solanaceae
Scientific Name: *Solanum erianthum*
Habit: Shrubs or small trees 10' or taller.
Leaves: Alternate; blades ovate to obovate, up to 7 ½" long, brownish with dense star-shaped hairs.
Flowers: Radial; corollas white, up to ¾" broad.
Fruit: Berries roundish, yellow when mature, up to ¾" broad.
Bloom Period: Spring, summer, fall.
Distribution: Cameron and Hidalgo counties.
Comments: This attractive shrub or small tree is becoming popular with gardeners. The hairy leaves give it an unusual color. It is attractive to wildlife and is a host plant for the Creamy Stripe-streak butterfly. The leaves emit an odor when rubbed.

Potato Tree *Solanum erianthum*

Potato Tree *S. erianthum* branch with young fruit

Tomato

Family: Solanaceae
Scientific Name: *Solanum lycopersicum* [*Lycopersicon esculentum* var. *cerastiforme*]
Habit: Annuals or perennials with sprawling stems, hairy and glandular and with a strong tomato scent.
Leaves: Alternate; blades deeply lobed or compound, ovate to oblong, up to 16" long.
Flowers: Radial; corollas yellow, 5-lobed, the lobes about ⅜" long.
Fruit: Berries roundish, red or yellow at maturity, about ¾" broad.
Bloom Period: Spring, summer.
Distribution: Cameron and Hidalgo counties.
Comments: This species was introduced from South America and has become naturalized. It is often cultivated for its edible fruit. The neighborhood of Los Tomates near the University of Texas Brownsville campus was named for this plant. While this plant, other tomatoes, potatoes, and eggplants are edible members of the Solanaceae family, it must be remembered that the deadly nightshade is in this same family.

Tomato *Solanum lycopersicum*

Jerusalem Cherry

Family: Solanaceae
Scientific Name: *Solanum pseudocapsicum*
Habit: Entirely hairless shrubs up to 4' or taller.
Leaves: Alternate; blades lanceolate to oblanceolate, up to 4" long, the margins slightly wavy.
Flowers: Radial, single or few; corollas 5-lobed, white, up to 5/16" broad.
Fruit: Berries round, red to yellow when ripe, up to 3/4" broad.
Bloom Period: Spring, summer, fall.
Distribution: Cameron and Hidalgo counties.
Comments: This species was introduced from the Old World and has become naturalized in warmer regions. The ripe berries of *S. triquetrum* are red, but it is herbaceous, and its young stems are triangular. Our other *Solanum* species are spiny or hairy, or their ripe fruits are black. The yellow anthers of Jerusalem cherry distinguish it from chile piquín, *Capsicum annuum,* which has bluish anthers.

Jerusalem Cherry *Solanum pseudocapsicum*

Buffalo Bur

Family: Solanaceae
Scientific Name: *Solanum rostratum*
Habit: Spiny annuals up to 28" tall.
Leaves: Alternate; blades deeply lobed, with spines and star-shaped hairs, up to 4 3/4" long.
Flowers: Radial; corollas 5-lobed, yellow, about 1" broad.
Fruit: Berries enclosed by the very spiny calyx lobes.
Bloom Period: Spring, summer, fall.
Distribution: Cameron and Starr counties.
Comments: This species is easily recognized by its spines and yellow corollas. *Shinners & Mahler's Illustrated Flora of North Central Texas* (Diggs et al. 1999) suggests that the common name originated from the burlike fruit becoming entangled in the hair of buffalos or cattle. The leaves and fruit are toxic to livestock.

Buffalo Bur *Solanum rostratum*

Sterculiaceae

Texas Nightshade

Family: Solanaceae
Scientific Name: *Solanum triquetrum*
Habit: Perennials, mostly hairless, with spindly, clambering stems; younger stems 3-cornered; older stems woody.
Leaves: Alternate; blades triangular, sometimes with basal lobes that turn outward, up to 1 ⅛" long.
Flowers: Radial; corollas 5-lobed, white, sometimes purplish, up to ⅝" broad.
Fruit: Berries roundish, red when mature, up to ⅝" broad.
Bloom Period: Spring, summer, fall.
Distribution: Cameron, Hidalgo, Willacy, and Starr counties.
Comments: This species is easily recognized by its spindly stems, which are 3-cornered on the young parts. The specific epithet *triquetrum* is Latin for "3-angled."

Texas Nightshade *Solanum triquetrum*

Texas Nightshade *S. triquetrum* plant in fruit

Rio Grande Ayenia

Family: Sterculiaceae
Scientific Name: *Ayenia limitaris*
Habit: Hairy shrubs up to 5' tall, usually less. Older stems more or less maroon in color.
Leaves: Alternate; blades ovate, up to 3" long, the margins toothed.
Flowers: Inflorescences from the leaf axils each with usually 3 flowers; petals 5, yellowish, bent backward, about ⅛" long, the bases threadlike.
Fruit: Capsules roundish, compartmentalized, spiny, about 5/16" tall.
Bloom Period: All seasons.
Distribution: Cameron, Hidalgo, and Willacy counties.
Comments: This species is federally listed as endangered. Plants usually grow in thickets on loamy soils. The younger plants have the appearance of herbs rather than shrubs. Because the plants resemble mallows, they are probably often overlooked. The Turk's Cap White-skipper butterfly has been observed using this species as a host plant.

Rio Grande Ayenia *Ayenia limitaris*

Rio Grande Ayenia *A. limitaris* spiny fruit

Dwarf Green-Flowered Ayenia

Family: Sterculiaceae
Scientific Name: *Ayenia* sp.
Habit: Herbs, much branched, up to 14" tall.
Leaves: Alternate; blades roundish on lower stems, lanceolate on upper stems, up to 2 1/8" long.
Flowers: In groups from the leaf axils; corollas greenish, bent backward, about 1/8" long or less, the bases threadlike.
Fruit: Capsules roundish, compartmentalized, spiny and hairy, about 1/8" long.
Bloom Period: Spring, summer, winter.
Distribution: Cameron, Hidalgo, and Starr counties.
Comments: This species has tentatively been identified as *A. filiformis,* but that species has purple flowers and is reported only from West Texas. *Ayenia* sp. has also been identified as *A. pilosa,* which has been reported for our area, but *A. pilosa* has purple flowers, and *Ayenia* sp. has green flowers. We have not been successful in finding *A. pilosa* with purple flowers. Possibly plants of panicled sand mallow, *Sidastrum paniculatum,* were mistaken for *A. pilosa,* although the spiny capsules of *A. pilosa* are unlike those of the Malvaceae family. A new revision of the genus *Ayenia* is needed.

Dwarf Green-Flowered Ayenia *Ayenia* sp.

Mexican Mallow

Family: Sterculiaceae
Scientific Name: *Hermannia texana*
Habit: Hairy plants, woody basally and sometimes higher, up to 20" tall.
Leaves: Alternate; blades ovate to elliptic, up to 1 3/4" long, the margins toothed.
Flowers: Inflorescences carrying several pendulous flowers; petals orange to red-orange, up to 1/2" long.
Fruit: Spiny, roundish, about 3/8" tall, splitting into 5 parts.
Bloom Period: Spring, summer, fall.
Distribution: Hidalgo, Willacy, and Starr counties.
Comments: This is a plant of sandy or calcareous soils. The flowers resemble candy corn. The corollas usually do not open fully.

Mexican Mallow *Hermannia texana*

Mexican Mallow *H. texana* spiny fruit

Sterculiaceae

Pyramid Flower

Family: Sterculiaceae
Scientific Name: *Melochia pyramidata*
Habit: Usually herbaceous, 24" tall or less.
Leaves: Alternate; blades ovate with toothed margins, up to 2 ⅜" long.
Flowers: Inflorescences from the axils of the leaves; petals 5, pink, rarely white, up to ⁵⁄₁₆" long.
Fruit: Capsules papery, pyramid shaped, spiny, about ⁵⁄₁₆" broad.
Bloom Period: Spring, summer, fall.
Distribution: Cameron, Hidalgo, Willacy, and Starr counties.
Comments: The specific epithet *pyramidata* is derived from the shape of the capsule. A white-flowered population was seen in Roma, Starr County.

Pyramid Flower *Melochia pyramidata*

Woolly Pyramid Bush

Family: Sterculiaceae
Scientific Name: *Melochia tomentosa*
Habit: Hairy shrubs up to 6½' tall. Mature stems dark maroon colored.
Leaves: Alternate; blades ovate, hairy, with toothed margins, up to 2 ⅜" long
Flowers: Inflorescences showy, from the axils of the leaves; flowers slightly bilateral; petals 5, purple to pink or white, up to ⅝" long.
Fruit: Capsules hairy and spiny, pyramid shaped, about ⁵⁄₁₆" broad.
Bloom Period: Spring, summer, fall.
Distribution: Cameron, Hidalgo, and Starr counties.
Comments: The specific epithet *tomentosa* comes from a Latin word meaning "densely woolly." In Texas this species is found only in our area. It is a beautiful plant, and it does well in the garden. It is an excellent butterfly nectar plant. Growers have been propagating woolly pyramid bush from seeds and also cuttings.

Woolly Pyramid Bush *Melochia tomentosa*

Woolly Pyramid Bush *M. tomentosa* fruit

Hierba Del Soldado

Family: Sterculiaceae
Scientific Name: *Waltheria indica*
Habit: Hairy shrubs, sometimes herbs, up to 6' tall.
Leaves: Alternate; blades ovate with toothed margins, up to 2" long.
Flowers: In tight clusters in the leaf axils; petals 5, yellow, up to ¼" long.
Fruit: Capsules club shaped, about ⅛" tall.
Bloom Period: All seasons.
Distribution: Cameron, Hidalgo, Willacy, and Starr counties.
Comments: This is a plant of sandy and caliche soils. It is a host plant for the larvae of the Hieroglyphic moth. A small population at Boca Chica Beach bore a large number of larvae; Dr. Gail McClain at the University of Texas at Brownsville identified them. The Spanish common name translates as "soldier plant."

Hierba Del Soldado *Waltheria indica*

Salt Cedar

Family: Tamaricaceae
Scientific Name: *Tamarix ramosissima*
Habit: Trees up to 26' tall.
Leaves: Alternate, tiny and scalelike.
Flowers: Tiny, in cylindrical clusters up to 2 ¾" long
Fruit: Capsules small, separating into 3–5 valves.
Bloom Period: All seasons.
Distribution: Cameron and Starr counties.
Comments: These plants were introduced and have become naturalized. Farmers have used the large branches for fence posts that later rooted and produced trees. The young stems are slender and threadlike. The species athel, *T. aphylla* (not pictured), also occurs here. Its tiny leaves clasp completely around the stems; those of *T. ramosissima* do not clasp the stems. Both species can be very invasive, growing along the Rio Grande, resacas, arroyos, and places where water stands for a time.

Salt Cedar *Tamarix ramosissima*

Turneraceae

Orinoco Jute

Family: Tiliaceae
Scientific Name: *Corchorus hirtus* var. *glabellus*
Habit: Herbs, sometimes shrubby, 24" or sometimes taller.
Leaves: Alternate; blades ovate to narrowly lanceolate with toothed margins, up to 2 ⅜" long.
Flowers: Radial, about ¼" broad, 1, 2, or 3 arising opposite the leaves; petals yellowish, usually 5.
Fruit: Capsules up to 2½" long.
Bloom Period: All seasons.
Distribution: Cameron, Hidalgo, and Starr counties.
Comments: This species prefers wet areas and is rarely seen, or else is overlooked. Some Asian species of *Corchorus* are cultivated for their coarse fibers (jute), which are used for sacking, carpeting, and twine.

Orinoco Jute *Corchorus hirtus* var. *glabellus*

Hierba Del Venado, Damiana

Family: Turneraceae
Scientific Name: *Turnera diffusa* var. *aphrodisiaca*
Habit: Shrubs up to 6' tall, usually smaller and appearing to be herbaceous.
Leaves: Alternate; blades ovate to elliptic with toothed margins, up to ¾" long.
Flowers: Radial; petals 5, yellow, up to ⁵⁄₁₆" long.
Fruit: Capsules about ³⁄₁₆" tall.
Bloom Period: All seasons.
Distribution: Cameron, Hidalgo, and Starr counties.
Comments: The plants are rare in Cameron and Hidalgo County. Most of the reports of this species in Texas have been from Starr County. They grow on gravelly or caliche hilltops. When bruised, the leaves emit a pleasant scent. This species is a host plant for the Mexican Fritillary butterfly. As the variety name implies, *T. diffusa* var. *aphrodisiaca* has long been considered an aphrodisiac; however, its effectiveness is doubtful (Correll and Johnston 1970). *Hierba del venado* translates as "deer's plant."

Hierba Del Venado, Damiana *Turnera diffusa* var. *aphrodisiaca*

Hackberry, Palo Blanco

Family: Ulmaceae

Scientific Name: *Celtis laevigata*

Habit: Trees up to 52' tall; bark light gray, warty appearing.

Leaves: Alternate; blades asymmetric, lanceolate to ovate, with 3 main veins from the base, up to 4" long; sometimes rough on the lower surface.

Flowers: Radial, tiny, perfect or unisexual on the same plant.

Fruit: Drupes globose, orange color, up to 5/16" broad.

Bloom Period: Spring.

Distribution: Cameron, Hidalgo, Willacy, and Starr counties.

Comments: Hackberry is widespread, occurring from Florida to Oklahoma and Texas, and in northeastern Mexico (Correll and Johnston 1970). It grows throughout Texas except for West Texas (Turner et al. 2003). The drupes have a thin layer of sweet material covering the comparatively large seeds. Extremely important for wildlife, these trees provide food, shelter, and nesting sites for a broad range of species.

Hackberry, Palo Blanco *Celtis laevigata*

Spiny Hackberry, Granjeno

Family: Ulmaceae

Scientific Name: *Celtis pallida*

Habit: Shrubs or small trees up to 20' tall, with unequal paired spines.

Leaves: Alternate; blades ovate to elliptic, apically usually toothed, with 3 main veins from the base, up to 1 1/8" long.

Flowers: Radial, tiny, perfect or unisexual on the same plant.

Fruit: Drupes roundish, about 1/4" broad, orange colored.

Bloom Period: Spring.

Distribution: Cameron, Hidalgo, Willacy, and Starr counties.

Comments: Granjeno occurs in South and West Texas, into Arizona and northern Mexico (Turner et al. 2003). The drupes are eaten by birds and mammals. They are also considered a treat by some of the human population. This species is a host plant for the American Snout butterfly. The wood was formerly used for tool handles.

Spiny Hackberry, Granjeno *Celtis pallida*

Urticaceae

Cedar Elm, Olmo

Family: Ulmaceae
Scientific Name: *Ulmus crassifolia*
Habit: Trees up to 80' tall; the younger branches sometimes have corky ridges.
Leaves: Alternate; blades ovate with prominent veins and toothed margins, up to 2 ⅜" long.
Flowers: Radial, tiny, perfect or unisexual on the same plant.
Fruit: Dry, winged, called a samara, about ⅜" long.
Bloom Period: Summer.
Distribution: Cameron, Hidalgo, and Starr counties.
Comments: This beautiful tree is usually found in moist places such as banks of resacas or ponds or along the Arroyo Colorado. It is a favorite growing site for Spanish moss, *Tillandsia usneoides,* in the LRGV.

Cedar Elm, Olmo *Ulmus crassifolia*

Florida Pellitory

Scientific Name: *Parietaria floridana*
Family: Urticaceae
Habit: Hairy herbs with juicy stems, leaning or prostrate.
Leaves: Alternate; blades mostly ovate, up to 1 ¼" long.
Flowers: Tiny, greenish, in clusters in the leaf axils.
Fruit: Achenes very small.
Bloom Period: Spring, summer.
Distribution: Cameron and Hidalgo counties.
Comments: This species is much like Pennsylvania pellitory, *P. pensylvanica,* but is not as commonly seen. It has a greater tendency toward erect or leaning growth than that species. Also, its calyx more or less equals the length of the bracts around it, while the calyx of Pennsylvania pellitory is shorter than the subtending bracts.

Florida Pellitory *Parietaria floridana*

Pennsylvania Pellitory

Family: Urticaceae
Scientific Name: *Parietaria pensylvanica*
Habit: Herbs with juicy stems, erect when young, leaning or prostrate when older, with curved or hooked hairs.
Leaves: Alternate; blades lanceolate, up to about 2" long.
Flowers: Tiny, greenish, in clusters in the leaf axils.
Fruit: Achenes very small.
Bloom Period: Winter, spring.
Distribution: Cameron and Hidalgo counties.
Comments: This small plant is commonly found in the garden. Once established, its seedlings are continually found in flower pots and various places in the garden.

Pennsylvania Pellitory *Parietaria pensilvanica*

Stinging Weed, Stinging Nettle

Family: Urticaceae
Scientific Name: *Urtica chamaedryoides*
Habit: Erect annuals with stinging hairs.
Leaves: Opposite; blades lanceolate to ovate, usually about 1 ½–2" long.
Flowers: Tiny, in clusters in the leaf axils.
Fruit: Achenes very small.
Bloom Period: Winter, spring, occasionally early summer.
Distribution: Cameron, Hidalgo, Willacy, and Starr counties.
Comments: Members of this genus have been cooked and eaten as greens. However, a collector of the plants would have to be extremely careful (wearing gloves, etc.) with these plants. The stinging hairs are like hundreds of miniature hypodermic needles ready to inject irritating material into hands or arms. This is a host plant for the Red Admiral butterfly.

Stinging Weed, Stinging Nettle *Urtica chamaedryoides*

Verbenaceae

Common Bee Brush, White Brush

Family: Verbenaceae
Scientific Name: *Aloysia gratissima*
Habit: Shrubs up to 10' tall, usually less.
Leaves: Opposite; blades narrowly elliptic, the margins sometimes toothed, up to ⅞" long.
Flowers: Slightly bilateral, fragrant; corollas 5-lobed, white or very pale pink, about ⅛" broad.
Fruit: Dry, of 2 nutlets (mericarps).
Bloom Period: Spring, summer, fall
Distribution: Cameron, Hidalgo, Willacy, and Starr counties.
Comments: White brush is planted for its pleasant aroma and as an ornamental to attract nectaring butterflies. We have two varieties of this species. Var. *gratissima* leaf margins are entire or with a few teeth; var. *schulzae* leaf margins are toothed, at least on some of the leaves. This plant can be toxic to livestock.

Common Bee Brush, White Brush *Aloysia gratissima*

Sweet Stem, Vara Dulce

Family: Verbenaceae
Scientific Name: *Aloysia macrostachya*
Habit: Shrubs to 6½' tall.
Leaves: Opposite; blades hairy, ovate, with toothed margins, up to 1 ⅝" long, aromatic when bruised.
Flowers: Slightly bilateral, 5-lobed, purple, 3⁄16" broad.
Fruit: Dry, of 2 nutlets (mericarps).
Bloom Period: Spring, summer.
Distribution: Hidalgo and Starr counties.
Comments: Sweet stem is usually found growing on gravelly hills or caliche soils. A beautiful shrub, it does well in the garden and is an excellent butterfly nectar plant. The Spanish common name translates as "sweet stem."

Sweet Stem, Vara Dulce *Aloysia macrostachya*

Berlandier's Fiddlewood

Family: Verbenaceae
Scientific Name: *Citharexylum berlandieri*
Habit: Shrubs, occasionally growing to almost 20' tall but usually around 8' tall.
Leaves: Opposite; blades ovate, up to 3 ½" long.
Flowers: Inflorescences from the leaf axils or the stem tips; corollas white, slightly bilateral, 5-lobed, ³⁄₁₆" broad.
Fruit: Roundish drupes, orange-red, turning black when mature, about ³⁄₁₆" broad.
Bloom Period: Spring, summer.
Distribution: Cameron and Hidalgo counties.
Comments: This is a common species on clay dunes near the coast. It is rarely seen in Hidalgo County. When the plants are stressed by drought, the leaves turn yellow to orange to red-orange in color. In Texas this species is found only in the Lower Rio Grande Valley. This is a beautiful shrub for the garden, with its bright green leaves, white flowers, and later the reddish fruit, which eventually turn black. An occasional plant will bloom but never produce fruit. This seems to enable it to produce more blooms over a longer period. *C. berlandieri* was named in honor of Jean Louis Berlandier, who made important plant collections in Mexico and Texas in the 1820s.

Berlandier's Fiddlewood *Citharexylum berlandieri*

Berlandier's Fiddlewood *C. berlandieri* inflorescence

Mexican Fiddlewood

Family: Verbenaceae
Scientific Name: *Citharexylum brachyanthum* [*C. spathulatum*]
Habit: Shrubs up to 8' tall, the young stems square.
Leaves: Opposite or in clusters; blades spatulate, up to 1" long.
Flowers: One or two produced at the leaf axils; corollas white, slightly bilateral, 5-lobed, about ¼" broad.
Fruit: Roundish drupes, orange-red when mature, about ¼" broad.
Bloom Period: Spring, summer.
Distribution: Hidalgo and Starr counties.
Comments: In Texas this species is found only in Starr, Zapata and Webb counties. Some sources include western Hidalgo County. It also occurs in adjacent Mexico. Mission fiddlewood, *C. spathulatum*, is sometimes recognized as a separate species.

Mexican Fiddlewood *Citharexylum brachianthum*

Verbenaceae

Dakota Vervain

Family: Verbenaceae
Scientific Name: *Glandularia bipinnatifida* [*Verbena bipinnatifida*]
Habit: Stems prostrate or occasionally leaning.
Leaves: Opposite; blades ovate in outline, many of them divided and redivided (bipinnate), up to 1 ⅜" long.
Flowers: Inflorescences showy; corollas purple, slightly bilateral, up to ⅝" broad but often less.
Fruit: Dry, 4 black nutlets (mericarps).
Bloom Period: All seasons.
Distribution: Cameron, Hidalgo, Willacy, and Starr counties.
Comments: Dakota vervain is quite widespread in Texas, and several varieties are recognized. Ours is var. *bipinnatifida*. It is usually recognized by its twice divided leaves. This leaf characteristic and the usually bluish purple flowers distinguish it from alfombrilla, *G. delticola*, which has pinkish purple flowers and leaf margins that are toothed or shallowly cut.

Dakota Vervain *Glandularia bipinnatifida*

Dakota Vervain *G. bipinnatifida* pinkish form

Dakota Vervain *G. bipinnatifida* (upper left, bluish purple flowers) growing with Alfombrilla *G. delticola* (lower foreground, pinkish purple flowers)

Alfombrilla

Family: Verbenaceae
Scientific Name: *Glandularia delticola* [*Verbena delticola*]
Habit: Stems usually sprawling.
Leaves: Opposite; blades triangular to ovate with toothed or shallowly cut margins, up to 2 ⅜" long.
Flowers: Inflorescences showy; corollas purple or pinkish, slightly bilateral, 5-lobed, about ⅜" broad.
Fruit: Dry, 4 nutlets (mericarps).
Bloom Period: Spring.
Distribution: Cameron County.
Comments: This species is known in Texas only from Cameron County. It also grows in adjacent Mexico. The illustration (above, right) is a mixed population of Alfombrilla *G. delticola*, and Dakota Vervain, *G. bipinnatifida*. The Spanish common name is derived from *alfombra*, for "carpet."

Alfombrilla *Glandularia delticola*

Rio Grande Mock Vervain

Family: Verbenaceae
Scientific Name: *Glandularia polyantha* [*Verbena polyantha*]
Habit: Stems sprawling.
Leaves: Opposite; blades ovate to triangular in outline with toothed and lobed margins, up to ¾" long.
Flowers: Inflorescences showy; corollas purple, slightly bilateral, 5-lobed, about ½" broad.
Fruit: Dry, 4 nutlets (mericarps). The scars usually are almost as broad as the nutlets.
Bloom Period: All seasons.
Distribution: Cameron, Hidalgo, and Willacy counties.
Comments: In Texas this species is known almost exclusively from our area. The specific epithet *polyantha* comes from Greek words for "many" and "flowers."

Rio Grande Mock Vervain *Glandularia polyantha*

Beaked Vervain

Family: Verbenaceae
Scientific Name: *Glandularia quadrangulata* [*Verbena quadrangulata*]
Habit: Stems prostrate.
Leaves: Opposite; blades ovate in outline, with toothed and lobed margins, up to 1 ⅛" long.
Flowers: Inflorescences not showy; corollas white or whitish, slightly bilateral, 5-lobed, about 3⁄16" broad.
Fruit: Dry, 4 nutlets (mericarps), more or less dumbbell shaped.
Bloom Period: Spring, summer, fall.
Distribution: Cameron, Hidalgo, Willacy, and Starr counties.
Comments: This species is easy to recognize with its very small white flowers and dumbbell-shaped nutlets.

Beaked Vervain *Glandularia quadrangulata*

Brushland Lantana

Family: Verbenaceae
Scientific Name: *Lantana achyranthifolia* [*L. macropoda*]
Habit: Shrubs usually about 2' tall, rarely up to 6' tall.
Leaves: Opposite; blades ovate, up to 3" long; margins with small teeth, the tips leaning toward the apices of the leaves.

Brushland Lantana *Lantana achyranthifolia*

Verbenaceae

Flowers: In heads about ⅝" broad; floral bracts broadly ovate, the tips attenuate; corollas white with yellow centers, fading to bluish; slightly bilateral, 5-lobed, about ³⁄₁₆" broad. Some forms have pinkish to purple flowers.
Fruit: Dry, enclosed in the fruiting calyx.
Bloom Period: Spring, summer, fall.
Distribution: Cameron, Hidalgo, Willacy, and Starr counties.
Comments: Desert lantana, *L. macropoda,* was recently "lumped" with *L. achyranthifolia,* even though the two taxa are readily distinguished.

Brushland Lantana *L. achyranthifolia* form with pinkish flowers

West Indian Lantana

Family: Verbenaceae
Scientific Name: *Lantana camara*
Habit: Shrubs up to 6' tall, usually less, often with stiff hairs.
Leaves: Opposite, ovate, 2" of longer, the margins finely toothed.
Flowers: In heads up to 1 ⅛" broad, the corollas slightly bilateral, mostly opening yellow, then turning pink.
Fruit: Drupes roundish, turning black when mature.
Bloom Period: Spring, summer, fall, sometimes winter.
Distribution: Cameron and Hidalgo County.
Comments: There is confusion in the identification of some of our *Lantana* species.
L. camara has leaves with fine-toothed margins, without noticeable white hairs along the veins on the lower leaf surfaces. The similar Texas lantana, *L. urticoides,* has leaves with coarse-toothed margins and usually with white hairs along the veins on the lower surfaces. According to Roger Sanders (2006), much hybridization has occurred between the two species, obscuring some of the distinguishing characteristics. In our area, it seems to be commonly agreed that the pink-flowered species is *L. camara,* and the yellow to orange- or red-flowered species is *L. urticoides. L. camara* is native to the West Indies but has been known in South Texas for over 100 years. A pathogen has recently infected local populations, causing proliferations of abnormal growth.

West Indian Lantana *Lantana camara*

Hammock Lantana

Family: Verbenaceae
Scientific Name: *Lantana canescens* [*L. microcephala*]
Habit: Shrubs up to 6' tall, usually shorter.
Leaves: Opposite; blades lanceolate, up to 2 ⅛" long, with small marginal teeth.
Flowers: In heads about ⅝" broad; corollas white with yellow centers, slightly bilateral, 5-lobed, about 3/16" broad.
Fruit: Dry, enclosed in the fruiting calyx.
Bloom Period: All seasons.
Distribution: Cameron and Hidalgo counties.
Comments: The very small marginal teeth on the leaves distinguish this species from brushland lantana, *L. achyranthifolia*.

Hammock Lantana *Lantana canescens*

Texas Lantana, Calico Bush

Family: Verbenaceae
Scientific Name: *Lantana urticoides* [*L. horrida*]
Habit: Shrubs up to 6' tall, usually less.
Leaves: Opposite; blades ovate, up to 2 ¼" long; margins with coarse teeth.
Flowers: In heads up to 1 ⅛" broad; floral bracts lanceolate; corollas opening yellow, usually turning orange, or red; slightly bilateral, 5-lobed, up to ½" broad.
Fruit: Drupes roundish, black when mature.
Bloom Period: Spring, summer, fall.
Distribution: Cameron, Hidalgo, Willacy, and Starr counties.
Comments: Texas lantana is a popular ornamental. A pathogen has recently infected local populations, causing proliferations of abnormal growth. The variation in flower color can make this a difficult species to identify. The lanceolate floral bracts and the leaf margins with coarse teeth are identifying characteristics. Texas lantana is widespread in South Texas, often seen blooming along ranch fences where birds have deposited the seeds. Lantanas are toxic to livestock, causing damage to the liver.

Texas Lantana, Calico Bush *Lantana urticoides*

Verbenaceae

Velvet Lantana

Family: Verbenaceae
Scientific Name: *Lantana velutina*
Habit: Shrubs up to 6' tall, usually less.
Leaves: Opposite; blades ovate to elliptic, up to 1 5/8" long, the surface with blisterlike bumps; margins with small teeth.
Flowers: In heads about 5/8" broad; floral bracts broadly ovate; corollas white with yellow centers (turning pinkish with age), slightly bilateral, 5-lobed, about 3/16" broad.
Fruit: Drupes roundish, pinkish, turning black at maturity.
Bloom Period: Spring, summer.
Distribution: Cameron, Willacy, and Starr counties.
Comments: Although listed in the *Manual of the Vascular Plants of Texas* (Correll and Johnston 1970), this species is not included in the *Atlas of the Vascular Plants of Texas* (Turner et al. 2003). A population was found by Mike Heep of Harlingen, who has been working at the University of Texas–Pan American for many years and who specializes in native shrubs. Other populations have since been seen in the LRGV. One population was seen with purple flowers.

Velvet Lantana *Lantana velutina*

Velvet Lantana *L. velutina* fruiting branch

Brushy Lippia

Family: Verbenaceae
Scientific Name: *Lippia alba*
Habit: Shrubs up to 6' tall.
Leaves: Opposite; blades ovate to lanceolate, up to 2 3/8" long, aromatic when bruised.
Flowers: In heads in the leaf axils; corollas purple to white, slightly bilateral, 5-lobed, about 1/8" broad.
Fruit: Small and dry, 2 nutlets enclosed in the flowering calyxes.
Bloom Period: All seasons.
Distribution: Cameron, Hidalgo, and Willacy counties.
Comments: The specific epithet *alba*, which usually is interpreted as "white," is somewhat confusing, since most of our *Lippia* species have purple or pinkish flowers. The *Atlas of the Vascular Plants of Texas* (Turner et al. 2003) lists our area as the only locality in Texas where this species is known. It frequently grows in low places. The erect stems sometimes fall over and take root, forming new plants.

Brushy Lippia *Lippia alba*

Mexican Oregano, Redbrush Lippia

Family: Verbenaceae
Scientific Name: *Lippia graveolens*
Habit: Shrubs up to 10' tall, or even trees up to 30' tall, but usually less in the LRGV.
Leaves: Opposite; blades ovate to elliptic, up to 1 ⅜" long, aromatic when bruised.
Flowers: In groups arising from the leaf axils; corollas whitish with yellow centers, slightly bilateral, 5-lobed, about 3/16" broad.
Fruit: Dry, football shaped, less than ⅛" long.
Bloom Period: Spring, summer, fall.
Distribution: Cameron, Hidalgo, and Starr counties.
Comments: Mexican oregano grows abundantly on gravelly hills. The leaves emit a strong scent of oregano when bruised. They are used in cooking in the same way as oregano.

Mexican Oregano, Redbrush Lippia *Lippia graveolens*

Texas Frog Fruit, Sawtooth Frog Fruit

Family: Verbenaceae
Scientific Name: *Phyla nodiflora* [*P. incisa. Lippia nodiflora*]
Habit: Prostrate perennials.
Leaves: Opposite, more or less oblanceolate, the length 5 or more times the width, with toothed margins, up to 1½" long.
Flowers: Inflorescences erect, flowers in heads arising from the leaf axils; corollas slightly bilateral, white with yellow or purplish centers, about ⅛" broad.
Fruit: Dry, separating into 4 nutlets (mericarps) less than ⅛" tall.
Bloom Period: Spring, summer, fall.
Distribution: Cameron, Hidalgo, and Willacy counties.
Comments: This is a useful species, producing an attractive ground cover without being invasive. It is both nectar plant and host plant for the Phaon Crescent butterfly.

Texas Frog Fruit, Sawtooth Frog Fruit *Phyla nodiflora*

Silky Leaf Frog Fruit

Family: Verbenaceae
Scientific Name: *Phyla strigulosa* [*Lippia strigulosa*]
Habit: Prostrate perennials.
Leaves: Opposite, obovate to ovate, the length 4 or less times the width, with toothed margins, up to 1 ⅝" long.
Flowers: Inflorescences erect, flowers in heads arising from the leaf axils; corollas slightly bilateral, white with yellow or purplish centers, about ⅛" broad.
Fruit: Dry, separating into 4 nutlets (mericarps) less than ⅛" tall.
Bloom Period: Spring, summer, fall.

Verbenaceae

Distribution: Cameron, Hidalgo, Willacy, and Starr counties.

Comments: The two *Phyla* species are similar but are easy to distinguish by the length/width ratio of the leaves. Texas frog fruit, *P. nodiflora,* has leaves with a 5:1 ratio or greater; *P. strigulosa* has leaves with a 4:1 ratio or less. Both species make an excellent ground cover. This species is nectar plant and host plant for the Phaon Crescent butterfly.

Silky Leaf Frog Fruit *Phyla strigulosa*

Common Velvet Bur

Family: Verbenaceae
Scientific Name: *Priva lappulacea*
Habit: Herbs up to 39" tall, usually much less, with erect to leaning or prostrate stems.
Leaves: Opposite; blades ovate, up to 1 ¼" long.
Flowers: Inflorescences slender and stretched out; corollas slightly bilateral, bluish to pinkish or white, about ⅛" broad.
Fruit: Dry, separating into 2 spiny nutlets (mericarps) about ⅛" tall.
Bloom Period: Spring, summer, fall.
Distribution: Cameron and Hidalgo counties.
Comments: This species is distinguished by the fruit, which separates into 2 spiny nutlets.

Common Velvet Bur *Priva lappulacea*

Coulter's Wrinklefruit

Family: Verbenaceae
Scientific Name: *Tetraclea coulteri*
Habit: Perennials; stems erect or leaning, up to 16" tall.
Leaves: Opposite; blades ovate to lanceolate, up to 1 ⅝" long.
Flowers: Slightly bilateral, bad-smelling, from the axils of the leaves; corollas cream colored, reddish on the outer parts, up to 1 ¾" long and ½" broad.
Fruit: Dry, dividing into 4 nutlets (mericarps) about 3/16" tall.
Bloom Period: Spring, summer, fall.
Distribution: Hidalgo and Starr counties.
Comments: With its opposite leaves, slightly bilateral flowers, and fruit of 4 nutlets, this species is a typical member of the Verbenaceae family. It is easily recognized by its fairly large, bad-smelling, cream-colored flowers.

Coulter's Wrinklefruit *Tetraclea coulteri*

Brazilian Vervain

Family: Verbenaceae
Scientific Name: *Verbena brasiliensis*
Habit: Plants robust, up to 5' or taller, usually less.
Leaves: Opposite; blades elliptic, with toothed margins, up to 2 1/8" long
Flowers: Inflorescences from the stem ends, the open flowers not touching; corollas purplish, slightly bilateral, 5-lobed, about 1/8" broad.
Fruit: Dry, 4 nutlets (mericarps) about 1/16" tall.
Bloom Period: Spring, summer, fall.
Distribution: Cameron County.
Comments: The small flowers plus the toothed, nonlobed leaves, separate this *Verbena* from the others. It is native to South America.

Brazilian Vervain *Verbena brasiliensis*

Gray Vervain

Family: Verbenaceae
Scientific Name: *Verbena canescens*
Habit: Plants hairy, up to 16" tall.
Leaves: Opposite; blades lanceolate, with toothed or lobed margins, up to 2" long.
Flowers: Inflorescences usually single, the flowers well separated; corollas purplish, slightly bilateral, 5-lobed, up to 7/16" broad.
Fruit: Dry, 4 nutlets about 1/16" tall.
Bloom Period: Spring, summer.
Distribution: Cameron, Hidalgo, Willacy, and Starr counties.
Comments: Rough leaf New Mexico vervain, *V. neomexicana,* is also hairy, but its bracts curl around the fruiting calyxes; those of *V. canescens* lean outward slightly. Clover's vervain, *V. cloverae,* has much larger flowers.

Gray Vervain *Verbena canescens*

Clover's Vervain

Family: Verbenaceae
Scientific Name: *Verbena cloverae*
Habit: Plants hairy, up to 20" tall.
Leaves: Opposite; blades ovate, with margins toothed and sometimes lobed, up to 1 3/16" long
Flowers: Inflorescences from the stem ends; corollas pink to purple, slightly bilateral, 5-lobed, 1/2" or broader.
Fruit: Dry, 4 nutlets (mericarps) less than 1/8" tall.
Bloom Period: Spring.
Distribution: Hidalgo and Starr counties.

Clover's Vervain *Verbena cloverae*

Verbenaceae

Comments: The larger flowers of this showy species separate it from the other *Verbena* species. They are more like those of the larger-flowering *Glandularia* species. Glandularias were at one time included in the genus *Verbena*. In general, they tend to have prostrate growth whereas the verbenas tend to be erect growing. This species was named in honor of Elzada Clover, a professor of botany at the University of Michigan. In 1937 she produced an extensive vegetational survey of the Lower Rio Grande Valley of Texas that is still used today.

Texas Vervain
Family: Verbenaceae
Scientific Name: *Verbena halei* [*V. officinalis*]
Habit: Plants slightly hairy, up to 40" tall.
Leaves: Opposite; blades ovate with toothed and lobed margins, up to 2 ¾" long.
Flowers: Inflorescences usually several or many; corollas blue, slightly bilateral, 5-lobed, up to 5/16" broad.
Fruit: Dry, 4 nutlets (mericarps) about 1/16" tall.
Bloom Period: Spring, summer, fall.
Distribution: Cameron, Hidalgo, Willacy, and Starr counties.
Comments: This species is similar to Rio Grande vervain, *V. runyonii,* but is smaller.

Texas Vervain *Verbena halei*

Rough Leaf New Mexico Vervain
Family: Verbenaceae
Scientific Name: *Verbena neomexicana* var. *hirtella*
Habit: Plants hairy, up to 40" tall, usually less.
Leaves: Opposite; blades ovate to elliptical, with toothed or lobed margins, up to 2" long.
Flowers: Inflorescences several or many from the stem ends; corollas blue to purple, slightly bilateral, 5-lobed, up to 5/16" broad.
Fruit: Dry, 4 nutlets (mericarps) about 1/16" tall.
Bloom Period: Spring, summer.
Distribution: Hidalgo County.
Comments: The floral bracts curl around the fruiting calyxes in this species. In gray vervain, *V. canescens,* a similar species, the bracts point slightly outward.

Rough Leaf New Mexico Vervain *Verbena neomexicana* var. *hirtella*

Rio Grande Vervain

Family: Verbenaceae
Scientific Name: *Verbena runyonii*
Habit: Robust plants up to 56" tall.
Leaves: Opposite; blades ovate with toothed and lobed margins, up to 2 3/4" long.
Flowers: Inflorescences several from the stem ends; corollas blue, slightly bilateral, 5-lobed, about 3/16" broad.
Fruit: Dry, 4 nutlets (mericarps) about 1/16" tall.
Bloom Period: Spring, summer, fall.
Distribution: Cameron and Hidalgo counties.
Comments: This species was named for Robert Runyon, photographer, former mayor of Brownsville, and self-taught botanist, who made many important collections of plants from this area. It looks like an extremely robust form of Texas vervain, *V. halei*. the inflorescences of *V. halei,* including the open flowers, are less than 1/8" in diameter; those of *V. runyonii* are 3/16" or broader.

Rio Grande Vervain *Verbena runyonii*

Nodding Green Violet

Family: Violaceae
Scientific Name: *Hybanthus verticillatus* var. *platyphyllus*
Habit: Low-growing herbs.
Leaves: Opposite; blades elliptic to linear, up to 1½" long.
Flowers: Radial or almost so, from the axils of the leaves, nodding downward; petals 5, up to 3/16" long, the lower one green and white or tipped with lavender, the upper ones violet or white.
Fruit: Capsules about 3/16" tall.
Bloom Period: All seasons.
Distribution: Cameron, Hidalgo, Willacy, and Starr counties.
Comments: This small species escapes notice because of its small size and the small flowers, which hang downward. It is widespread in Texas, especially in disturbed areas. It is a host plant for the Variegated Fritillary butterfly.

Nodding Green Violet *Hybanthus verticillatus* var. *platyphyllus*

Vitaceae

Mistletoe

Family: Viscaceae
Scientific Name: *Phoradendron tomentosum*
Habit: Woody parasites on trees, especially mesquite, with green stems; hairy on younger parts.
Leaves: Opposite, green, more or less elliptic, up to 4 ⅞" long.
Flowers: Very tiny, male and female flowers on separate plants.
Fruit: A white drupe about ¼" broad; inner part sticky.
Bloom Period: Fall, winter, spring.
Distribution: Cameron and Hidalgo counties.
Comments: Seven species of mistletoe are known to grow in Texas. *P. tomentosum* is the only one that grows in our area. It is widespread in Texas. Although the stems are green and photosynthetic, mistletoe cannot produce all the complex molecules it needs, and so it extracts them from its host tree. Continued growth can eventually kill the tree. This species is a host plant for the Great Purple Hairstreak butterfly.

Mistletoe *Phoradendron tomentosum*

Pepper Vine

Family: Vitaceae
Scientific Name: *Ampelopsis arborea*
Habit: Woody vines with tendrils.
Leaves: Alternate, once or twice compound, about 6" long.
Flowers: Inflorescences arising opposite the leaves; flowers tiny.
Fruit: Roundish berries up to ⅜" broad, black when mature.
Bloom Period: Spring.
Distribution: Cameron and Hidalgo counties.
Comments: The genus name *Ampelopsis* is from the Greek words *ampolos* (vine) and *opsis* (likeness). The specific epithet *arborea* means "tree," referring to the vine's habit of growing up onto trees.

Pepper Vine *Ampelopsis arborea*

Marine Ivy, Possum Grape

Family: Vitaceae
Scientific Name: *Cissus incisa*
Habit: Woody vines with tendrils, from a large fleshy tuber.
Leaves: Alternate, fleshy, bad-smelling when bruised, up to 3 ⅛" long; blades deeply lobed or compound, the divisions toothed. Occasionally, the leaves may be simple.
Flowers: Inflorescences arising opposite the leaves; flowers tiny.
Fruit: Roundish berries up to 5/16" broad, black when mature.
Bloom Period: Spring.
Distribution: Cameron, Hidalgo, Willacy, and Starr counties.
Comments: The fleshy, bad-smelling leaves on this vine are major identifying characteristics of this species. When flowering, possum grape attracts large numbers of nectaring insects.

Marine Ivy, Possum Grape *Cissus incisa*

Marine Ivy, Possum Grape *C. incisa* plant in fruit

Princess Vine

Family: Vitaceae
Scientific Name: *Cissus sicyoides*
Habit: Robust, perennial hairless vines with tendrils.
Leaves: Alternate, simple; blades ovate, up to 4" long.
Flowers: Inflorescences emerge opposite the leaves; flowers greenish white to reddish, about 5/16" broad.
Fruit: Roundish berries about ¼" broad, black when mature.
Bloom Period: Spring.
Distribution: Hidalgo County.
Comments: This vine is native to tropical America. It is extremely invasive, growing rampantly in two citrus orchards (French et al. 2003), a plant nursery, and a home, all in the Weslaco area of Hidalgo County.

Princess Vine *Cissus sicyoides*

Zygophyllaceae 425

Mustang Grape

Family: Vitaceae
Scientific Name: *Vitis mustangensis*
Habit: Woody vines with tendrils.
Leaves: Alternate; blades palmately lobed, up to 7½" broad.
Flowers: Tiny, greenish, in clusters.
Fruit: Berries up to ⅝" broad, in clusters, purple to black when mature.
Bloom Period: Spring.
Distribution: Kenedy County
Comments: Mustang grape is included because it is often seen along Highway 77 a short distance north of Willacy County. It is often cultivated, although more in the past than today. The grapes are used to make excellent jelly, and the vines are much more vigorous and productive than imported vines.

Mustang Grape *Vitis mustangensis*

Mustang Grape *V. mustangensis* inflorescence

Guayacán

Family: Zygophyllaceae
Scientific Name: *Guaiacum angustifolium* [*Porlieria angustifolia*]
Habit: Shrubs or small trees up to 23' tall, usually appearing grayish green or blue-green in color.
Leaves: Opposite, compound; leaflets about ⅝" long.
Flowers: Radial; petals 5, purplish, rarely white, up to ⅜" long.
Fruit: Winged and usually 2-lobed with 2 red seeds, up to ¾" broad.
Bloom Period: Spring, summer.
Distribution: Cameron, Hidalgo, and Starr counties.
Comments: This woody plant would be beautiful in the garden. It vaguely resembles the appearance of the northern evergreens sometimes used here. Although slow growing, it has attractive leaves, flowers, and fruit. It is a host plant for the Lyside Sulphur butterfly and is heavily browsed by deer. The wood is green, dense, and sinks in water. It is prized for articles such as knife handles.

Guayacán *Guaiacum angustifolium*

California Caltrop

Family: Zygophyllaceae
Scientific Name: *Kallstroemia californica*
Habit: Prostrate annuals, the young growth hairy.
Leaves: Opposite, compound; leaflets up to ⅝" long.
Flowers: Radial; petals 5, yellow, up to ¼" long.
Fruit: Dry and knobby, separating into 10 one-seeded parts.
Bloom Period: Summer, fall.
Distribution: Cameron, Hidalgo, Willacy, and Starr counties.
Comments: This plant is fairly easy to recognize by its prostrate habit, compound leaves, yellow radial flowers, and fruit without spines.

California Caltrop *Kallstroemia californica*

Big Caltrop

Family: Zygophyllaceae
Scientific Name: *Kallstroemia maxima*
Habit: Prostrate annuals with succulent stems, forming mats. Young stems somewhat hairy; older stems up to 3' long, without hairs.
Leaves: Opposite, compound with 3–4 pairs of leaflets; leaflets oblong to elliptical, ¼–1 ⅛" long.
Flowers: Radial, emerging from the leaf axils, on erect stalks. Petals yellow, about ⅜" long and ¼" broad or more; anthers orange colored.
Fruit: Dry, without spines, separating into 10 seeds.
Bloom Period: Spring, summer, fall.
Distribution: Cameron County.
Comments: This species is much more attractive than California caltrop, *K. californica*. The flowers are larger and are held above the plant. The flowers open around noon and close by 6:00 P.M. The anthers are prominent, with variable color. Different collections of this species have been reported with yellow anthers and with coral red anthers. This is the first record of *K. maxima* in the LRGV. In Texas it has been reported for Harris, Washington, and Austin counties. It is described as a weedy species and also grows in coastal Florida, Georgia, Alabama, and South Carolina. It is widespread in the Caribbean, tropical Mexico, and South America.

Big Caltrop *Kallstroemia maxima*

Zygophyllaceae

Creosote Bush

Family: Zygophyllaceae
Scientific Name: *Larrea tridentata*
Habit: Shrubs up to 10' tall, with many individual slender stems arising from the base, with odor of creosote, especially after rains.
Leaves: Opposite, compound with only 2 leaflets about ⅜" long.
Flowers: Radial; petals 5, yellow, up to ⅜" long.
Fruit: Capsule roundish, 5-lobed, woolly with red or white hairs, about ¼" broad.
Bloom Period: Spring, summer.
Distribution: Starr County.
Comments: The common name for this species is quite appropriate. After rains, where these shrubs are abundant, the air is filled with the pleasant, clean scent of creosote, which is not strong enough to be offensive. The plants are more abundant in Zapata County and farther west; probably a few exist in western Starr County.

Creosote Bush *Larrea tridentata*

Goat Head, Puncture Weed

Family: Zygophyllaceae
Scientific Name: *Tribulus terrestris*
Habit: Prostrate annuals, the young growth hairy.
Leaves: Opposite, compound; leaflets up to ⁷⁄₁₆" long.
Flowers: Radial; petals 5, yellow, up to ³⁄₁₆" long.
Fruit: Dry, up to ³⁄₁₆" broad, with spines about ¼" long, separating into 5 parts.
Bloom Period: Spring, summer, fall.
Distribution: Cameron, Hidalgo, Willacy, and Starr counties.
Comments: This species is often found in disturbed places. It is recognized by its prostrate nature, compound leaves, yellow flowers, and spiny fruit. It is introduced, native of the Mediterranean area. The common name goat head is appropriate, since the fruit resembles the head of a goat.

Goat Head, Puncture Weed *Tribulus terrestris*

APPENDIX
Butterflies and Moths

Butterflies

American Snout	*Libytheana carinenta*	Lacy's Scrub-hairstreak	*Strymon alea*
		Large Orange Sulphur	*Phoebis agarithe*
		Laviana White-skipper	*Heliopetes laviana*
Black Swallowtail	*Papilio polyxenses*	Leaf Wing	*Anaea species*
Blue-eyed Sailor	*Dynamine dyonis*	Legume Caterpillar	*Selenisa sueroides*
Blue metalmark	*Lasaia sula*	Little Yellow	*Euramea lisa*
Bordered Patch	*Chlosyne lacinia*	Lyside Sulphur	*Kricogonia lyside*
Brown-banded Skipper	*Timochares ruptifasciatus*		
		Malachite	*Syproeta stelenes*
		Mallow Scrub-hairstreak	*Strymon istapa*
		Manfreda Giant Skipper	*Stallingsia maculosa*
Cassius Blue	*Leptotes cassius*	Mexican Silverspot	*Dione moneta*
Cloudless Sulphur	*Phoebis sennae*	Mexican Fritillary	*Euptoieta hegesia*
Common Mestra	*Mestra amymone*	Mexican Bluewing	*Myscelia ethusa*
Common Buckeye	*Junonia coenia*	Monarch	*Danaus plexippus*
Creamy Stripe-streak	*Arawacus jada*		
Crimson Patch	*Chlosyne janais*	Pale-banded Crescent	*Phyciodes tulcis*
Curve-Wing Metalmark	*Emesis emisia*	Phaon Crescent	*Phyciodes phaon*
		Pipevine Swallowtail	*Battus philenor*
Definite Patch	*Chlosyne definita*		
		Queen	*Danaus gilippus*
Elada Checkerspot	*Texola elada*		
Elf	*Microtia elva*	Reakirt's Blue	*Hemiargus isola*
		Red Rim	*Beblis hyperia*
Funereal Dusky Wing	*Erynnis funeralis*	Red Admiral	*Vanessa atalanta*
		Rounded Metalmark	*Calephelis perditalis*
Giant Swallowtail	*Papilio cresphontes*		
Gray Ministreak	*Ministrymon azia*	Silver-banded Hairstreak	*Chlorostrymon simaethis*
Gray Cracker	*Hamadryas februa*		
Great Purple Hairstreak	*Atlides halesus*	Soldier	*Danaus eresimus*
Great Southern White	*Ascia monuste*	Southern Dogface	*Colias cesonia*
Gulf Fritillary	*Agraulis vanillae*	Sulphur	Family Pieridae
Heliconia		Texan Crescent	*Phyciodes texana*
Family Nymphalidae, Subfamily Heliconiinae		Texas Powdered Skipper	*Systasea pulverulenta*
		Theona Checkerspot	*Thessalia theona*

Tiny Checkerspot	*Dymasia dymas*	**Moths**	
Turk's Cap White-skipper	*Heliopetes macaira*		
Two-barred Flasher	*Astraptes fulgerator*	Hawk moths	Family Sphingidae
		Hieroglyphic moth	*Diphthera festiva*
Variegated Fritillary	*Euptoieta claudia*		
Vesta Crescent	*Phyciodes vesta*	Saucy Beauty moth	*Phaloesia sausia*
		Shinx moth	Family Sphingidae
Western Pigmy Blue	*Brephidium exile*		
White Peacock	*Anartia jatrophae*		
White-patched Skipper	*Chiomara asychis*		
Xami Hairstreak	*Callophrys xami*		

GLOSSARY

ACHENE. A small, dry one-seeded fruit, typical of the Asteraceae family. At the base of the seed (point of attachment), the ovary wall is grown to the seed coat.

AGGREGATE FRUIT. Fruit such as mulberry or blackberry in which the small fruits of many close-packed flowers grow together to form what appears to be a single fruit.

ALTERNATE LEAVES. Leaves appearing singly at any given point on a stem.

ANNUAL. A plant that completes its life cycle in one season, then dies.

ANTHER. The pollen-producing part of the stamen.

APEX (plural apices). The tip end of a stem or inflorescence.

APPRESSED. Flatly pressed against.

ATYPICAL. Not typical.

AWN. A very small bristle or scale.

AXIL; AXILLARY. The angle formed where the leaf joins the stem; in the location of the axil.

BERRY. A juicy fruit with many seeds.

BILATERAL FLOWER. Only one line through the center of a flower will divide it into two mirror images. Example, an iris flower.

BIENNIAL. A plant that requires two years to complete its life cycle, then dies.

BLADE. The flat, broad part of a leaf.

BRACT. A modified leaf, associated with some inflorescences.

BRISTLES. Stiff hairs.

BULB. An underground bud made of thickened modified leaves and a small amount of stem. Example, an onion.

CALYX (plural calyces). The sepals of a flower, collectively; see sepal.

CAPSULE. A dry fruit made of more than one carpel, which usually splits open.

CARPEL. The units that make up the ovary with its style and stigma.

CATKIN. A small inflorescence with tiny flowers without petals, all the same sex.

CHAFF. Dry scales, in the Asteraceae sometimes found interspersed among the disk flowers.

CLAWED PETAL. A petal with a narrow stalk at the base.

COLUMN. The style surrounded by the anther filaments.

COMPOUND LEAF. A leaf divided into two or more leaflets.

COROLLA. The petals of a flower, collectively.

DEHISCENT FRUIT. A fruit that splits open.

DISJUNCT POPULATIONS. Populations that are widely separated geographically, with none of that species occurring in the areas between.

DISK FLOWER. In the Asteraceae, the radially symmetrical flowers in the central portion.

DISTAL. Away from the point of attachment.

DRUPE. A juicy, one-seeded fruit, with the seed enclosed in a hard covering. Example, a peach.

ELLIPSOID FIGURE. A three-dimensional figure broadest about the middle, tapering to both ends.

ELLIPTIC FIGURE. A two-dimensional figure broadest about the middle, tapering to both ends.

ENDEMIC. Limited to a given region.

ENTIRE MARGIN. A leaf margin that is smooth (not toothed or lobed).

EPIPHYTE. A plant that grows on another plant but is not a parasite.

FLORET. A small flower.

FOLLICLE. A dry fruit that originates from one carpel and splits on one side.

FRUIT. A matured ovary. The structure that encloses the seeds.

GENUS. A group of related species.

GLOCHIDS. Barbed spines, especially in the cactus family.

GYMNOSPERM. A seed plant, usually woody, that does not have flowers but produces seeds in its cones.

HEAD. A cluster of flowers originating from a common receptacle.

HERB, HERBACEOUS. A non-woody plant.

HOST PLANT. A plant on which an insect lays its eggs and on which the developing larvae feed.

INDEHISCENT FRUIT. A fruit that does not split open.

INFERIOR OVARY. The other floral parts are attached to the top or sides of the ovary.

INFLORESCENCE. A flowering stalk.

INVOLUCRE. A cluster of bracts below the flowers, especially in the Asteraceae.

KEELED. Having a prominent longitudinal ridge, as in the keel of a boat.

LANCEOLATE. Two dimensional, narrowly ovate, in the shape of a lance.

LEAFLET. A leaflike division of a compound leaf.

LEGUME. A fruit made of one carpel, breaking open on two sides. Also, a member of the legume family (Fabaceae).

LINEAR. Long and narrow, with the sides more or less parallel.

LOBED. Moderately to deeply indented toward the midrib or base.

LOMENT. A legume constricted between the seeds and breaking into individual segments at maturity.

MEMBRANOUS. Thin and soft.

MERICARP. Unit of a dry fruit that splits into usually four one-seeded parts. Especially found in the Boraginaceae, Lamiaceae, and Verbenaceae.

NODE. The point on a stem where a leaf emerges

NON-VASCULAR. Without a specialized conducting system for water and nutrients.

NUTLET. Another name for a mericarp, found in a dry fruit that splits into usually four one-seeded parts. Especially found in the Boraginaceae, Lamiaceae, and Verbenaceae families.

OBCONIC. Shaped like an upside-down cone.

OBCORDATE. Shaped like an upside-down heart.

OBLANCEOLATE. The reverse of lanceolate. The broadest part is toward the apex.

OBLONG. Longer than broad, with the sides more or less parallel. Refer to drawings, Introduction.

OBOVATE. Two dimensional, the reverse of ovate, with the broadest part toward the apex. Refer to drawings, Introduction.

ONCE COMPOUND LEAF. A leaf divided into two or more leaflets that are not further subdivided.

OPPOSITE. Two leaves, branches, or flowers emerging from the stem at the same point, on opposite sides.

OVARY. The swollen innermost part of a flower that will eventually grow to become the fruit containing the seeds.

OVATE. A two-dimensional figure, broader below the middle.

OVOID. A three-dimensional figure, broader below the middle. Example, an egg.

PALMATE LEAF. A compound leaf with the leaflets all originating from the end of the petiole.

PAPPUS. In the Asteraceae family, a modification of the calyx, which can be small scales to plumose bristles as seen in the dandelion.

PEDICEL. The stalk of a single flower.

PEDUNCLE. The main stalk of an inflorescence.

PELTATE LEAF. A leaf with the petiole attached at or near the center rather than at the margin of the leaf.

PERENNIAL. A plant that persists from year to year, producing flowers and seeds each year.

PERFECT FLOWER. A flower that has both male and female parts.

PETAL. One of the second set of appendages on a flower, usually the showy part.

PETIOLE. The stalk of a leaf.

PHYLLARY. One of the bracts that make up the involucre of a member of the Asteraceae, encircling the base of the flower cluster.

PINNA. Technically, the first division of a compound leaf. Further divisions of the leaf are called leaflets. In common usage, divisions of a once-compound leaf are usually called leaflets.

PINNATE LEAF. A compound leaf with the leaflets arising along the axis.

PNEUMATOPHORE. In mangroves, an erect, aerial root arising from a horizontal, underground root.

PRICKLE. A sharp outgrowth produced by the bark or epidermis. Example, dewberry and some mimosas.

RACHIS. In a compound leaf, the continuation of the petiole where the leaflets are attached.

RADIAL FLOWER. A flower in which any of several lines through the center will result in two mirror images.

RAY FLOWER. In the Asteraceae family, the strap-shaped flower found around the edge of the disk. Example, the yellow "petals" of a sunflower.

RECEPTACLE. The enlarged end of a stem to which the floral parts are attached.

REFLEXED. Bent backward or downward.

REGULAR FLOWER. Radial, a flower shape in which any of several lines through the center will produce two mirror images; also called radial.

RESACA. An oxbow lake, left when a river changes its course.

RHIZOME. A horizontal underground stem.

RHOMBIC. More or less diamond shaped.

ROSETTE. A circular cluster of leaves crowded at the base of the plant.

SCHIZOCARP. A fruit that splits into two separate parts.

SEDGE. A member of the Cyperaceae family.

SEPAL. The outermost series of flower parts, usually green but sometimes colored.

SESSILE. Without a stalk, attached directly to the axis.

SILIQUE. An elongate capsule with two seed chambers, in the Brassicaceae family.

SINUATE. Wavy in and out.

SPATULATE. Shaped like a spatula.

SPECIFIC EPITHET. The second part of the name of a species. Example: in the name *Ruellia nudiflora*, the specific epithet is *nudiflora*.

SPIKE. An unbranched inflorescence, the flowers without pedicels.

SPIKELET. A secondary spike of flowers in a compound inflorescence, as in the Cyperaceae (sedges) and Poaceae (grasses).

SPINE. A hard, sharp-pointed modification of a leaf or leaf part.

SPORANGIUM. A body that produces spores.

SPORE. In ferns, a one-celled reproductive body, produced instead of a seed.

SPOROCARP. A body that produces sporangia.

STAMEN. The "male" part of a flower, made up of filament and anther.
STIGMA. The sticky portion at the top of a flower, where the pollen adheres.
STIPULE. One of a pair of appendages found at the base of some leaves.
STOLON. A modified creeping stem, usually underground.
SUBTENDED, SUBTENDING. Placed directly below.
SUPERIOR OVARY. The other flower parts are attached below the ovary.
SYMBIOTIC RELATIONSHIP. A close, permanent relationship between individuals of two different species. In a mutualistic relationship, both benefit; in a parasitic relationship, one benefits, and the other is harmed.

TENDRIL. In some vines, a wiry appendage of a stem or leaf that wraps around a support.
TEPALS. A term for the sepals and petals together, used when they look alike or when one set is missing but it is not obvious which is present.
THORN. A stem branch, usually short, that ends in a hard point.
TUBER. A thickened, solid underground stem.
TUBERCLES. Small bumps on a surface.
TWICE COMPOUND. In a compound leaf or inflorescence, the first division is divided again into smaller segments.

UMBEL. An inflorescence with the flowers all arising from more or less the same point.
UTRICLE. A fruit consisting of a single seed surrounded by an inflated papery material.

VASCULAR. Provided with a specialized conducting system for water and nutrients.
VISCID. Sticky.

WHORLED. Three or more leaves (sometimes other parts) arising in a circle from the same point on a stem.
WINGED FRUIT. A fruit with one or more dry, flat, extensions that may function as wings.
WOOLLY. Covered with long, matted hairs.

BIBLIOGRAPHY

Ajilvsgi, Geyata. 2002. *Wildflowers of Texas.* Rev. ed. Shearer Publishing. 414 pp.

Brock, Jim P., and Kenn Kaufman. 2003. *Butterflies of North America.* Houghton Mifflin. 383 pp.

Carr, Bill, and David Diamond. 1995. *Vascular Plants Endemic to Texas.* Working draft.

Cassell's Spanish Dictionary. 1968. Ed. Edgar A. Peers, J. V. Barragan, F. A. Vinyals, and J. A. Mora. Funk and Wagnalls Publishing Company. 1477 pp.

Cheatham, Scooter, and M. C. Johnston, with L. Marshall. 1995. *The Useful Wild Plants of Texas, the Southeastern and Southwestern United States, the Southern Plains, and Northern Mexico.* 2 vols. Useful Wild Plants, Inc. 1: 568 pp., 2: 599 pp.

Clover, Elzada U. 1937. Vegetational survey of the Lower Rio Grande Valley, Texas. *Madroño* 4: 41–66.

Correll, Donovan S., and Helen B. Correll. 1972. *Aquatic and Wetland Plants of the Southwestern United States.* Environmental Protection Agency. 1777 pp.

Correll, D. S., and M. C. Johnston. 1970. *Manual of the Vascular Plants of Texas.* Texas Research Foundation. 1881 pp.

Davis, Anna May Tarrence. 1942. *A Study of Boscaje de la Palma in Cameron County, Texas, and of Sabal Texana.* MA thesis, University of Texas. 111 pp.

Diggs, George M. Jr., B. Lipscomb, and R. O'Kennon. 1999. *Shinners & Mahler's Illustrated Flora of North Central Texas.* Botanical Research Institute of Texas. 1626 pp.

Diggs, George M. Jr., B. L. Lipscomb, M. D. Reed, and R. J. O'Kennon. 2006. *Illustrated Flora of East Texas.* Botanical Research Institute of Texas. 1594 pp

Enquist, Marshall. 1987. *Wildflowers of the Texas Hill Country.* Lone Star Botanical. 275 pp.

Everitt, James H., D. L. Drawe, and R. I. Lonard. 1999. *Field Guide to the Broad-Leaved Herbaceous Plants of South Texas.* Texas Tech University Press. 277 pp.

———. 2002 . *Trees, Shrubs and Cacti of South Texas.* Rev. ed. Texas Tech University Press. 213 pp.

Everitt, James H., Robert I. Lonard, and Christopher R. Little. 2007. *Weeds in South Texas and Northern Mexico.* Texas Tech University Press. 222 pp.

Flora of North America Editorial Committee, eds. 1993–2007. *Flora of North America,* vols. 1–5, 19–27. Oxford University Press.

French, Victor J., R. Lonard, and J. Everitt. 2004. *Cissus sicyoides* C. Linnaeus (Vitaceae), a potential exotic pest in the Lower Rio Grande Valley, Texas. *Subtropical Plant Science* 55 (1):72–74.

Fryxell, Paul A. 2007. Two new species of Malvaceae from Sonora, Mexico and Texas. *Lundellia* 10:1–6.

———. 1988. *Malvaceae of Mexico.* Systematic Botany Monographs 25. American Society of Plant Taxonomists. 522 pp.

Gould, Frank W. 1975. *The Grasses of Texas.* Texas A&M University Press. 653 pp.

Handbook of Texas Online. S.v. "Berlandier, Jean Louis." http://www.tshaonline.org/handbook/online/articles/BB/fbe56.html (Accessed May 4, 2009).

———. S.v. "Drummond, Thomas."

http://www.tshaonline.org/handbook/online/articles/DD/fdr8.html (Accessed May 4, 2009).

———. S.v. "Lindheimer, Ferdinand Jacob." http://www.tshaonline.org/handbook/online/articles/LL/fli4.html (Accessed May 4, 2009).

———. S.v. "Runyon, Robert." http://www.tshaonline.org/handbook/online/articles/RR/fru37_print.html (Accessed May 5, 2009).

———. S.v. "Wright, Charles." http://www.tshaonline.org/handbook/online/articles/WW/fwr1_print.html (Accessed May 5, 2009).

Hart, Charles R., Tam Garland, A. Catherine Barr, Bruce B. Carpenter, and John C. Reagor. 2003. *Toxic Plants of Texas.* Texas Cooperative Extension, the Texas A&M University System. 243 pp.

Hatch, Stephan L., J. L. Schuster, and D. L. Drawe. 1999. *Grasses of the Texas Gulf Prairies and Marshes.* Texas A&M University Press. 355 pp.

Ideker, Joe. 1996. *Pereskia aculeata* (Cactaceae) in the Lower Rio Grande Valley of Texas. *Sida* 17 (2): 527–28.

Johnston, M. C. 1990. *The Vascular Plants of Texas: A List, Up-dating the Manual of the Vascular Plants of Texas.* 2nd ed. Marshall C. Johnston. 107 pp.

Jones, Fred B. 1975. *Flora of the Texas Coastal Bend.* Mission Press. 262 pp.

Jones, Stanley D., and J. K. Wipff. 2003. *A 2003 Updated Checklist of the Vascular Plants of Texas.* CD-ROM. Botanical Research Center. 712 pp.

Lehman, Roy L., Ruth O'Brien, and Tammy White. 2005. *Plants of the Texas Coastal Bend* Texas A&M University Press. 352 pp.

Liggio, Joe, and Ann O. Liggio. 1999. *Wild Orchids of Texas.* University of Texas Press. 228 pp.

Lonard, Robert I. 1993. *Guide to the Grasses of the Lower Rio Grande Valley, Texas.* University of Texas–Pan American Press. 240 pp.

Lonard, Robert I., James H. Everitt, and Frank W. Judd. 1991. *Woody Plants of the Lower Rio Grande Valley, Texas.* Miscellaneous Publications no. 7. Texas Memorial Museum. 179 pp.

Morey, Roy. 2008. *Little Big Bend.* Texas Tech University Press. 329 pp.

NOAA Southern Regional Climate Center. http://www.srcc.lsu.edu/ (Accessed October 12, 2004).

Poole, Jackie M. 1975. Taxonomic Study of *Acleisanthes* (Nyctaginaceae). MA thesis, The University of Texas at Austin.

Poole, Jackie M., William R. Carr, Dana M. Price, and Jason R. Singhurst. 2007. *Rare Plants of Texas.* Texas A&M University Press. 640 pp.

Powell, A. Michael. 1998. *Trees and Shrubs of the Trans-Pecos and Adjacent Areas.* University of Texas Press. 498 pp.

Powell, A. Michael, and James F. Weedin. 2004. *Cacti of the Trans-Pecos and Adjacent Areas.* Texas Tech University Press. 509 pp.

Powell, A. Michael, James F. Weedin, and Shirley A. Powell. 2008. *Cacti of Texas: A Field Guide*. Texas Tech University Press. 383 pp.

Richardson, Alfred. 1995. *Plants of the Rio Grande Delta.* University of Texas Press. 332 pp.

———. 2002. *Wildflowers and Other Plants of Texas Beaches and Islands.* University of Texas Press. 247 pp.

Richardson, Alfred, and Ken King. 2006. *Capraria mexicana* (Scrophulariaceae) in Cameron County, Texas: rediscovered in the United States. *Sida* 22(2): 1237–38.

———. 2008. *Neptunia plena* (Fabaceae: Mimosidae) rediscovered in Texas. *Journal of the Botanical Research Institute of Texas* 2 (2):1491–93.

———. 2009. *Tournefortia hirsutissima* (Boraginaceae) new to the flora of Texas. *Journal of the Botanical Research Institute of Texas.* In press.

Sanders, Roger. 2006. Taxonomy of Lantana sect. *Lantana* (Verbenaceae), I: Correct application of *Lantana camara* and associated names. *Sida* 22 (1): 381–421.

Schery, Robert W. 1972. *Plants for Man.* Prentice-Hall. 657 pp.

Silverthorne, Elizabeth. 1996. *Legends and Lore of Texas Wildflowers.* Texas A&M University Press. 240 pp.

Standley, Paul C. 1920–26. *Trees and Shrubs of Mexico.* Contributions from the U.S. National Herbarium 23. Smithsonian Institution. 1721 pp.

Stutzenbaker, Charles D. 1999. *Aquatic and Wetland Plants of the Western Gulf Coast.* Texas Parks and Wildlife Press. 465 pp.

Taylor, R. B., J. Rutledge, and J. C. Herrera. 1999. *A Field Guide to Common South Texas Shrubs.* Texas Parks and Wildlife Press. 106 pp.

Tull, Delena. 1987. *A Practical Guide to Edible and Useful Plants.* Texas Monthly Press. 518 pp.

Turner, B.L. 2009. *Convolvulus carrii*, a Localized Endemic from Southernmost Texas. *Phytologia* 91 (3): 394–400

———. 2009. Taxonomy of *Iva angustifolia* and *I. asperifolia* (Asteraceae). *Phytologia* 91 (1).

Turner, Billie L., H. Nichols, G. Denny, and O. Doron. 2003. *Atlas of the Vascular Plants of Texas.* Botanical Research Institute of Texas. 888 pp.

Vines, Robert A. 1960. *Trees, Shrubs, and Woody Vines of the Southwest.* University of Texas Press. 1104 pp.

Wagner, David L. 2005. *Caterpillars of Eastern North America*. Princeton University Press. 512 pp.

Ward, Daniel B. 2006. A name for a hybrid Kalanchoe now naturalized in Florida. *Cactus and Succulent Journal* 78 (2): 92–95.

Warnock, Barton H. 1970. *Wildflowers of the Big Bend Country Texas.* Sul Ross State University. 158 pp.

———. 1974. *Wildflowers of the Guadalupe Mountains and the Sand Dune Country, Texas.* Sul Ross State University. 176 pp.

———. 1977. *Wildflowers of the Davis Mountains and the Marathon Basin, Texas.* Sul Ross State University. 274 pp.

Weniger, Del. 1970. *Cacti of the Southwest.* University of Texas Press. 249 pp.

Wunderlin, Richard, and Bruce F. Hansen. 2003. *Guide to the Vascular Plants of Florida*. 2nd ed. University Press of Florida. 787 pp.

INDEX

Abronia ameliae, 325
abutilon, 301–6
Abutilon abutiloides, 301
Abutilon berlandieri, 302
Abutilon fruticosum, 302, 304
Abutilon hulseanum, 303
Abutilon hypoleucum, 303
Abutilon parvulum, 304
Abutilon theophrasti, 304
Abutilon trisulcatum, 305
Abutilon umbellatum, 305
Abutilon wrightii, 306
acacia, 239–41
Acacia angustissima, 239
Acacia berlandieri, 239
Acacia farnesiana, 240
Acacia greggii var. *wrightii,* 240
Acacia rigidula, 241
Acacia schaffneri, 241
Acacia smallii, 240
Acacia wrightii, 240
Acalypha monostachya, 205
Acalypha poirettii, 206
Acalypha radians, 206
Acanthaceae family, 48–54
Acanthocereus pentagonus, 160
Acanthocereus tetragonus, 160
Achatocarpaceae family, 55
Achyranthes aspera, 57
Achyranthes philoxeroides, 59
Acleisanthes anisophylla, 325
Acleisanthes longiflora, 326
Acleisanthes obtusa, 326
Acourtia runcinata, 82
Acourtia wrightii, 82
Adelia vaseyi, 207
Aeschynomene indica, 253
afinador, 179
agalinis, 382–83
Agalinis heterophylla, 382
Agalinis maritima, 382
Agalinis maritima var. *grandiflora,* 383

Agalinis strictifolia, 383
Agavaceae family, 16–20
agave, 16–18
Agave americana, 16, 17
Agave lophantha, 17
Agave scabra, 16, 17
Agave variegata, 18
Aizoaceae family, 55–57
Alabama lip fern, 13
Alamo vine, 198
alfombrilla, 413
algae, 4, 11
alicoche, 163, 164
Alismataceae family, 21–23
alkali bulrush, 31
alkaliweed, silky, 190
Alliaceae family, 23–24
alligator weed, 59
Allionia choisyi, 327
Allionia incarnata, 327
Allium drummondii, 23, 24
Allium runyonii, 24
Allowissadula lozanii, 306
allthorn, 176
Alophia drummondii, 40
Aloysia gratissima, 411
Aloysia macrostachya, 411
Alternanthera caracasana, 58
Alternanthera paronychioides, 58
Alternanthera philoxeroides, 59
Alternanthera pungens, 58, 59
alternate leaves, 6
amapola del campo, 341
Amaranthaceae family, 57–66
Amaranthus arenicola, 60
Amaranthus berlandieri, 61
Amaranthus greggii, 60
Amaranthus palmeri, 61
Amaranthus polygonoides, 61
Amaranthus spinosus, 62
amargosa, 390
Amaryllidaceae family, 16–19, 25

ambiguous green thread, 130
Amblyolepis setigera, 83
Ambrosia chieranthifolia, 83
Ambrosia confertiflora, 84
Ambrosia psilostachya, 84
Ambrosia trifida var. *texana,* 85
Amelia's sand verbena, 325
American bluehearts, 385
American bulrush, 39
American germander, 291
American nightshade, 399
Ammannia coccinea, 297, 298
Ammannia robusta, 298
Ammannia teres, 297
Ammi majus, 68
Ammoselinum butleri, 68
Amoreuxia wrightii, 141
Ampelopsis arborea, 423
Amphyachyris dracunculoides, 85
Amyris madrensis, 375
Amyris texana, 375
anacahuita (anacahuite), 141
Anacardiaceae family, 67
anacua, 143
Anagallis arvensis, 360
Anagallis minima, 361
Ancistrocactus scheeri, 161
angel trumpet, 393
angel trumpets, 326
annual saltwort, 184
annual seepweed, 186
annual waterwort, 173
annual Western umbrella grass, 37
Anoda pentaschista, 307
Anredera leptostachys, 139
anther (flower part), 7
Anthericaceae family, 26
Anthericum chandleri, 26
Antigonon leptopus, 354
Aphanostephus ramosissimus var. *ramosissimus,* 86

Aphanostephus skirrhobasis var. *kidderi,* 86
Aphanostephus skirrhobasis var. *thalassius,* 86–87
aphid vine, 77
Apiaceae family, 68–72
Araceae family, 27
Arecaceae family, 27
Argemone aenea, 344
Argemone mexicana, 344–45
Argemone sanguinea, 345
Argythamnia humilis, 220
Argythamnia mercurialina var. *pilosissima,* 221
Argythamnia neomexicana, 221
Aristolochiaceae family, 73
Aristolochia erecta, 73
Aristolochia longiflora, 73
Aristolochia pentandra, 73
arrowhead, 22
arrow leaf milkvine, 81
arrow leaf morning glory, 198
Artemisia ludoviciana subsp. *mexicana,* 87
Asclepiadaceae family, 73–81
Asclepias curassavica, 73
Asclepias emoryi, 74
Asclepias linearis, 74
Asclepias obovata, 75
Asclepias oenotheroides, 75
Asclepias prostrata, 76, 80
ash, Mexican, 334
ashy dogweed, 132
ashy sand mat, 209
Asian ponyfoot, 191
Asiatic hawkweed, 137
Asteraceae (sunflower) family, 8, 81–138
Aster spinosus, 91
Aster subulatus, 127
Astolepis sinuata, 13
Astragalus nuttallianus var. *austrinus,* 254
Astrophytum asterias, 161
Atriplex acanthocarpa, 180
Atriplex arenaria, 181
Atriplex canescens, 180
Atriplex matamorensis, 181
Atriplex pentandrra, 181
Audubon's water lily, 333
Avicenniaceae family, 138

Avicennia germinans, 138, 367
ayenia, 403–4
Ayenia filiformis, 404
Ayenia limitaris, 403
Ayenia pilosa, 404
Ayenia sp., 404
Azolla caroliniana, 12
Azollaceae family, 12
Azolla mexicana, 12

baby bonnets, 255
baby's breath, pink, 360
baccharis, Texas, 89
Baccharis halimifolia, 87
Baccharis neglecta, 88
Baccharis salicifolia, 88
Baccharis texana, 89
Bacopa monnieri, 384
Bacopa rotundifolia, 384
Bahia absinthifolia var. *dealbata,* 89
Bailey's ball moss, 28
bald cypress, Montezuma, 15
ball moss, 28–29
balloon vine, 379–80
banana water lily, 333
Baptisia bracteata, 254
Baptisia leucophaea, 254
Barbados cherry, 301
barbas de chivato, 362
barbed-wire cactus, 160
barreta, 376
Basellaceae family, 139
basket flower, 117
Bastardia viscosa, 307
Bataceae family, 139
Batis maritima, 139
beach croton, 219
beach evening primrose, 339
beach ground cherry, 397
beach morning glory, 196
beaked burhead, 21
beaked vervain, 414
beans, 253, 255, 261, 266, 268, 270
———. *See also* Leguminosae family
bearded dalea, 260
bedstraw, 371
beebalm, 286–87
bee brush, common, 411

bee wild bundleflower, 243
bequilla, 269
Berlandier croton, 217
Berlandier's alicoche, 163
Berlandier's fiddlewood, 412
Berlandier's Indian mallow, 302
Berlandier's woolly pod bladderpod, 154
Berlandier wolfberry, 394
Bernardia myricifolia, 208
betony, shade, 290
betony leaf mistflower, 93
bicolored greggia, 153
Bidens odorata, 89
Bidens pilosa, 89
big caltrop, 426
big flower bladderpod, 154
big foot water clover, 12–13
Bignoniaceae family, 140
Billieturnera helleri, 308
bindweed, 188–90
bishop's weed, large, 68
Bixaceae family, 141
black brush, 241
black mangrove, 138
black mimosa, 246
blackroot, 120
black willow, 379
bladder mallow, 309
bladderpod, 154–56
bladderpod sida, 314–15
bladderwort, 293
blade (leaf part), 5
Blank's alicoche, 163
bloodleaf, Palmer's, 66
bluebell gentian, 277
blue boneset, 127
bluebonnet, Texas, 265
blue creeping sage, 288
blue curls, 282
blue-eyed grass, 41
blue mistflower, 91
blue morning glory, 196
blue mud plantain, 45
bluets, prairie, 371
blue water lily, 332
blue weed sunflower, 106
Blutaparon vermiculare, 62
Boerhavia coccinea, 328
Boerhavia erecta, 328

INDEX

Boerhavia intermedia, 329
Boerhavia scandens, 329
Bogenhardia crispa, 309
Bolboschoenus maritimus, 31
Boraginaceae family, 141–50
Bordas Escarpment, 2
border bonebract, 123
Borrichia frutescens, 90
botanical terms, 5–8
Boussingaultia leptostachya, 139
Bowlesia incana, 69
bracted sida, 315
bracted zornia, 273
bractless brookweed, 361
branched tooth cup, 300
brasil, 365
Brassicaceae family, 150–59
Brassica juncea, 150
Brazilian pepper, 67
Brazilian vervain, 420
Brazoria arenaria, 284
bristly tropic croton, 216
brittlewort, 11
Bromeliaceae family, 28–29
brookweed, 361–62
broom groundsel, 124
broomrape, 341–42
broomweed, 85, 102, 114
broomwort, sweet, 388–89
Browne's savory, 285
brown-eyed Susan, 122
Brown's flaveria, 98
brush holly, 274
brushland lantana, 414–15
brush noseburn, 229
brushy lippia, 417
Bryophyllum daigremontianum, 200
Bryophyllum tubiflorum, 199
Buchnera americana, 385
Buchnera floridana, 385
Buckley's yucca, 20
buckwheat, 354–55
Buddlejaceae family, 159–60
Buddleja sessiliflora, 159
buffalo bur, 402
bulb lip fern, 13
bull nettle, 213
bulrush, 31, 38, 39
Bumelia celastrina, 382

bur clover, 266
burhead, 21–22
bush morning glory, 194
bush sunflower, 124
Butler's sand parsley, 68
buttercup, 341, 363
butterfly bush, Rio Grande, 159
butterfly pea, 255
butterweed, 104, 114
buttonbush, 368–69
button mangrove, 188
buttonweed, 370, 374

Cactaceae (cactus) family, 160–73
caesalpinia, 230–32
Caesalpinia bonduc, 230–31
Caesalpinia caudata, 231
Caesalpinia mexicana, 232
Caesalpinoideae family, 230–38
Cakile geniculata, 151
calderona, 283
Calibrachoa parviflora, 390
calico bush, 416
California caltrop, 426
California loosestrife, 300
Calliandra conferta, 242
Callirhoe incolucrata var. *lineariloba,* 308
Callisia micrantha, 29
Callitrichaceae family, 173
Callitriche terrestris, 173
caltrop, 426
Calylophus australis, 335
Calylophus serrulatus, 335
Calymmandra candida, 95
Calyptocarpus vialis, 90
calyx (flower part), 7
Cameron County, 2
Campanulaceae family, 173–74
camphor daisy, 122
camphor weed, 108
Canavalia maritima, 255
Canavalia rosea, 255
candelilla, 222
Capparaceae (Capparidaceae) family, 174–77
Capparis incana, 174
Capparis spinosa, 174
Capraria mexicana, 385

Capsella bursa-partoris, 151
Capsicum annuum, 402
Capsicum annuum var. *aviculare,* 391
cardinal feather, 206
Cardiospermum dissectum, 379
Cardiospermum halicacabum, 380
Carlowrightia parviflora, 48
Carlowrightia texana, 48
Carolina crane's bill, 279
Carolina ponyfoot, 191
Carolina water fern, 12
Carolina wolfberry, 394
Carr's bindweed, 189
Caryophyllaceae family, 177–78
Cassia aristellata, 232
Cassia bauhinioides, 236
Cassia colunteoides, 238
Cassia durangensis var. *iselyi,* 236
Cassia fasciculata var. *ferrisiae,* 233
Cassia lindheimeriana, 237
Cassia occidentalis, 237
Cassia pumilio, 238
Cassia texana, 233
Cassytha filiformis, 292
Castela erecta subsp. *texana,* 390
Castela texana, 390
Castilleja indivisa, 386
castor bean, 228
cat brier, 46
catchweed, 371
catclaw vine, 140
catnip noseburn, 230
cat's eye, 142
cattail, 47
cebolleta, 25
cedar elm, 409
Celastraceae family, 178–79
celery leaf buttercup, 363
Celosia nitida, 63
Celosia palmeri, 63
Celtis laevigata, 408
Celtis pallida, 408
cenicilla, 56
cenizo, 386
Centaurea americana, 117

Centaurium breviflorum, 276
Centrosema virginianum, 255
Centunculus minimus, 361
century plant, 16
Cephalanthus occidentalis var. *californicus,* 368–69
Cephalanthus salicifolius, 369
Ceratophyllaceae family, 180
Ceratophyllum demersum, 180
Cercidium macrum, 235
Cereus pentagonus, 160
Cevallia sinuata, 295
chaff flower, 57–58, 59
chaffweed, 361
Chamaecrista calycioides, 232
Chamaecrista fasciculata, 233
Chamaecrista flexuosa var. *texana,* 233
Chamaesaracha conioides, 391
Chamaesaracha coronopus, 391–92
Chamaesaracha sordida, 391, 392
Chamaesaracha villosa, 393
Chamaesyce albomariginata, 208
Chamaesyce cinerascens, 209
Chamaesyce cordifolia, 209
Chamaesyce glyptosperma, 210
Chamaesyce hypericifolia, 210
Chamaesyce laredana, 211
Chamaesyce nutans, 211
Chamaesyce prostrata, 212
Chamaesyce serpens, 212
chapote, 205
Characeae family, 11
Chara sp., 11
Cheaetopappa asteroides, 90–91
Cheilanthes alabamensis, 13
Chenopodium berlanderi, 182
Chenopodium carinatum, 183
Chenopodium desiccatum var. *leptophylloides,* 183
Chenopodium murale, 182
Chenopodium praetericola, 183
Chenopoliaceae family, 180–86
cherry, 301, 396–99, 402
chickweed, 178, 323
chiggery grapes, 149
Chihuahuan balloon vine, 379

chile piquín (chilipiquín), 391
chinaberry tree, 321
Chiococca alba, 369
chisme, 359
Chlorocantha spinosa, 91
Chloroleucon ebano, 28, 242
chomonque, 101
Christmas senna, 238
Chromolaena odorata, 91
Cienfuegosia drummondii, 309
Cirsium texanum, 92
Cissus incisa, 424
Cissus sicyoides, 424
Cistaceae family, 186–87
Citharexylum berlandieri, 412
Citharexylum brachyanthum, 412
Citharexylum spathulatum, 412
clammy weed, 176
clapdaisy, 92
clappia, 92
Clappia suaedifolia, 92
clasping false pimpernel, 387
classification system, plant, 3–4
Clematis drummondii, 362
Cleome gynandra, 175
Cleomella angustifolia, 175
cliff morning glory, 197
climbing devil's claw, 332
climbing hemp weed, 113
climbing milkweed, 79
climbing senna, 238
climbing wortclub, 329
clover, 12–13, 259, 266–67, 373
Clover's vervain, 420–21
Clusiaceae family, 187
Cnidoscolus texanus, 213
Cnidoscolus urens, 213
coastal germander, 292
coastal lazy daisy, 86–87
coastal marsh fimbry, 37
coastal plain heliotrope, 146
coastal sands, 2
coastal water hyssop, 384
coast globe amaranth, 65
coat buttons, 133
Cocculus diversifolius, 46, 322
Cochleospermaceae family, 141
Cochorus hirtus var. *glabellus,* 407

cocklebur, 137
cock's comb, 63
coffee senna, 237
Coldenia canescens, 149
colima, 377
Colombian waxweed, 298
Colubrina greggii, 364
Colubrina texensis, 364
coma, 382
Combretaceae family, 188
Commelinaceae family, 29–31
Commelina diffusa, 30, 31
Commelina elegans, 30–31
Commelina erecta, 30, 31
Commicarpus scandens, 329
common balloon vine, 380
common bee brush, 411
common buttonbush, 368–69
common chickweed, 178
common dandelion, 128
common jimmyweed, 109
common least daisy, 90–91
common purslane, 358
common silverbush, 221
common sow thistle, 126
common sunflower, 105
common velvet bur, 419
common watermeal, 42
common water nymph, 42–43
Compositae family, 8, 81–138
compound leaf structures, 5
comsumption weed, 87
Condalia hookeri, 365
Condalia spathulata, 365
coneflower, long headed, 121
Conocarpus erectus, 188
Conoclinium betonicifolium, 93
Convolvulaceae family, 188–98
Convolvulus arvensis, 188, 189
Convolvulus carrii, 189
Convolvulus equitans, 189, 190
Conyza canadensis var. *glasbrata,* 93
Conyza coulteri, 111
Cooperia drummondii, 25
Cooperia smallii, 25
copperleaf, 205–6
copper mallow, 308
coppery false fanpetals, 308
coral bean, 261
Cordia boissieri, 141

INDEX

Coreopsis basalis, 94
Coreopsis nuecensoides, 94
Coreopsis tinctoria, 94–95
corky stemmed passion flower, 347
corolla (flower part), 7
corona de Cristo, 346
corona de reina, 354
Cortés croton, 215
Corydalis micrantha subsp. *texensis,* 275
Coryphantha macromeris var. *runyonii,* 162
Coryphantha pottsiana, 162, 169
Coryphantha robertii, 162
Coryphantha runyonii, 162
Coryphantha sunderlandii, 162
Cory's croton, 216
cotton batting false cudweed, 119
Coulter's laennecia, 111
Coulter's wrinklefruit, 419
Coursetia axillaris, 255
cowpea, yellow, 273
cow pen daisy, 134
coyotillo, 366
Crassulaceae family, 199–201
creeping burhead, 22
creeping lady's-sorrel, 342
crenate leaf snake herb, 49
creosote bush, 427
Cressa depressa, 190
Cressa nudicaulis, 190
Cressa truxillensis, 190
crest-rib morning glory, 195
Croptilon divaricatum, 95
Croptilon rigidifolium, 95
crotalaria, 256–57
Crotalaria incana, 256
Crotalaria retusa, 256
Crotalaria spectabilis, 257
croton, 213–20
Croton argenteus, 213
Croton argyranthemus, 214
Croton capitatus var. *lindheimeri,* 214
Croton ciliatoglandulifer, 215
Croton cortesianus, 215
Croton coryi, 216
Croton glangulosus var. *pubentissimus,* 216

Croton humilis, 217
Croton incanus, 217
Croton leucophyllus, 218
Croton lindheimerianus, 218
Croton parksii, 219
Croton punctatus, 219
Croton texensis, 220
Croton torreyanus, 217
crowded heliotrope, 144
crow poison, 24
Cruciferae family, 150–59
crucillo, 372
crusita, 91
Cryptantha mexicana, 142
Cryptantha texana, 142
Cryptostegia grandiflora, 76
Cucurbitaceae family, 201–2
cudweed, 100–101, 119
Cuphea carthagenensis, 298
cure-for-all, 118
Cuscutaceae family, 202–4
Cuscuta cuspidata, 202, 203
Cuscuta glabrior, 203
Cuscuta leptantha, 202, 203
Cuscuta pentagona var. *glasbrior,* 203
Cuscuta runyonii, 204
cusp dodder, 202
cut leaf gilia, 352
cut leaf stickleaf, 296
cut leaved evening primrose, 340
cylindrical yellow cress, 156
cylindric sedge, 35
Cylindropuntia leptocaulis, 171
Cynanchum angustifolium, 77
Cynanchum barbigerum, 77
Cynanchum racemosum var. *unifarium,* 78
Cyperaceae family, 31–39
Cyperus alternifolius, 33
Cyperus articulatus, 32
Cyperus digitatus, 32
Cyperus hermaphroditus, 36
Cyperus involucratus, 33
Cyperus iria, 33
Cyperus macrocephalus var. *eggarsii,* 34
Cyperus ochraceus, 34
Cyperus odoratus, 34
Cyperus odoratus subsp. *engelmannii,* 35

Cyperus polystachyos, 35
Cyperus retrorsus, 35
Cyperus tenuis, 36
Cyperus thyrsiflorus, 36
cypress, 15

dahlia cactus, 165
daisy, 83, 86–87, 90–91, 95, 97, 100, 113, 117, 122, 128, 134, 136
Dakota vervain, 413
dalea, 257–61
Dalea aurea, 257
Dalea austrotexana, 258
Dalea compacta, 258
Dalea emarginata, 259, 260
Dalea lanata, 258
Dalea nana, 259
Dalea obovata, 260
Dalea pogonathera var. *walkerae,* 260
Dalea scandens var. *pauciflora,* 261
Dalea thyrsiflora, 261
damiana, 407
dandelion, 120–21, 128
Datura wrightii, 393
Daucus pusillus, 69
David's milkberry, 369
dayflower, spreading, 30
delta arrowhead, 23
depressed wand-like bundle flower, 244
depression weed, 88
Descurainia pinnata, 152
desert Christmas cactus, 171
desert goosefoot, 183
desert mint, 284
desert peony, pink, 82
desert spikes, 363
desert tobacco, 395–96
desert yaupon, 179
desert zinnia, 138
Desmanthus bicornutus, 243
Desmanthus virgatus var. *depressus,* 244
devil's backbone, 200
devil's bouquet, 331
devil's claw, 332, 348
devil's head, 163
devil weed, Mexican, 91

dewberry, 368
Diaperia candida, 95, 96
Diaperia verna, 95, 96
Dichondra carolinensis, 191
Dichondra micrantha, 191
Dichromena colorata, 38
Dicliptera sexangularis, 49
Dicliptera vahliana, 49
dicots (dicotyledoneae), 4, 48–427
Diodia teres, 370
Diodia tricocca, 373
Diodia virginiana, 370
Diospyros texana, 205
disc flower (flower part), 8
disc water hyssop, 384
ditaxis, 222
Ditaxis humilis, 220
Ditaxis neomexicana, 221
Ditaxis pilosissima, 221
Ditaxis sp., 222
dodder, 202–4
dog cholla, 167
dogweed, 131–32
Dolicothele sphaerica, 169
dollar weed, 70–71
downy ground cherry, 398
Doxantha unguis-cati, 140
Draba cuneifolia, 152
Drosera annua, 204
Drosera brevifolia, 204
Droseraceae family, 204
Drosera leucantha, 204
Drummond's rattlebush, 268
Drummond's snake cotton, 64
Drummond's wood sorrel, 343
duckweed, lesser, 41
duraznillo, 368
Dutchman's breeches, 377
dwarf dalea, 259
dwarf green-flowered ayenia, 404
dwarf Indian mallow, 304
dwarf senna, 238
dwarf spike rush, 36
dwarf sundew, 204
Dyschoriste crenulata, 49
Dysphania carinata, 183
Dyssodia pentachaeta, 131
Dyssodia tenuiloba, 131
Dyssodia tephroleuca, 132

Eastern sleepy daisy, 136
Ebenaceae family, 205
ebony, Texas, 242
Echeandia chandleri, 26
Echeandia texensis, 26
Echinocactus asterias, 161
Echinocactus scheeri, 161
Echinocactus sinuatus, 166
Echinocactus texensis, 163
Echinocereus angusticeps, 164
Echinocereus berlandieri, 163, 165
Echinocereus blankii, 163
Echinocereus enneacanthus, 164
Echinocereus fitchii, 166
Echinocereus papillosus var. *angusticeps,* 164
Echinocereus pentalophus, 163, 165
Echinocereus poselgeri, 165
Echinocereus reichenbachii var. *fitchii,* 166
Echinodorus berteroi, 21
Echinodorus cordifolius, 22
Echinodorus rostratus, 21
Echinodorus rostratus var. *lanceolatus,* 21
Eclipta alba, 96
Eclipta prostrata, 96
Ehretia anacua, 143
Eichhornia crassipes, 44
elbow bush, narrow leaf, 333
elegant gayfeather, 112
Eleocharis interstincta, 36
Eleocharis parvula, 36
Eleocharis quadrangulata, 37
elm, cedar, 409
elm leaved urvillea, 381
Elytraria bromoides, 50
Emerus herbacea, 269
Emilia fosbergii, 96–97
Emilia sonchifolia, 97
Emory's milkweed, 74
encino, 274
Engelmannia peristenia, 97
Engelmannia pinnatifida, 97
Engelmann's daisy, 97
Ephedra antisiphylitica, 14
Ephedraceae family, 14
Ephedra pedunculata, 14

erect joint fir, 14
erect leaf agalinis, 383
erect spiderling, 328
Ericameria austrotexana, 114
Erigeron geiseri, 98
Erigeron ortegae, 91
Erigeron procumbens, 97
Erigeron versicolor, 98
Eriogonum greggii, 354
Eriogonum multiflorum, 355
Eriogonum riograndensis, 355
Erodium cicutarium, 278
Eryngium naturtiifolium, 70
Erythrina herbacea, 261
Escobaria runyonii, 162
Esenbeckia berlandieri, 376
Esenbeckia runyonii, 376
espanta vaqueros, 66
Eucnide bartonioides, 295
Eupatorium azureum, 127
Eupatorium betonicifolium, 93
Eupatorium incarnatum, 98–99
Eupatorium odoratum, 91
Euphorbia albomarginata, 208
Euphorbia antisiphyllitica, 222
Euphorbiaceae family, 205–30
Euphorbia cinerascens, 209
Euphorbia cordifolia, 209
Euphorbia cyathophora, 223
Euphorbia glyptosperma, 210
Euphorbia graminea, 223
Euphorbia helleri, 224
Euphorbia heterophylla var. *cyathophora,* 223
Euphorbia hypericifolia, 210
Euphorbia innocua, 224
Euphorbia laredana, 211
Euphorbia nutans, 211
Euphorbia prostrata, 212
Eurystemon mexicanum, 45
Eurytaenia texana, 70
Eustoma exaltatum, 277
Evax candida, 95
Evax multicaulis, 96
Evax verna, 96
evening primrose, 339–40, 341
evolvulus, 192–93
Evolvulus alsinoides var. *angustifolius,* 192
Evolvulus nuttallianus, 192
Evolvulus sericeus, 192, 193

INDEX

eyebane, 211
Eysenhardtia texana, 262

Fabaceae family, 230–73
Fagaceae family, 274
false broomweed, 114
false dandelion, 120–21
false mesquite calliandra, 242
false ragweed, 116
fanpetals, 308, 316
Fatuoua villosa, 323
Fendler's ivy leaf ground cherry, 397
fern, 4, 12–13
Ferocactus hamatacanthus var. *sinuatus,* 166
Ferocactus setispinus, 167
few-flowerd climbing dalea, 261
few flowered St. John's wort, 187
fiddle dock, 357
fiddleleaf, 280
fiddle leaf tobacco, 396
fiddlewood, 412
field anoda, 307
field bindweed, 188
field ragweed, 84
filament (flower part), 7
Fimbristylis castanea, 37
fine leaf fumitory, 276
finger flat sedge, 32
fire wheel, 100
fishhook cactus, 161
Fitch's rainbow cactus, 166
five needle dogweed, 131
five wing spiderling, 329
Flacourtiaceae family, 274
flagrant flat sedge, 35
flamethrower, yellow, 358
Flaveria brownii, 98
Flaveria oppositifolia, 98
flax, 293–94
fleabane, 97–98, 118
flecha de agua, 22
Fleishmannia incarnata, 98–99
flor de San Juan, 72
Florestina tripeteris, 99
Florida buttonweed, 374
Florida pellitory, 409
flower parts, 7

Forestiera angustifolia, 55, 333
four o'clock, 326, 330–31
four wing saltbush, 180
fragrant begga ticks, 89
Frankeniaceae family, 275
Frankenia johnstonii, 275
Fraxinus berlandieriana, 334
fresno, 334
fringed sunflower, 106
Froelichia drummondii, 64
Froelichia gracilis, 64
frog fruit, 418–19
frostweed, 134–35
Fuirena simplex, 37
Fumariaceae family, 275–76
Fumaria parviflora, 276
Funastrum clausum, 78
Funastrum cynanchoides, 79

gaillardia, 99–100
Gaillardia aestivalis, 99
Gaillardia pulchella var. *australis,* 100
Gaillardia suavis, 100
Galactia canescens, 262
Galactia marginalis, 263
Galactia texana, 262, 263
Galium aparine, 371
Galium virgatum, 371
Galphimia angustifolia, 300
Gamochaeta pensilvanica, 100–101
garlic guinea weed, 349
garlic weed, 349
garrapatilla, 327
gaura, 335–37
Gaura brachycarpa, 335
Gaura drummondii, 336
Gaura mckelveyae, 336
Gaura odorata, 336
Gaura parviflora, 337
Gaura sinuata, 337
Gentianaceae family, 276–78
Georgia rockrose, 186
Geraniaceae family, 278–79
Geranium carolinianum, 279
Geranium texanum, 279
Gerardia heterophylla, 382
Gerardia maritima, 383
Gerardia strictifolia, 383
germander, 291, 292

giant bulrush, 38
Gilia incisa, 352
girdlepod, white, 372
Glandularia bipinnatifida, 413
Glandularia delticola, 413
Glandularia polyantha, 414
Glandularia quadrangulata, 414
glandular milkwort, 353
Glinus radiatus, 322
globe berry, 201
glory of Texas, 173
Gnaphalium chilense, 119
Gnaphalium pensulvanicum, 100–101
goat bush, 390
goat head, 427
Gochnatia hypoleuca, 101
golden dalea, 257
golden eye, skeleton leaf, 135
golden fruited dock, 357
golden prickly poppy, 344
goldenrod, 125
golden tick seed, 94
golden wave, 94–95
goldenweed, 110, 114, 136
golden zephyr lily, 25
Gomphrena nealleyi, 65
googly-eyed vine, 150
goosefoot, 182–83
Gossypianthus lanuginosus var. *lanuginosus,* 65
graceful sand mat, 210
granjeno, 408
grapes, 149, 425
grasses, 11, 37–38, 41, 44
grassleaf spurge, 223
gray golden aster, 107
gray nicker, 230–31
gray vervain, 420
great braided sedge, 39
green false nightshade, 392
Green Island echeandia, 26
Greggia camporum, 153
Gregg's buckwheat, 354
Gregg's colubrina, 364
Gregg's keelpod, 158
Gregg's pigweed, 60
Grindelia adenodonta, 101
Grindelia microcephala, 101
Grindelia microcephala var, *adenodonta,* 101

Grindelia oolepis, 102
gromwell, rough, 148
ground cherry, 396–99
groundsel, 123–24
Grusonia schottii, 167
Guaiacum angustifolium, 425
guajillo, 239
guapilla, 28
guayacán, 425
Guilleminea lanuginosa, 65
Gulf of Mexico, 2
gulf sea rocket, 151
gumhead, 103
gumweed, 101–2
Gutierrezia glutinosa, 103
Gutierrezia sphaerocephala, 102
Gutierrezia texana var. *glutinosa,* 103
guttapercha, 178
Gymnosperma glutinosum, 103
gymnosperms, 4, 14–15

Habernaria repens, 43
habitats, 2
hachinal, 299
hackberry, 408
hair covered cactus, 169
hairs, plant, 4
hairy evolvulus, 192
hairy false nightshade, 392
hairy silverbush, 221
hairy tickle tongue, 378
hairy wedelia, 135
Hamatocactus bicolor, 166, 167
Hamatocactus setispinus, 167
hammock lantana, 416
Haplopappus spinulosus, 136
Havardia pallens, 244
hawkweed, Asiatic, 137
heart leaf fanpetals, 316
heart leaf hibiscus, 310
heart leaf sand mat, 209
Hebbronville Plain, 2
Hechtia glomerata, 28
Hedeoma drummondii, 284
hedgehog cactus, 167
hedge parsley, 71
Hedyotis nigricans, 160, 371
Heimia salicifolia, 299
Helenium amarum, 104

Helenium linifolium, 104
Helenium microcephalum, 105
Helianthemum georgianum, 186
Helianthus annuus, 105
Helianthus argophyllus, 106
Helianthus ciliaris, 106
Helianthus praecox subsp. *runyonii,* 107
Helietta parvifolia, 376
heliotrope, 144–47
Heliotropium angiospermum, 144
Heliotropium confertifolium, 144
Heliotropium curassavicum, 145
Heliotropium indicum, 145
Heliotropium procumbens, 146
Heliotropium racemosum, 146, 147
Heliotropium texanum, 146, 147
Heliotropium torreyi, 147
Heller's spurge, 224
hemp, Indian, 317
hemp weed, climbing, 113
henbit, 285
Herissantia crispa, 309
Hermannia texana, 404
Heteranthera dubia, 44, 45
Heteranthera liebmannii, 44
Heteranthera limosa, 45
Heteranthera mexicana, 45
Heteranthera mexicanum, 45
Heteranthera multiflora, 45
Heterotheca canescens, 107
Heterotheca latifolia, 108
Heterotheca subaxillaris, 108, 122
Hibiscus cardiophyllus, 310
Hibiscus martianus, 310
Hidalgo County, 2, 3
hierba de alacrán, 352
hierba de la golondrina, 212
hierba de la hormiga, 327
hierba del sapo, 70
hierba del soldado, 406
hierba del venado, 407
hoary black foot, 113
hoary caper, 174

hoary milkpea, 262
hoary pea, 271
hoary rattlebox, 256
Hoffmanseggia glauca, 234
hog plum, 364
Homalocephala texensis, 163
honey mesquite, 252
Hooker's plantain, 350
hornwort, submerged, 180
horse crippler, 163
horsemint, 286
horse purslane, 57
horse weed, 93
huisache, 240
huisache daisy, 83
huisachillo, 241
hyacinth, water, 44
Hybanthus verticillatus var. *platyphyllus,* 422
hydrilla, 40
Hydrilla verticillata, 40
Hydrocharitaceae family, 40
Hydrocotyle bonariensis, 70–71
Hydrophyllaceae family, 280–82
Hymenopappus artemisiifolius var. *riograndensis,* 108
Hymenoxys linearifolia, 129
Hymenoxys odorata, 109
Hypericaceae family, 187
Hypericum pauciflorum, 187

Ibervillea lindheimeri, 201
Indian blanket, Southern, 100
Indian chickweed, 323
Indian heliotrope, 145
Indian hemp, 317
Indian mallow, 301–6
Indian rush pea, 234
indigo, 254, 264
Indigofera miniata, 264
Indigofera suffruticosa, 264
involucre (flower part), 8
Ipomoea alba, 193
Ipomoea amnicola, 194
Ipomoea carnea subsp. *fistulosa,* 194
Ipomoea cordatotriloba var. *torreyana,* 195
Ipomoea costellata var. *costellata,* 195

INDEX

Ipomoea costellata var. *edwardsensis,* 195
Ipomoea fistulosa, 194
Ipomoea hederacea, 196
Ipomoea imperati, 196
Ipomoea pes-caprae subsp. *brasiliensis,* 197
Ipomoea rupicola, 197
Ipomoea sagittata, 198
Ipomoea sinuata, 198
Ipomoea stolonifera, 196
Ipomoea trichocarpa var. *torreyana,* 195
Iresine palmeri, 66
Iridaceae family, 40–41
Isely's durango senna, 236
Isocarpha oppositifolia var. *achyranthes,* 109
Isocoma coronopifolia, 109
Isocoma drummondii, 110
Iva angustifolia, 110
Iva annua var. *caudata,* 110–11

jaboncillo, 380–81
Jamaican weed, 280
Jann's Indian mallow, 303
Jatropha berlandieri, 225
Jatropha cathartica, 225
Jatropha dioica, 226
Jefea brevifolia, 111
Jerusalem cherry, 402
jicamilla, 225
jimmyweed, common, 109
jimson weed, 393
Johnston's frankenia, 275
jointed flat sedge, 32
jointed spike rush, 36
joint fir, 14
joint vetch, 253
Julocroton argenteus, 213
junco, 176
junior Tom Thumb cactus, 162
Justicia pilosella, 50
Justicia runyonii, 51
Justicia turneri, 51
jute, orinoco, 407

Kalanchoe daigremontiana, 200
Kalanchoe delagoensis, 199
Kalanchoe tubiflora, 199

Kalanchoe xhoughtonii, 200
Kallstroemia californica, 426
Kallstroemia maxima, 426
Karwinskia humboldtiana, 366
keeled goosefoot, 183
keelpod, Gregg's, 158
Kidder's lazy daisy, 86
kidneywood, Texas, 262
knotted hedge parsley, 72
knotweed leaf-flower, 227
Koeberliniaceae family, 176
Koeberlinia spinosa, 176
Kosteletzkya depressa, 311
Kosteletzkya virginica, 310–11
Krameriaceae family, 283
Krameria lanceolata, 283
Krameria ramosissima, 283
Kunth's evening primrose, 340

Labiatae family, 284–92
Lactuca intybacea, 112
lady finger cactus, 165
lady's thumb, 356
Laennecia coulteri, 111
Lamiaceae family, 284–92
Lamium amplexicaule, 285
lampazos, 332
lance-leaf cotton flower, 65
lance leaf loosestrife, 299
lantana, 414–17
Lantana achyranthifolia, 414–15
Lantana camara, 415
Lantana canescens, 416
Lantana horrida, 416
Lantana macropoda, 414–15
Lantana microcephala, 416
Lantana urticoides, 415, 416
Lantana velutina, 417
Laredo flax, 294
Laredo sand mat, 211
large bishop's weed, 68
large flower broomrape, 341
large-head flat sedge, 34
large redstem, 298
Larrea tridentata, 427
La Sal del Rey, 2
Launaea intybacea, 112
Lauraceae family, 292
lawn orchid, 43
lazy daisy, 86–87

leafless cressa, 190
leaf margins, 4, 7
leaf mustard, 150
leaf types and characteristics, 4–7
least snout bean, 268
leatherleaf, 178
leather stem, 226
Lechea san-sabeana, 187
Lechea tenuifolia, 187
Leguminosae family, 230–73
Lemna aequinoctialis, 41
Lemnaceae family, 41–42
Lemna obscura, 41
lemon vine, 172
Lenophyllum texanum, 201
Lentibulariaceae family, 293
Lepidium austrinum, 153
Lepidium lasiocarpum var. *orbiculare,* 153
Lesquerella argyraea, 155
Lesquerella grandiflora, 154
Lesquerella lasiocarpa var. *berlandieri,* 154
Lesquerella lindheimeri, 155
Lesquerella thamnophila, 156
lesser duckweed, 41
lettuce, 27, 69, 112
Leucaena leucocephala, 245
Leucaena pulverulenta, 245
Leucophyllum frutescens, 386
Liatris elegans, 112
lilac tasselflower, 97
lila de los llanos, 26
Liliaceae family, 20, 23–24, 26
lily, 19, 25–26, 297, 332–33
lily of the lomas, 26
Limnosciadium pumilum, 71
limoncillo, 116–17, 376
Limonium carolinianum, 351
Limonium nashii, 351
Linaceae family, 293–94
Linaria texana, 388
Lindernia dubia var. *anagallidea,* 387
Lindheimer's bladderpod, 155
Lindheimer's senna, 237
Lindheimer's sida, 316
Lindheimer's stickleaf, 296
linear leaf four o'clock, 331
Linum alatum, 293

Linum elongatum, 294
Linum imbricatum, 294
Linum lundellii, 294
lippia, 417–19
Lippia alba, 417
Lippia graveolens, 418
Lippia nodiflora, 418
Lippia strigulosa, 418–19
Lithospermum incisum, 148
Lithospermum matamorense, 148
little chiles, 168
little head gumweed, 101
little mallow, 311
live oak, 274
liverspot lily, 19
lizard tail, 337
Loasaceae family, 295–97
lobed leaves, 7
Lobelia berlandieri, 173
Loganiaceae family, 159–60
long headed coneflower, 121
Longhorn passion flower, 347
long leaf pondweed, 46
loosestrife, 299–300
Lophophora williamsii, 161, 168
Lorna's savory, 285
lotebush, 366
lotus, yellow, 324
love vine, 292
low amaranth, 61
low croton, 217
Lower Rio Grande Valley, 1–4
Lower Rio Grande valley barrel cactus, 166
low growing prickly pear, 171
low wild mercury, 220
Lozano's false Indian mallow, 306
Ludwigia decurrens, 338
Ludwigia octovalvis, 338
Ludwigia peploides, 338–39
Lundell's flax, 294
Lupinus subcarnosus, 265
Lupinus texensis, 265
Lycium berlandieri, 394
Lycium carolinianum var. *quadrifidum,* 394
Lycopersicon esculentum var. *cerastiforme,* 401
lyreleaf parthenium, 115

Lythraceae family, 297–300
Lythrum alatum var. *lanceolatum,* 299, 300
Lythrum californicum, 299, 300
Lythrum lanceolatum, 299

Macfadyena unguis-cati, 140
Machaeranthera phyllocephala, 122
Machaeranthera pinnatifida, 136
Macroptilium atropurpureum, 266
Macrosiphonia lanuginosa, 72
Macrosiphonia macrosiphon, 72
Madiera vine, 139
Malachra capitata, 311
mallow, 301–6, 308, 309, 310–13, 314, 317–20, 321, 404
Malpighiaceae family, 300–301
Malpighia glabra, 301
Malvaceae family, 301–21
malva de caballo, 311
malva loca, 312
Malva parviflora, 311
Malvastrum americanum, 312
Malvastrum aurantiacum, 312
Malvastrum coromandelianum, 312–13
Malvaviscus arboreus var. *drummondii,* 313
Malvaviscus drummondii, 313
Malvaviscus penduliflorus, 313
Mammillaria heyderi, 168
Mammillaria longimamma, 169
Mammillaria multiceps, 169
Mammillaria prolifera var. *texana,* 169
Mammillaria robertii, 162
Mammillaria runyonii, 162
Mammillaria sphaerica, 169
Manfreda longiflora, 18
Manfreda maculosa, 18, 19
Manfreda sileri, 18, 19
Manfreda variegata, 19
mangrove, 138, 188, 367
Manihot walkerae, 226
many spiked flat sedge, 35
many stem evax, 96
manzanita, 301

Margaranthus solanaceus, 395
margin (leaf part), 5, 7
margined milkpea, 263
marine ivy, 424
mariola, 116
marsh fimbry, coastal, 37
Marsileaceae family, 12
Marsilea macropoda, 12
Martyniaceae family, 348
Matamoros saltbush, 181
Matelea brevicoronata, 79, 80
Matelea parviflora, 79, 80
Matelea retuculata, 80
Matelea sagittifolia, 81
Matt chaff flower, 58
Maurandya antirrhiniflora, 387
mauve Indian mallow, 303
Maytenus phyllanthoides, 178
Maytenus texana, 178
McKelvey's beeblossom, 336
meadow pink, 278
Mecardonia procumbens, 388
Mecardonia vandellioides, 388
Medicago polymorpha, 266
Melampodium cinereum, 113
Melia azedarach, 321
Meliaceae family, 321
Melilotus albus, 267
Melilotus indicus, 267
Melochia pyramidata, 405
Melochia tomentosa, 405
meloncito, 202
Melothria pendula, 202
Menispermaceae family, 322
Menodora heterophylla, 334
Mentzelia incisa, 296
Mentzelia lindheimeri, 296
Mentzelia nuda, 296
Mentzelia nuda var *stricta,* 297
Merremia dissecta, 189, 198
mescal bean, 270
mesquite, 252–53
mesquite calliandra, false, 242
mesquite plains milkvine, 80
Mexican ash, 334
Mexican bastardia, 307
Mexican buttonbush, 369
Mexican caesalpinia, 232
Mexican capraria, 385
Mexican cat's eye, 142
Mexican croton, 215

INDEX 449

Mexican devil weed, 91
Mexican fiddlewood, 412
Mexican hat, 121
Mexican mallow, 404
Mexican mud plantain, 45
Mexican oregano, 418
Mexican persimmon, 205
Mexican poppy, 344–45
Mexican tournefortia, 150
Mexican urvillea, 381
Meximalva filipes, 314
Micromeria brownei var. *pilosiuscula,* 285
midrib (leaf part), 5
Mikania scandens, 113
milkpea, 262–63
milkvine, 79–81
milkweed, 74–76, 79
milkwort, 353
Mimosa asperata, 246
Mimosa latidens, 247
Mimosa malacophylla, 248
Mimosa pigra var. *berlandieri,* 246
Mimosa strigillosa, 249
Mimosa texana, 250
Mimosa wherryana, 250
Mimosoideae family, 239–53
mint, 284, 286, 291
Mirabilis albida, 330
Mirabilis austrotexana, 330
Mirabilis linearis, 331
mirasol, 105
mistflower, 91, 93, 98–99, 127
mistletoe, 423
Mitracarpus breviflorus, 372
mock pennyroyal, 284
Modiola caroliniana, 314
Molluginaceae family, 322–23
Mollugo verticillata, 323
Monarda citriodora, 286
Monarda fruticulosa, 286
Monarda punctata var. *lasiodonta,* 287
monocots (monocotyledoneae), 4, 16–47
Montezuma bald cypress, 15
moonflower, 193, 393
moonlight cactus, 172
Moraceae family, 323–24
Mormon tea, 14

morning glory, 194–98
mortonia, 179
Mortonia greggii, 179
Morus alba, 324
Morus rubra, 324
moss, 28–29
mottled tuberose, 19
mountain laurel, Texas, 270
mud babies, 45
mulberry, 324
mulberry weed, 323
muskgrass, 11
Mustang grape, 425
mustard, 150, 152, 158

nailwort, South Texas, 177
Najadaceae family, 42–43
Najas guadalupensis, 42–43
nama, 280–81
Nama hispidum, 280
Nama jamaicense, 280
Nama parvifolium, 281
Nama stenocarpum, 280, 281
Nama undulatum, 281
naming of plants, 3
narrow leaf elbow bush, 333
narrowleaf globe mallow, 318
narrow leaf goldshower, 300
narrow leaf heliotrope, 147
narrow leaf kalanchoe, 199
narrow leaf puccoon, 148
narrow leaf rhombopod, 175
narrow leaf shrubby wood sorrel, 343
narrow leaf sumpweed, 110
narrow leaf swallow wort, 77
narrow leaved sandbar willow, 378
necklace pod, 270
Nelumbo lutea, 324
Nelumbonaceae family, 324
Neonesomia palmeri, 114
Neptunia plena, 251
Neptunia pubescens, 252
Nerisyrenia camporum, 153
net leaf rabbit's ears, 273
nettle, 213, 295, 410
nettled globe berry, 395
nettle leaf goosefoot, 182
Nicotiana glauca, 395
Nicotiana repanda, 396

Nicotiana trigonophylla, 396
night-blooming cereus, 160
nightshade, 392–93, 399–400, 403
nipple cactus, 168
nitella, 11
Nitella sp., 11
nodding green violet, 422
noseburn, 229–30
notched rattleweed, 256
Notholaena sinuata, 13
Nothoscordum bivalve, 24
Nueces green thread, 130
Nuttallianthus texanus, 388
Nyctaginaceae family, 325–32
Nyctaginea capitata, 331
Nymphaeaceae family, 332–33
Nymphaea elegans, 332
Nymphaea mexicana, 333

oak, live, 274
Oenothera drummondii, 339
Oenothera grandis, 339
Oenothera kunthiana, 340
Oenothera laciniata, 340, 341
Oenothera rosea, 340–41
Oenothera speciosa, 341
Oenothera tetraptera, 340
ojo de víbora, 192
old man's beard, 362
Oleaceae family, 333–34
Oligomeris linifolia, 363
olive, wild, 141
olmo, 409
Onagraceae family, 335–41
once compound leaf, 5
onion *(Allium),* 23–24
opposite leaves, 6
Opuntia engelmannii var. *lindheimeri,* 170, 171
Opuntia leptocarpa, 171
Opuntia leptocaulis, 171
Opuntia lindheimeri var. *lindheimeri,* 170
Opuntia macrorhiza, 171
Opuntia schottii, 167
Orchidaceae family, 43
oregano, Mexican, 418
oreja de perro, 149
oreja de ratón, 208
orinoco jute, 407

Orobanchaceae family, 341–42
Orobanche ludoviciana subsp. *ludoviciana,* 341
Orobanche ludoviciana subsp. *multiflora,* 342
ovary (flower part), 7
Oxalidaceae family, 342–44
Oxalis berlandieri, 343
Oxalis corniculata var. *wrightii,* 342
Oxalis dichondrifolia, 342
Oxalis dillenii, 344
Oxalis drummondii, 343
Oxalis frutescens subsp. *angustifolia,* 343
Oxalis stricta, 344
Oxalis tuberosa, 343

Packera tampicana, 114
Padre Island mistflower, 93
paintbrush, Texas, 386
palafoxia, 114–15
Palafoxia hookeriana, 114–15
Palafoxia texana var. *ambigua,* 115
pale blue-eyed grass, 41
palillo, 215
Palmae family, 27
palma pita, 20–21
palmately compound leaf, 5
Palmer's bloodleaf, 66
Palmer's cock's comb, 63
Palmer's goldenweed, 114
Palmer's wigweed, 61
palmetto, Rio Grande, 27
palm grove passion vine, 346
palm leaf globe mallow, 319
palo blanco, 408
palo verde, 235
panicled sand mallow, 318
Papaveraceae family, 344–45
paper flower, 120
Papilionoideae family, 253–73
papillosus, small, 164
Parietaria floridana, 409
Parietaria pensylvanica, 409, 410
Parkinsonia aculeata, 234
Parkinsonia texana, 234, 235
Parkinsonia texana var. *macra,* 235
Park's croton, 219

Paronychia jonesii, 177
parsley, 68, 71–72
Parthenium confertum, 115, 116
Parthenium hysterophorus, 116
Parthenium incanum, 116
partridge pea, 233
Passifloraceae family, 346–47
Passiflora filipes, 346
Passiflora foetida var. *gossypifolia,* 346
Passiflora pallida, 347
Passiflora suberosa, 347
Passiflora tenuiloba, 347
passion flower, 346–47
passion vine, palm grove, 346
Paysonia grandiflora, 154, 155
Paysonia lasiocarpa subsp. *berlandieri,* 154
pea, 232–33, 234, 255, 264, 267, 271
peach bush, 368
pearlhead, Rio Grande, 109
pearl net leaf milkvine, 80
Pectis angustifolia var. *tenella,* 116–17
Pedaliaceae family, 348
Pediomelum rhombifolium, 267
pellitory, 409–10
pencil cactus, 165
pencilflower, sticky, 271
Pennsylvania cudweed, 100–101
Pennsylvania pellitory, 410
penny leaf wood sorrel, 342
peonia, 82
peony, pink desert, 82
Peperomia pellucida, 350
pepper, 67
pepper vine, 423
pepperweed, Southern, 153
perennial saltwort, 185
Pereskia aculeata, 172
Perezia runcinata, 82
Perezia wrightii, 82
perianth (flower part), 7
Perityle microglossa var. *microglossa,* 117
Persicaria amphibia, 355
Persicaria maculosa, 356
Persicaria pensylvanica, 356

persimmon, 205
Peruvian pepper, 67
petal (flower part), 7
Petalostemum emarginatum, 259
Petalostemum obovata, 260
petiole (leaf part), 5
Petiveria alliacea, 349
petunia, 52–53, 390
Petunia parviflora, 390
peyote, 168
Phacelia congesta, 282
Phacelia patuliflora var. *austrotexana,* 282
Phaseolus caracalla, 272
Phaulothamnus spinescens, 55
Phemeranthus aurantiacus, 358, 360
Phemeranthus parviflorus, 358, 360
Philoxerus vermicularis, 62
Phlox drummondii subsp. *drummondii,* 352–53
Phlox drummondii var. *peregrina,* 352–53
Phorandendron tomentosum, 423
Phyla incisa, 418
Phyla nodiflora, 418, 419
Phyla strigulosa, 418–19
Phyllanthus polygonoides, 227
Phyllanthus tenellus, 227
phyllary (flower part), 8
Physalis cinerascens var. *cinerascens,* 396–97
Physalis cinerascens var. *spathulifolia,* 397
Physalis hederifolia var. *fendleri,* 397
Physalis lobata, 399
Physalis pubescens, 398
Physalis virginiana var. *texana,* 398
Physalis viscosa var. *cinerascens,* 396–97
Physaria argyrea, 155
Physaria lindheimeri, 155
Physaria thamnophila, 156
Phytolaccaceae family, 55, 349
pidgeon berry, 349
pie print, 304

INDEX

pigweed, 60, 62
pincushion cactus, 168
pincushion daisy, 100
pink baby's breath, 360
pink desert peony, 82
pink evening primrose, 341
pink mint, 291
pinks, 277–78
pink smartweed, 356
pinnately compound leaf, 5
pinweed, san saba, 187
Piperaceae family, 350
Pisonia aculeata, 332
Pistia stratiotes, 27
pistil (flower part), 7
Pithecellobium ebano, 242
Pithecellobium flexicaule, 242
Pithecellobium pallens, 244
pitseed goosefoot, 182
plains black foot daisy, 113
plains gaura, 335
plains gumweed, 102
Plantaginaceae family, 350–51
Plantago hookeriana, 350
Plantago rhodosperma, 351
plantain, 45, 350–51
Plectocephalus americanus, 117
Pluchea carolinensis, 118
Pluchea odorata, 118
Pluchea purpurascens, 118
Plumbaginaceae family, 351–52
Plumbago scandens, 352
plume tooth beebalm, 287
poinsettia, wild, 223
Poinsettia helleri, 224
pointed watermeal, 42
Poiret's copperleaf, 206
poison bitterweed, 109
poison ivy, 67
Polanisia dodecandra subsp. *riograndensis,* 176
Polanisia erosa subsp. *breviglandulosa,* 177
Polemoniaceae family, 352–53
Polianthes maculosa, 18
Polianthes runyonii, 18
Polly prim, 160
Polygala alba, 353
Polygalaceae family, 353
Polygala glandulosa, 353

Polygonaceae family, 354–57
Polygonum amphibium, 355
Polygonum pensylvanicum, 356
Polygonum persicaria, 356
Polypremum procumbens, 160
Pomaria austrotexana, 235
pondweed, long leaf, 46
pong flat sedge, 34
Pontederiaceae family, 44–45
ponyfoot, 191
popinac, 245
poppy, 344–45
Porlieria angustifolia, 425
Portulacaceae family, 358–60
Portulaca mundula, 359
Portulaca oleracea, 358
Portulaca pilosa, 359
Portulaca retusa, 358
Portulaca umbraticola subsp. *coronata,* 359
Portulaca umbraticola subsp. *lanceolata,* 359
Potamogetonaceae family, 46
Potamogeton diversifolius, 46
Potamogeton latifolius, 46
Potamogeton nodosus, 46
Potamogeton pulcher, 46
Potamogeton pusillus, 46
potato tree, 401
powderpuff, 249
prairie acacia, 239
prairie agalinis, 382
prairie bluets, 371
prairie broomweed, 85
prairie bur, 283
prairie clover, 259
prairie dalea, 258
prairie dog sunshade, 71
prairie gaillardia, 99
prairie Mexican clover, 373
prairie milkweed, 75
prairie onion, 23
prickles on stem, defined, 4
prickly caesalpinia, 230–31
prickly mallow, 317
prickly pear, 170, 171
primrose, 335, 338–40, 341
primrose willow, 338
Primulaceae family, 360–62
princess vine, 424
Priva lappulacea, 419

Proboscidea fragrans, 348
Proboscidea louisianica subsp. *fragrans,* 348
Proboscidea parviflora, 348
Prosopis glandulosa, 252
Prosopis reptans var. *cinerascens,* 253
prostrate fleabane, 97
prostrate ground cherry, 391
prostrate lawn weed, 90
prostrate milkweed, 76
prostrate sand mat, 212
prostrate starwort, 178
prostrate water hyssop, 388
Prunus texana, 368
Pseudabutilon lozanii, 306
Pseudognaphalium austrotexanum, 119
Pseudognaphalium stramineum, 119
Psilostrophe gnaphalioides, 120
Psoralea rhombifolia, 267
Pteridaceae family, 13
Pterocaulon virgatum, 120
puncture weed, 427
purple allamanda, 76
purple bean, 266
purple ground cherry, 399
purple marsh fleabane, 118
purple pleat leaf, 40
purslane, 55–57, 358
pussyfoot, 260
pyramid bush, woolly, 405
pyramid flower, 405
Pyrrhopappus pauciflorus, 120–21
queen of the night, 172
queen's delight, 228–29
queen's wreath, 354
quelite, 61
Quercus virginiana, 274
Quincula lobata, 399

rabbit lettuce, 69
rabbit tobacco, 95
ragweed, 83–85, 116
railroad vine, 197
rainfall, average, 3
rain lily, 25
Randia rhagocarpa, 372
Ranunculaceae family, 362–63
Ranunculus sceleratus, 363

raspilla, 248
Ratibida columnaris, 121
Ratibida columnifera, 121
Ratibida peduncularis, 121
rattlebush, Drummond's, 268
rattlesnake weed, 69
rattleweed, notched, 256
ray flower (flower part), 8
Rayjacksonia phyllocephala, 122
Raynal's upright braided sedge, 38–39
receptacle (flower part), 8
red berry nightshade, 400
redbrush lippia, 418
redbud, 334
red center morning glory, 194
red mangrove, 367
red poppy, 345
red sage, 288
red seeded plantain, 351
redstem, large, 298
red stem stork's bill, 278
Resedaceae family, 363
retama, 234
Rhamnaceae family, 364–66
rhizome, defined, 4
Rhizophoraceae family, 367
Rhizophora mangle, 367
rhombopod, narrow leaf, 175
Rhynchosia americana, 268
Rhynchosia minima, 268
Rhynchosida physocalyx, 314–15
Rhynchospora colorata, 38
rice field flat sedge, 33
Richardia brasiliensis, 373
Richardia scabra, 373
Richardia tricocca, 373
Ricinus communis, 228
ridge seed sand mat, 210
Rio Grande abutilon, 303
Rio Grande ayenia, 403
Rio Grande butterfly bush, 159
Rio Grande Delta, 2
Rio Grande mock vervain, 414
Rio Grande palmetto, 27
Rio Grande pearlhead, 109
Rio Grande phlox, 352–53
Rio Grande plains milkvine, 79
Rio Grande ragweed, 83

Rio Grande River, 2
Rio Grande skullcap, 290
Rio Grande tick seed, 94
Rio Grande Valley selenia, 157
Rio Grande vervain, 422
river sage, 288
Rivina humilis, 349
rocket mustard, 158
rocktrumpet, woolly, 72
Roosevelt weed, 88
root cactus, 161
Rorippa teres, 156
Rosaceae family, 368
roseling, Southern coastal, 29
rose sundrops, 340–41
rosita, 276
Rotala ramosior, 300
rouge plant, 349
rough buttonweed, 370
rough gromwell, 148
rough leaf New Mexico vervain, 421
rough leaved agave, 17
round copperleaf, 205
round head broomweed, 102
round leaf scurf pea, 267
rubber vine, 76
Rubiaceae family, 368–74
Rubus riograndis, 368
Rubus trivialis, 368
Rudbeckia hirta, 122
Ruellia nudiflora var. *nudiflora,* 52
Ruellia nudiflora var. *runyonii,* 52
Ruellia occidentalis, 53
Ruellia runyonii, 52
Ruellia yucatana, 53
Rumex chrysocarpus, 357
Rumex pulcher, 357
Runyon's coryphantha, 162
Runyon's dodder, 204
Runyon's esenbeckia, 376
Runyon's huaco, 18
Runyon's onion, 24
Runyon's pincushion cactus, 162
Runyon's premature sunflower, 107
Runyon's skullcap, 289
Runyon's violet wild petunia, 52

Runyon's water willow, 51
rushpea, 231, 235
Russian thistle, 184
Rutaceae family, 375–78

Sabal mexicana, 27
Sabal texana, 27
Sabatia arenicola, 277, 278
Sabatia campestris, 277, 278
sabino, 15
sacasil, 165
sacasile, 139
sage, 87, 287–88, 386
Sagittaria longiloba, 22
Sagittaria platyphylla, 23
saladillo, 134
Salicaceae family, 378–79
Salicornia bigelovii, 184, 185
Salicornia utahensis, 184, 185
Salicornia virginica, 184, 185
saline soils, 2
Salix exigua subsp. *interior,* 378
Salix interior, 378
Salix nigra, 379
Salsola australis, 184
Salsola kali, 184
Salsola tragus, 184
saltbush, 180–81
salt cedar, 406
saltmarsh mallow, 310–11
saltmarsh morning glory, 198
saltmarsh pink, 277
saltmarsh sandspurry, 177
saltwort (annual and perennial), 184, 185
Salvia ballotiflora, 287
Salvia coccinea, 288
Salvia misella, 288
Samolus ebracteatus subsp. *alysoides,* 361
Samolus ebracteatus subsp. *cuneatus,* 361
Samolus ebracteatus subsp. *ebracteatus,* 361
Samolus parviflorus, 362
Samolus valerandi subsp. *parviflorus,* 362
sand atriplex, 181
sandbell, 280
sand brazoria, 284

INDEX

sand deposits, 2
sand dollar cactus, 161
sandhills amaranth, 60
sand lily, 297
sand mat, 208–11, 212
sand scorpionweed, South Texas, 282
sandspurry, saltmarsh, 177
sandy land bluebonnet, 265
sangre de drago, 226
san saba pinweed, 187
Sanvitalia ocymoides, 122–23
Sapindaceae family, 379–81
Sapindus saponaria var. *drummondii,* 380–81
Sapotaceae family, 382
Sarcocornia utahensis, 184, 185
Sarcostemma clausum, 78
Sarcostemma cynanchoides, 79
sauz, 379
savannah milkweed, 75
savory, 285
sawtooth frog fruit, 418
scabrum, 17
scarlet musk flower, 331
scarlet pea, 264
scarlet pimpernel, 360
scarlet sage, 288
scarlet spiderling, 328
Schaefferia cuneifolia, 179
Schinus terebinthifolius, 67
Schoenoplectus californicus, 38
Schoenoplectus erectus subsp. *raynalii,* 38–39
Schoenoplectus pungens var. *longispicatus,* 39
Schoenoplectus tabernaemontani, 39
Schott's twintip, 389
Schrankia latidens, 247
Scirpus californicus, 38
Scirpus maritimus, 31
Scirpus pungens var. *longispicatus,* 39
Sclerocactus scheeri, 161
Sclerocarpus uniserialis var. *austrotexanus,* 123
Scoparia dulcis, 388–89
scorpion's tail, 144
scrambled eggs, 275
scratch daisy, 95

screw bean mesquite, 253
Scrophulariaceae family, 382–89
Scutellaria drummondii var. *runyonii,* 289
Scutellaria muriculata, 290
sea lavender, 351
sea ox eye, 90
seaside agalinis, 383
seaside goldenrod, 125
seaside heliotrope, 145
sea urchin cactus, 161
sedge family, 31–39
Sedum texanum, 201
seepweed, 186
seepwillow, 88
Selenia grandis, 157
Selenicereus spinulosus, 172
Senecio ampullaceus, 123
Senecio imparipinnatus, 114
Senecio riddellii, 124
Senecio spartioides, 124
Senecio tampicanus, 114
senna, 233, 236–38
Senna bauhinioides, 236
Senna durangensis var. *iselyi,* 236
Senna lindheimeriana, 237
Senna occidentalis, 237
Senna pendula, 238
Senna pumilio, 238
sensitive brier, 247
sensitive plant, 249
sepal (flower part), 7
serjania, 381
Serjania brachycarpa, 381
Sesbania drummondii, 268
Sesbania exaltata, 269
Sesbania herbacea, 269
Sesbania macrocarpa, 269
Sesuvium maritimum, 55
Sesuvium portulacastrum, 56
Sesuvium sessile, 56
Sesuvium verrucosum, 56
shade betony, 290
shaggy false nightshade, 393
shaggy tuft, sweet, 54
shakeshake, 256
shepherd's purse, 151
Shinners' rocket, 159
shiny bush, 350

shiny cock's comb, 63
short gland clammy weed, 177
shorthorn jefea, 111
short ray rock daisy, 117
showy crotalaria, 257
showy evening primrose, 339
showy palafoxia, 114–15
shrimp plant, Texas, 54
shrubby beebalm, 286
shrubby blue sage, 287
shrubby Indian mallow, 301
shrubby indigo, 264
Sibara runcinata, 157
Sibara viereckii, 157
sida, 308, 314–18
Sida abutifolia, 315
Sida ciliaris, 315
Sida cordifolia, 316
Sida filicaulis, 315
Sida filipes, 314
Sida helleri, 308
Sida lindheimeri, 316
Sida paniculata, 318
Sida physocalyx, 314–15
Sida rhombifolia, 317
Sida spinosa, 317
Sidastrum paniculatum, 318, 404
Sideroxylon celastrinum, 382
Sierra Madre torchwood, 375
Siler's tuberose, 19
silky alkaliweed, 190
silky evolvulus, 193
silky leaf frog fruit, 418–19
silver bladderpod, 155
silverbush, 221
silver croton, 214
silverhead, 62
silver july croton, 213
silver leaf nightshade, 400
silverleaf sunflower, 106
Simaroubaceae family, 390
simple leaf structure, 5
Simsia calva, 124
Siphonoglossa greggii, 50, 51
Siphonoglossa pilosella, 50
Sisymbrium irio, 158
Sisyrinchium biforme, 41
Sisyrinchium langloisii, 41
six angle fold wing, 49
skeleton leaf golden eye, 135

skullcap, 289–90
slender dodder, 203
slender evolvulus, 192
slender leaf fournerve, 129
slender lobe passion flower, 347
slender passion flower, 346
slender sea purslane, 55
slender snake cotton, 64
slender stalked Mexican mallow, 314
slim leaf sneezeweed, 104
slim milkweed, 74
slim vetch, 272
small flowered devil's claw, 348
small flowered wrightwort, 48
small flower ground cherry, 396–97
small flower milk vetch, 254
small headed sneezeweed, 105
small-leaf nama, 281
small-leaved wissadula, 320
small papillosus, 164
small Venus' looking glass, 174
smartweed, 355, 356
Smilacaceae family, 46
Smilax bona-nox, 46
smooth buttonweed, 374
smooth chaff flower, 58
smooth five angled dodder, 203
smooth senna, 238
smooth stem wild indigo, 254
smooth umbrella wort, 327
snailseed, variable leaf, 322
snail vine, 272
snake cotton, 64
snake eyes, 55
snapdragon vine, 387
sneezeweed, 104–5
snoutbean, 268
soapberry, 380–81
soils, 2
Solanaceae family, 390–403
Solanum americanum, 399
Solanum campechiense, 400
Solanum eleagnifolium, 400
Solanum erianthum, 401
Solanum lycopersicum, 401
Solanum nodiflorum, 399
Solanum pseudocapsicum, 402

Solanum ptychanthum, 399
Solanum rostratum, 400, 402
Solanum triquetrum, 402, 403
Solidago altissima var. *altissima,* 125
Solidago altissima var. *pluricephala,* 125
Solidago canadensis, 125
Solidago sempervirens, 125
sombrerillo, 70–71
Sonchus asper, 126
Sonchus oleraceus, 126
Sophora secundiflora, 270
Sophora tomentosa var. *occidentalis,* 270
sorrel, 342–44
sour clover, 267
Southern coastal roseling, 29
Southern flat sedge, 36
Southern Indian blanket, 100
Southern pepperweed, 153
South Texas false cudweed, 119
South Texas four o'clock, 330
South Texas globe mallow, 319
South Texas nailwort, 177
South Texas rushpea, 235
South Texas sand scorpionweed, 282
South Texas Sand Sheet, 2
Southwest bedstraw, 371
sow thistle, 126
Spanish dagger, 20–21
Spanish moss, 29
Spergularia marina, 177
Spergularia salina, 177
Spermacoce floridana, 374
Spermacoce glabra, 374
Sphaeralcea angustifolia var. *angustifolia,* 318
Sphaeralcea angustifolia var. *cuspidata,* 318
Sphaeralcea angustifolia var. *lobata,* 318
Sphaeralcea angustifolia var. *oblongifolia,* 318
Sphaeralcea hastulata, 319
Sphaeralcea lindheimeri, 319
Sphaeralcea pedatifida, 319
spiderling, 328–29
spider wisp, 175

spiderwort, stemless, 31
spike rush, 36–37
spine, leaf, defined, 4
spiny aster, 91
spiny fruit saltbush, 180
spiny goldenweed, 136
spiny hackberry, 408
spiny leaf chaff flower, 59
spiny pigweed, 62
spiny sow thistle, 126
Spiranthes vernalis, 43
spotted tuberose, 18
spreading dayflower, 30
spreading sida, 315
spreading sweetjuice, 322
spread wing, Texas, 70
spring ladies' tresses, 43
spring mistflower, 127
spurge, 223–24
square bud daisy, 128
square bud primrose, 335
square stem spike rush, 37
squaw bush, 365
Stachys crenata, 290, 291
Stachys drummondii, 290, 291
stamen (flower part), 7
star cactus, 161
Starr County, 2
star-shaped hair on leaf or stem, defined, 4
starwort, prostrate, 178
Stellaria cuspidata subsp. *prostrata,* 178
Stellaria media, 178
stellate hair on leaf or stem, defined, 4
stemless spiderwort, 31
Stemodia lanata, 389
Stemodia schottii, 389
Stemodia tomentosa, 389
Stenandrium dulce, 50, 54
Sterculiaceae family, 403–6
stickleaf, 296
sticky pencilflower, 271
sticky snakeweed, Texas, 103
stigma (flower part), 7
Stillingia sylvatica, 228
Stillingia treculiana, 229
stinging cevallia, 295
stinging nettle, 410
stinging weed, 410

INDEX

stinkweed, 182
stipule (leaf part), 5
St. John's wort, few flowered, 187
stolon, defined, 4
stonecrop, Texas, 201
stonewort, 11
straggler daisy, 90
strawberry cactus, 164
strawberry pitaya, 164
style (flower part), 7
Stylosanthes viscosa, 271
Suaeda conferta, 185
Suaeda linearis, 186
Suaeda tampicensis, 186
submerged hornwort, 180
sulphur mallow, 309
sumpweed, 110–11
sundew, dwarf, 204
sunflower (Asteraceae) family, 8, 81–138
swallow wort, narrow leaf, 77
swan flower, 73
sweet broomwort, 388–89
sweet gaura, 336
sweetjuice, spreading, 322
sweet shaggy tuft, 54
sweet stem, 411
swordbean, 255
Symphyotrichum divaricatum, 127
Symphyotrichum subulatum, 127
Synthlipsis greggii, 158

tailed rushpea, 231
tailed seacoast sumpweed, 110–11
talayote, 78
Talinum angustissimum, 358
Talinum aurantiacum, 358
Talinum paniculatum, 358, 360
tall goldenrod, 125
tallow weed, 350
Tamaricaceae family, 406
Tamarix aphylla, 406
Tamarix ramosissima, 406
Tamaulipa azurea, 127
tampico seepweed, 186
Taraxacum officinale, 128
tasajillo, 171

tassle flower, 96–97
tatalencho, 103
Taxodiaceae family, 15
Taxodium distichum var. *mexicanum*, 15
Taxodium mucronatum, 15
Telosiphonia lanuginosa, 72
temperatures, 2
tenaza, 244
tender leaf-flower, 227
tepeguaje, 245
tepozán, 159
Tephrosia lindheimeri, 271
Tetraclea coulteri, 419
Tetragonotheca repanda, 128
Tetragonotheca texana, 129
Tetramerium platystegium, 54
Tetraneuris linearfolia, 129
Teucrium canadense, 292
Teucrium cubense, 292
Texas baccharis, 89
Texas bindweed, 190
Texas bluebonnet, 265
Texas cat's eye, 142
Texas colubrina, 364
Texas croton, 220
Texas doubtful palafoxia, 115
Texas ebony, 242
Texas frog fruit, 418
Texas geranium, 279
Texas giant ragweed, 85
Texas ground cherry, 398
Texas groundsel, 123
Texas heliotrope, 147
Texas Indian mallow, 302
Texas kidneywood, 262
Texas lantana, 416
Texas milkpea, 263
Texas mimosa, 250
Texas mountain laurel, 270
Texas nerve ray, 129
Texas nightshade, 403
Texas paintbrush, 386
Texas persimmon, 205
Texas prickly pear, 170
Texas purple sage, 386
Texas sabal palm, 27
Texas senna, 233
Texas shrimp plant, 54
Texas spread wing, 70
Texas sticky snakeweed, 103

Texas stonecrop, 201
Texas thistle, 92
Texas toad flax, 388
Texas torchwood, 375
Texas vervain, 421
Texas wrightwort, 48
Thamnosma texanum, 377
Thelesperma ambiguum, 130
Thelesperma megapotamicum var. *ambiguum*, 130
Thelesperma nuecense, 130
Thelocactus bicolor, 173
Thelocactus setispinus, 167
Thelypodiopsis shinnersii, 159
thicket threadvine, 77
thin leaf brookweed, 362
thistle, 92, 126, 184
thorn, defined, 4
thorn crested agave, 17
threadvine, thicket, 77
three furrowed Indian mallow, 305
three lobes florestina, 99
three seed croton, 218
three times compound leaf, 5
Thryallis angustifolia, 300
Thymophylla pentachaeta, 131
Thymophylla tenuiloba, 131
Thymophylla tephroleuca, 132
tick seed, 94
Tidestromia lanuginosa, 66
tie vine, 195
Tiliaceae family, 407
Tillandsia baileyi, 28
Tillandsia recurvata, 28–29
Tillandsia usneoides, 29
Tiny Tim, 131
Tiquilia canescens, 144, 149
tomato, 401
tomillo, 253
tooth cup, 297
toothed leaves, 7
torchwood, 375
Torilis arvensis, 71
Torilis nodosa, 72
Torrey's croton, 217
tournefortia, 149–50
Tournefortia hirsutissima, 149
Tournefortia volubilis, 150
Toxicodendron radicans, 67
Tradescantia micrantha, 29

Tradescantia subacaulis, 31
Tragia glanduligera, 229, 230
Tragia ramosa, 229, 230
trailing heliotrope, 146
Trecul's queen's delight, 229
tree tobacco, 395
Trianthema portulacastrum, 57
Tribulus terrestris, 427
Trichocoronis wrightii, 132
tridax, 133
Tridax procumbens, 133
Triodanis biflora, 174
Triodanis perfoliata var. *biflora,* 174
trixis, 133
Trixis inula, 133
Trixis radialis, 133
tropical Mexican clover, 373
tropical puff, 252
tuber, defined, 4
tuberose, 18–19
tube tongue, 50
tufted flax, 294
tufted sea blite, 185
tule, 38, 47
tulipán del monte, 310
tumbleweed, 184
Turk's cap, 313
Turneraceae family, 407
Turnera diffusa var. *aphrodisiaca,* 407
Turner's tube tongue, 51
turnsole, 145
twice compound leaf, 5
twinevine, 78
twintip, Schott's, 389
twisted rib, 167
two-leaved senna, 236
Typhaceae family, 47
Typha domingensis, 47
Typha latifolia, 47

Ulmaceae family, 408–9
Ulmus crassifolia, 409
Umbelliferae family, 68–72
umbrella grass, 37–38
umbrella Indian mallow, 305
umbrella plant, 33
umbrella wort, smooth, 327
unequal-leaf trumpets, 325
Urticaceae family, 409–10

Urtica chamaedryoides, 410
Urvillea ulmacea, 381
Utricularia gibba, 293

vara blanca, 174
vara dulce, 411
variable leaf snailseed, 322
Varilla texana, 134
variously colored fleabane, 98
Vasey's adelia, 207
veintiunilla, 73
velvet bur, common, 419
velvet lantana, 417
velvet leaf Indian mallow, 304
velvet leaf mallow, 320
velvet spurge, 224
velvety leaf gaura, 337
Venus' looking glass, small, 174
Verbena bipinnatifida, 413
Verbena brasiliensis, 420
Verbena canescens, 420, 421
Verbenaceae family, 411–22
Verbena cloverae, 420–21
Verbena deltricola, 413
Verbena halei, 421
Verbena neomexicana var. *hirtella,* 421
Verbena officinalis, 421
Verbena polyantha, 414
Verbena quadrangulata, 414
Verbena runyonii, 421, 422
Verbesina encelioides, 134
Verbesina microptera, 134–35
vervain, 413–14, 420–22
vetch, 253–54, 272
Vicia ludoviciana, 272
vidrillos, 139
Viereck's winged rockcress, 157
Vigna caracalla, 272
Vigna luteola, 273
Viguiera stenoloba, 135
vine four o'clock, 326
vine joint fir, 14
Violaceae family, 422
violet petunia, 52
Virginia buttonweed, 370
Viscaceae family, 423
Vitaceae family, 423–25
Vitis mustangensis, 425
voucher specimens, 1

Walker's manihot (Walker's manioc), 226
Waltheria indica, 406
water hyacinth, 44
water hyssop, 384, 388
water lettuce, 27
water lily, 332–33
watermeal, 42
water nymph, common, 42–43
water primrose, 338–39
water sensitive, 251
water smartweed, 355
water stargrass, 44
water willow, Runyon's, 51
waterwort, annual, 173
wavy leaf nama, 281
wavy leaved gaura, 337
wavy twinevine, 78
wax plant, 222
waxweed, Colombian, 298
Wedelia acapulcensis var. *hispida,* 135
Wedelia hispida, 135
Wedelia texana, 135
wedge leaf draba, 152
Western ragweed, 84
Western tansy mustard, 152
Western umbrella grass, annual, 37
Western wild petunia, 53
West Indian lantana, 415
wheatspike scaly stem, 50
wheel mallow, 314
Wherry's mimosa, 250
white bristly acacia, 239
white brush, 411
white girdlepod, 372
white leaf croton, 218
white margined sand mat, 208
white milkwort, 353
white mistflower, 98–99
white mulberry, 324
whitened leaf bahia, 89
white plumbago, 352
white sage, 87
white sweet clover, 267
white topped umbrella grass, 38
white twinevine, 78
white velvet leaf mallow, 321
whitlow wort, 152

INDEX

whorled leaves, 6
widow's tears, 30–31
wigweed, 61
Wilcoxia poselgeri, 165
wild buckwheat, 355
wild lettuce, 112
wild mercury, low, 220
wild olive, 141
wild petunia, 390
wild poinsettia, 223
Willacy County, 2
willow (*Salix* sp.), 378–79
willow leaf buttonbush, 369
willow leaved heimia, 299
winecup, 308
winged flax, 293
winged sea purslane, 56
wing leaf primrose willow, 338
wing pod portulaca, 359
wireweed, 127
wissadula, 320–21
Wissadula amplissima, 320
Wissadula hernandioides, 320
Wissadula parvifolia, 320
Wissadula periplocifolia, 321
wolfberry, 394
Wolffia brasiliensis, 42
Wolffia columbiana, 42
woodland sensitive pea, 232
woodrose, 198
wood sorrel, 342–44
woolly cotton flower, 65
woolly croton, 214
woolly dalea, 258

woolly pyramid bush, 405
woolly rocktrumpet, 72
woolly stemodia, 389
woolly white, 108
wortclub, climbing, 329
Wright's bugheal, 132
Wright's catclaw, 240
Wright's false mallow, 312
Wright's Indian mallow, 306
wrightwort, 48
wrinkled globe mallow, 319

Xanthisma spinulosum var. spinulosum, 136
Xanthisma texanum var. orientale, 136
Xanthium strumarium, 137
Xanthocephalum dracunculoides, 85
Xanthocephalum sphaerocephalum, 102
Xylosma flexuosa, 274
Xylothamia palmeri, 114

yard mallow, 312–13
yaupon, desert, 179
Yeatesia platystegia, 54
yellow cowpea, 273
yellow flamethrower, 358
yellow flowered alicoche, 164
yellow flowered mecardonia, 388
yellow flowered pincushion cactus, 169

yellow lotus, 324
yellow prickly poppy, 344–45
yellow rain lily, 25
yellow rock nettle, 295
yellow sanvitalia, 122–23
yellow show, 141
yellow sophora, 270
yellow water lily, 333
yellow wood sorrel, 344
yerba de tago, 96
Youngia japonica, 137
Yucatan wild petunia, 53
Yucca constricta, 20, 21
Yucca treculeana, 20–21

Zanthoxylum fagara, 377
Zanthoxylum hirsutum, 378
Zapata bladderpod, 156
zarza, 246
zarzamora, 368
Zephyranthes pulchella, 25
Zeuxine strateumatica, 43
Zexmenia brevifolia, 111
Zexmenia hispida, 135
Zinnia acerosa, 138, 144
Ziziphus obtusifolia, 366
Zornia bracteata, 273
Zornia reticulata, 273
Zygophyllaceae family, 425–27